Physics of Neural Networks

Springer

New York
Berlin
Heidelberg
Barcelona
Hong Kong
London
Milan
Paris
Singapore
Tokyo

Physics of Neural Networks

J. Leo van Hemmen Jack D. Cowan
Eytan Domany (Eds.)

Models of
Neural Networks IV

Early Vision and Attention

With 139 Figures

Springer

Series and Volume Editors:

J. Leo van Hemmen
Institut für Theoretische Physik
Technische Universität München
D-85747 Garching bei München
Germany
lvh@physik.tu-muenchen.de

Eytan Domany
Department of Electronics
Weizmann Institute of Science
76100 Rehovot
Israel
fedomany@weizmann.ac.il

Jack D. Cowan
Department of Mathematics
University of Chicago
Chicago, IL 60637
USA
cowan@synapse.uchicago.edu

Library of Congress Cataloging-in-Publication Data
Models of neural networks IV / J.L. van Hemmen, J.D. Cowan, E. Domany,
    editors.
        p.   cm. — (Physics of neural networks)
    Includes bibliographical references and index.
    ISBN 0-387-95105-9 (alk. paper)
        1. Neural networks (Computer science) — Mathematical models.
I. Cowan, J.D. (Jack D.).   II. Domany, E. (Eytan).
III. Hemmen, J.L van (Jan Leonard).   IV. Series.
QA76.87.M59   2001
006.3—dc20   95-14288

Printed on acid-free paper.

Production managed by Allan Abrams; manufacturing supervised by Jeffrey Taub.
Photocomposed copy prepared from the editors' LATEX files.

9 8 7 6 5 4 3 2 1

ISBN 0-387-95105-9   SPIN 10774300

Springer-Verlag   New York Berlin Heidelberg
A member of BertelsmannSpringer Science+Business Media GmbH

Preface

Close this book for a moment and look around you. You scan the scene by directing your attention, and gaze, at certain specific objects. Despite the background, you discern them. The process is partially intentional and partially preattentive. How all this can be done is described in the fourth volume of *Models of Neural Networks* devoted to *Early Vision and Attention* that you are holding in your hands. Early vision comprises the first stages of visual information processing. It is as such a scientific challenge whose clarification calls for a penetrating review. Here you see the result. The Heraeus Foundation (Hanau) is to be thanked for its support during the initial phase of this project.

John Hertz, who has extensive experience in both computational and experimental neuroscience, provides in "Neurons, Networks, and Cognition" a theoretical introduction to neural modeling. John Van Opstal explains in "The Gaze Control System" how the eye's gaze control is performed and presents a novel theoretical description incorporating recent experimental results. We then turn to the relay stations thereafter, the *lateral geniculate nucleus* (LGN) and the primary visual cortex. Their anatomy, physiology, functional relations, and ensuing response properties are carefully analyzed by Klaus Funke et al. in "Integrating Anatomy and Physiology of the Primary Visual Pathway: From LGN to Cortex", one of the most comprehensive reviews that is available at the moment.

How do we discern patterns? That is to say, how do we perform scene segmentation? It has been shown that this process is partially *pre*attentive and, so to speak, done on the spot in the primary visual cortex. Reinhard Eckhorn explains the underlying "Neural Principles of Preattentive Scene Segmentation" while Esther Peterhans et al. analyze a neuronal model of "Figure-Ground Segregation and Brightness Perception at Illusory Contours".

Scene segmentation can also be performed by a feedback process that is called 'attention'. A glance suffices to convince every beholder that the eye catches megabytes of data. Through attention we reduce this data flood by singling out specific objects. Ernst Niebur et al. indicate how this can be done by "Controlling the Focus of Visual Selective Attention" while Julian Eggert and Leo van Hemmen elucidate the feedback mechanism proper in "Activity-Gating Attentional Networks".

Ever tried to smash a busy buzzing fly against the wall? Then you know how good it is in avoiding you. That is to say, you realize that also insects such as flies may perform highly efficient visual-information processing. In their essay "Timing and Counting Precision in the Blowfly Visual System" Rob de Ruyter van Steveninck and Bill Bialek explain how this is done in early vision and show what key role is played by spikes. Finally, Wolfgang Maass approaches "Paradigms for Computing with Spiking Neurons" from the point of view of a computer scientist who is concerned with biological information processing. Enjoy!

The Editors

Contents

3   Integrating Anatomy and Physiology of the Primary Visual Pathway: From LGN to Cortex   97

K. Funke, Z. F. Kisvárday, M. Volgushev, and F. Wörgötter

4   Neural Principles of Preattentive Scene Segmentation: Hints from Cortical Recordings, Related Models, and Perception   183

Reinhard Eckhorn

Contributors

REINHARD ECKHORN, Biophysik Department, Philipps Universität, Renthof 7, D-35032 Marburg, Germany

JULIAN EGGERT, Future Technology Research, HONDA R&D Europe, Carl Legien Strasse 30, D-63073 Offenbach/Main, Germany

KLAUS FUNKE, ZOLTÁN F. KISVÁRDAY, MAXIM VOLGUSHEV, Abt. für Neurophysiologie, Medizinische Fakultät, Ruhr-Universität Bochum, Universitätsstrasse 150, D-44801 Bochum, Germany

J. LEO VAN HEMMEN, Physik Department, TU München, D-85747 Garching bei München, Germany

CHRISTOF KOCH, Division of Biology, California Institute of Technology, Pasadena, CA 91125, USA

ERNST NIEBUR, Krieger Mind/Brain Institute, Johns Hopkins University, 3400 N. Charles Street, Baltimore, MD 21218, USA

JOHN HERZ, NORDITA, Blegdamsvej 17, DK-2100 Copenhagen. Denmark

LAURENT ITTI, Computer Science Department, University of Southern California, Hedco Neuroscience Building, Los Angeles, CA 90089-2520, USA

WOLFGANG MAASS, Institut für Informationsverarbeitung, TU Graz, Klosterwiesgasse 32/II, A-8010 Graz, Austria

JOHN VAN OPSTAL, Medical Physics and Biophysics, University of Nijmegen, P.O. Box 9101, NL-6500 HB Nijmegen, Netherlands

ESTHER PETERHANS, RICK VAN DER ZWAN, BARBARA HEIDER, FRIEDRICH HEITGER, Neurology Department, Zürich University Hospital, CH-8091 Zürich, Switzerland

ROB DE RUYTER VAN STEVENINCK, WILLIAM BIALEK, NEC Research Institute, 4 Independence Way, Princeton, NJ 08540, USA

FLORENTIN WÖRGÖTTER, Department of Psychology, University of Stirling, Stirling, FK9 4LA Scotland/UK

1

Neurons, Networks, and Cognition: An Introduction to Neural Modeling

J. A. Hertz

ABSTRACT This chapter deals with the modeling of neural systems at three levels: (1) single neurons, described by Hodgkin-Huxley equations and simpler integrate-and-fire models; (2) the dynamics of local cortical circuitry, in particular the observed irregularity of neuronal firing; and (3) cognitive computation, illustrated by the case of associative memory. The examples show how physically-based analysis, particularly methods from statistical mechanics, gives insight into the dynamics and function of the nervous system.

1.1 Introduction

The phrase "neural networks" covers a wide spectrum of models applied to an equally wide spectrum of tasks. This chapter will focus on three problems which are central to the whole field. First, we will study the dynamics of single neurons, as described by the Hodgkin-Huxley equations. We will also examine a simpler class of models – integrate-and-fire neurons – and how they are related to the more microscopic Hodgkin-Huxley description.

We will then turn to the dynamics of networks of neurons, focusing on generic features of local neocortical circuitry. Although the neocortex carries out a wide variety of processing tasks, it is remarkably uniform in structure, leading us to conjecture that it does its many computations with the same hardware and "in the same programming language," i.e., with the same underlying dynamics. A major characteristic of these dynamics is the high irregularity of the firing of the neurons, which has been seen in electrophysiological recordings for many decades. We will try to model the generic dynamics of extensively-connected networks of size around 1 mm^2 (the size of the region providing the major input to a single neuron), with emphasis on understanding the observed highly irregular firing. The implication of this analysis is that the network dynamics may be generically chaotic, and this may impose a basic constraint on neural computation.

And so we turn to the question of what and how a brain computes, which brings us into the complex realm of cognitive science. There are many problems in cognitive science for which we could study the underlying computational mechanisms, and to begin to make some progress we focus on associative memory. Any working brain needs a memory, and all evidence about animal brains points to some kind of distributed mechanism. We will examine some simple network models due to Willshaw, Hopfield, and others and see how they perform elementary tasks such as noise reduction and pattern completion. The models are examples of a general computational paradigm called "computing with attractors": Items in the memory are represented by stationary firing patterns across the network.

It will be obvious that there is a big gap between the formal model level of description in these associative memory models and the complexities of cortical network dynamics. It is hard to reconcile stationary memory patterns with intrinsically chaotic dynamics. We will try to fill a little of this gap, showing how one can extend the associative memory models to time-dependent attractors (and we will see there is some evidence that something like them are used in cortical processing). Finally, we will see some recent results that suggest how such meaningful temporal structure might grow as a result of self-organization.

In all these investigations we will meet some of the basic theoretical concepts and tools for studying neural networks more generally. Particularly important among these is mean field theory, a tool borrowed from statistical physics, and especially relevant for such stongly-interconnected systems as cortical matter.

1.1.1   A few neuroanatomical facts

It is useful to begin with a few basic facts and numbers from neuroanatomy [1, 2]. The brain (a good fraction of which consists of the neocortex in higher mammals) contains roughly 10^{11} neurons, with perhaps 10^{14} synapses between them. The neocortex is a sheet about 2 mm thick with an area of about 2000 cm^2 (crumpled up to fit inside the head). Its structure is remarkably uniform across brain regions. A typical mm^2 section contains up to 10^5 neurons, and these typically receive half their inputs from other neurons less than a centimeter away and half from outside this local neighbourhood.

This connectivity pattern suggests that it would be fruitful to understand the dynamics of subnetworks of this size on their own. Within such a region, the connectivity, though far complete, is much more extensive, so we expect strong collective effects. With luck, a proper study of this problem will also give us clues about how to link a number of these units together into networks on a larger scale, eventually making contact with cognitive phenomena.

Anatomists call such a piece of brain a "cortical column." The term is often used to suggest that there is a kind of clump-like structure to the cortex, with stronger connections within clumps than between them, but this is seldom actually seen. Nevertheless, it still seems useful to analyze cortical subnetworks of this size ($10^4 - 10^5$ neurons), and I will follow convention in calling them "columns."

The cortex has a systematic layered structure. Furthermore, in some parts of it there is systematic lateral structure within columns; for example, in visual cortex one speaks of "minicolumns" which respond strongly to a particular range of orientations of bar- or grating-like stimuli. However, in most respects the neocortex is remarkably uniform. It is a kind of general-purpose processor; its different areas perform different tasks largely because they receive different input during development, not because of intrinsic variations in circuitry (though such variations also do exist and have some functional role). An experiment done about 10 years ago showed that it was possible to redirect visual input fibers to the auditory cortical area of newborn ferrets, and the neurons there became responsive to visual stimuli. They arranged themselves to form a topographic map of the visual field the way the visual cortex normally does, and these animals could even see [3].

Here, I will not assume anything about any internal structure. My columns will be statistically homogeneous; there will be no meaning one can attach to "position" within the network. The models will be specified by a fixed probability of synaptic connections between neurons of particular types. This obviously ignores many interesting problems, but it permits a first step toward studying generic dynamical properties. It turns out that just understanding the observed stability of the resting state of cortical networks without specific external input, which is characterized by very low firing rates ($O(1)$ spike/sec), is a nontrivial problem.

Of course, activity in a cortical network is not completely unspecific, with all neurons firing equally rapidly. The unspoken assumption in most thinking about brain function is that certain groups of cells firing at more or less the same time are what encode the information being processed. These were called "cell assemblies" by Donald Hebb, who was probably the first to make this assumption explicit [4]. Therefore, the networks we will describe here might be taken as models of such assemblies (or at least localized parts of them).

Sometimes people think of the brain as a multiprocessor system, with cortical columns as the processors. In this analogy, the neurons play the role of the transistors on a chip. But we don't know what the signals between neuorns mean or how are they processed, so this hypothesis is incomplete. We can't really test it systematically yet. Instead, for now, we will focus on lower-level questions: What kind of dynamics do small networks with connection densities comparable to that within cortical columns have? Under what conditions is the behavior we find compatible with neurophysiological findings? Only after we can answer questions like these can we turn to asking questions about cortical function.

1.1.2   A few neurophysiological facts

We want to make our modelling consistent with neurophysiological findings, which is a big order. I recommend Abbott and Dayan's book [5] for further background. Here we will emphasize just three features:

- Neuronal firing is generally rather irregular. This applies not only to the resting state of the cortex, but also to the stronger firing evoked by external stimuli. For example, in experiments in visual cortex, repeated presentation of the same visual stimulus to an animal evoke similar total numbers of spikes (in, say, a 100-ms interval), but the timing of individual spikes varies widely (typically $\sim \pm 25$ ms) from trial to trial. Recent work (which we will discuss in sect. 4.3) has focused attention on the possibility that sometimes more precise temporal structure might be present, but irregularity is in any case very common, if not universal, in neuronal firing.

- The total spike count in, say, that 100-ms time window, varies systematically with salient features of the stimulus. For example, in primary visual cortex, the evoked firing rate is a systematic, continuous function of the orientation of a bar-shaped stimulus. That is, the spike count *carries information about the stimulus.* This statement has a precise meaning, which can be made quantitative, in terms of information theory [6, 7].

- Even the highest of these observed rates are much less than the maximum rate a neuron could fire at (interspike interval of the order of the refractory period, 1-2 ms, corresponding to rates of 500-1000 Hz).

These features have been seen in probably hundreds of thousands of experiments over many decades. Curiously, however, it is only recently that anyone has seriously tackled the question of how they emerge from the dynamics of networks. We will review some of this recent work, most of which is due to Amit and Brunel [8] and his collaborators and to van Vreeswijk and Sompolinsky [9]. I will mostly follow the latter treatment here, supplementing it with some recent simulation results from my own group. But first we need to take a look at ways of modeling the neurons these networks are made up of.

1.2   Neurons

One can model the neurons in a network at different levels of detail. The most microscopic level I will talk about here is generally called "Hodgkin-

Huxley neurons." This model is a nonlinear electrical circuit (the resistances are voltage-dependent) which describes spike generation explicitly. More phenomenologically, I will (more often) use "integrate-and-fire" neurons. This model is a linear electrical circuit that describes approximately the response of the membrane potential to synaptic input, augmented by *ad hoc* spike generation when the potential reaches a threshold. Finally, I will sometimes use a still more phenomenological description with binary neurons ($S_i = 1$ for a neuron that fires within an appropriately chosen time interval, $S_i = 0$ for one that doesn't). With cautious interpretation, even this simplest description can give insight into the dynamics of the network.

1.2.1 Hodgkin-Huxley neurons

Even this level of description is highly idealized. The spatial structure of real neurons is functionally important: Synaptic input is received at many different places in the dendritic tree and propagated to the soma. Spikes are generated at the end of the axon near the soma and propagate along the axon and its branchings. The present model ignores all of that. The neuron is treated as a point object, best thought of explicitly as the region (the "axon hillock") where spikes are generated. All effects of the propagation of synaptic input through the dendritic tree are described in terms of effective inputs, as felt at the spike-generating region. If everything in the dendrites were linear, this would be formally justified. Dendrites are not linear (at least not always), so this is an idealization. Similarly, we ignore the spike propagation in the axon and its arborization.

The Hodgkin-Huxley model has been treated extensively elsewhere [10]. Therefore here I will be rather sketchy, focusing on the basic physics. The cell membrane is taken to have a capacitance and, in parallel with it and each other, several kinds of conducting ion channels (Fig. 1.1). If V is the membrane potential,

$$\begin{aligned} C\dot{V} \quad & + \quad G_{\mathrm{l}}(V - V_{\mathrm{l}}) + G_{\mathrm{s}}(V - V_{\mathrm{s}}) \\ & + \quad G_{\mathrm{Na}}[V](V - V_{\mathrm{Na}}) + G_{\mathrm{K}}[V](V - V_{\mathrm{K}}) = I_{\mathrm{ext}}. \end{aligned} \quad (1.1)$$

Here C is the capacitance of the cell membrane, each G is the net conductance of a different kind of ion channel, and I_{ext} describes externally injected current (as in an experiment). With each conductance, the potential appears relative to the equilibrium potential for that kind of ion; these effective batteries are maintained by active molecular pumps which we do not describe here. The term with G_{s} describes the ion flow through synaptic channels when they are open (briefly after a presynaptic spike). (For simplicity we only consider a single kind of synaptic channel here; otherwise there would be several such terms.) G_{Na} and G_{K} are from Na and K channels involved in the spike generation mechanism. We write $G_{\mathrm{Na}}[V]$ and $G_{\mathrm{K}}[V]$ (with brackets instead of parentheses) to indicate that these conductances do not just depend on the instantaneous membrane potential,

Outside

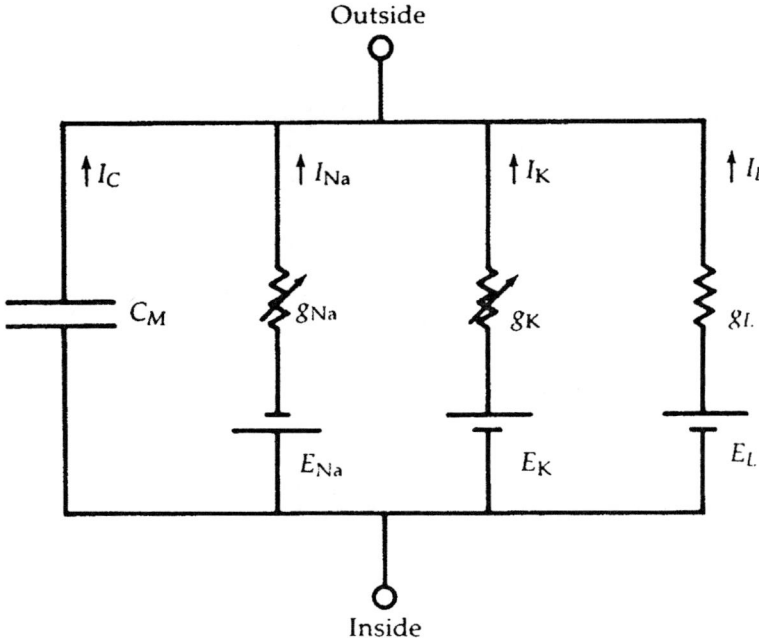

Inside

FIGURE 1.1. The cell membrane described as an electrical circuit. The capacitative branch represents the dielectric properties of the membrane. The three conducting branches represent active sodium and potassium and passive leakage conductances with their different equilibrium potentials. (The synaptic conductances are not shown.) (From B. Hille, *Ionic Channels of Excitable Membranes*, 2nd ed., Sinauer, 1992)

but rather on its (recent) history. Finally, all other channels are lumped into a nonspecific "leakage" conductance G_l.

The synaptic conductance G_s comes from synaptic channel openings set off by spiking of presynaptic neurons. If t_j^k is the time of the kth spike by neuron j,

$$G_s(t) = \sum_j \sum_k g_s(t - t_j^d - t_j^k), \qquad (1.2)$$

where $g_s(t)$ gives the time dependence of the postsynaptic conductance change produced by a single presynaptic spike, and t_j^d is the transmission delay between the spike in presynaptic neuron j and the beginning of the postsynaptic current. This transient conductance is a result of various processes involved in the opening and closing of the synaptic channel, but here we model it as a fixed function of t. For fast synaptic channels, this function is commonly taken to be proportional to $t \exp(-t/\tau_s)$, with a rise-

and-fall time of the order of 1 ms. (Some synaptic conductances can be voltage-dependent, but I won't consider those here.)

For a typical cortical neuron receiving 10^4 inputs, with each firing at, say, 5 spikes/s, there are roughly 50 incoming spikes overlapping in the 1-ms time associated with the opening and closing of the synaptic channel. If those inputs are independent, then the typical fluctuations in this input are of order $\sqrt{50} \approx 7$. Thus the input signal $G_s(t)$ has a DC component, but with significant fluctuations around it. Their correlation time is of the order of 1 ms.

All synaptic conductances are positive, of course, but the currents they produce can have either sign (excitatory or inhibitory), according to the sign of $V - V_s$. Fast excitatory channels are permeable to Na ions, whose equilbrium potential $V_s = V_{Na} \approx +55$ mV. Fast inhibitory channels are selective for Cl ions, for which $V_s = V_{Cl} \approx -70$ mV (typically below the equilibrium potential of the cell).

At a given constant I_{ext} and considering only the DC part of G_s, and for V not too large, we can find a stationary solution V_0 of (1.1). The fluctuations in synaptic input (individual presynaptic spikes) will cause the membrane potential to rise or fall in response; then it will relax back to V_0.

If all the conductances in the problem were constant, this would be all there was to the problem. It would just be a linear circuit. It is the voltage dependence of the Na and K conductances $G_{Na}[V]$ and $G_K[V]$ that makes it nontrivial. In particular, the voltage dependence of G_{Na} is such that when the potential is raised high enough, the conductance increases dramatically, permitting more Na ions to flow in and raise the potential even further. Above a threshold potential, this rise leads to a spike: a brief excursion to a very high membrane potential, followed by a fall, an overshoot, and finally a relaxation back to the original state.

Let us take a closer look at the spike generation mechanism, starting with the Na conductance G_{Na}. The protein forming the channel can assume various conformations with different probabilities (given by statistical mechanics) at different potentials. The Hodgkin-Huxley Na channel seems to behave as if four particular local binary degrees of freedom have to be in the right configuration ("open") for the channel to let Na ions through. Empirically, these degrees of freedom seem to be independent; thus, the probability that they are all open is just the product of their individual opening probabilities. Furthermore, three of these degrees of freedom are identical, in the sense that their opening probabilities are the same. Therefore the average conductance is just the total number of channels times the probability that a given one is open to Na ions, i.e.,

$$G_{Na}[V] = g_{Na}m^3[V]h[V], \tag{1.3}$$

where g_{Na} is a constant and m and h are the opening probabilities for the two kinds of degrees of freedom. (We are assuming that there are many

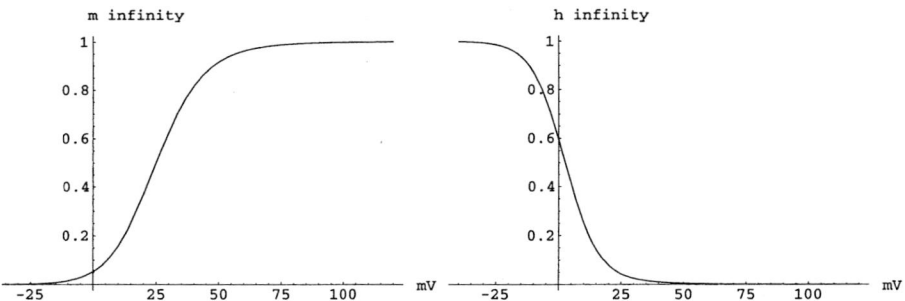

FIGURE 1.2. Plots of m_∞ (left) and h_∞ (right) as functions of the membrane potential.

channel molecules so we can use this average description and ignore fluctuations. This might not be completely accurate, but it describes the main features of spike generation.)

When the membrane potential V changes, the Boltzmann probabilities for the different local conformations change, and therefore so does G_{Na}. However, the actual numbers of molecules in the open and closed configurations for their m's and h's do not follow the changes in V instantaneously, since there are generally (free) energy barriers between their two local minima. The kinetics obey a first-order differential equation,

$$\frac{dm}{dt} = \alpha_m(V)(1 - m) - \beta_m(V)m. \tag{1.4}$$

α_m and β_m are the probabilities of the transitions closed→open and vice versa, respectively. In the simplest model, they would both exhibit activated behavior, e.g.

$$\alpha_m(V) \propto \exp[-b_\alpha(V - V_\alpha)/T], \tag{1.5}$$

with the barrier taken linear in V. Eqn. (1.4) can be written in the form

$$\tau_m(V)\frac{dm}{dt} = m_\infty(V) - m, \tag{1.6}$$

with

$$m_\infty(V) = \frac{\alpha_m(V)}{\alpha_m(V) + \beta_m(V)} \tag{1.7}$$

and

$$\tau_m^{-1}(V) = \alpha_m(V) + \beta_m(V). \tag{1.8}$$

Although we suppose that $\alpha_m(V)$ and $\beta_m(V)$ (or, equivalently, $m_\infty(V)$ and $\tau_m^{-1}(V)$)) depend instantaneously on $V(t)$, Eqns. (1.4) and (1.6) show that $m(t)$ depends on the past of $V(t)$. The same applies to h. Thus G_{Na} is a functional of $V(t)$. The same kind of argument applies to G_K.

The function $m_\infty(V)$ (Fig. 1.2, left panel) is a sigmoidal function of V. (This emerges naturally for the activated α_m and β_m (1.5).) It is small for

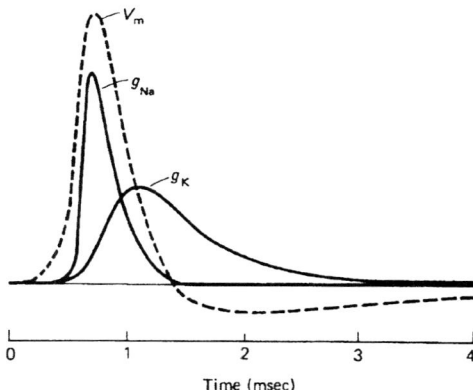

FIGURE 1.3. The shape of the action potential (dashed line) and the Na and K conductances as calculated from the changes in the Na and K given by the Hodgkin-Huxley equations. (From ref. 1)

V around the resting potential of the neuron, but increases rapidly once V gets above around -60 mV, saturating above $V \approx -20$ mV. The other function in G in Eqn. (1.3), h (Fig. 1.2, right panel), has an $h_\infty(V)$ sigmoid which goes the other way: It is high at low V and goes to zero at high V.

Spike generation is possible because the relaxation time for m, $\tau_m(V)$, is *much* smaller (by more than an order of magnitude) than that of h, $\tau_h(V)$. Thus initially, h does not have time to change from its resting value (about 0.6 for membrane potentials a little below threshold). To within $\tau_m(V) \ll 1$ ms, m follows V nearly instantaneously. The small rising V caused by excitatory synaptic input leads to an increasing G_{Na}, just following the sigmoidal rise in $m_\infty(V)$. This conductance increase leads to further Na inflow and a consequent further rise in membrane potential. If this increase is sufficient to outweigh the decrease due to leakage, the result will be a runaway rise in V. This is the way a "spike" is generated (Fig. 1.3). V rises to something near the Na equilibrium potential $\approx +55$ mV before the h variable can react. At such high voltages, $h_\infty(V)$ is nearly zero, so once h does relax to $h_\infty(V)$, the active Na channel closes.

On the same timescale the K conductance also comes into play. It can be described by an equation much like (1.3):

$$G_K[V] = g_K n^4[V]. \tag{1.9}$$

The n variable has a relaxation time $\tau_n(V)$ of the same order as $\tau_h(V)$ and an $n_\infty(V)$ with a shape something like that of $m_\infty(V)$. Thus at high V, the K channels open and let K ions out, bringing the cell eventually back to its resting state.

Many other kinds of channels with similar but different kinetics are known to exist, so this description is only a kind of minimal model. However, it does exhibit the basic physics of spiking neurons.

One can simulate large networks (thousands) of Hodgkin-Huxley or similar model neurons. However, this is costly because the full set of highly nonlinear equations has to be integrated numerically with time steps smaller than the fastest processes going on (the linetics of m). Furthermore, simplified models (when we can have confidence in their validity) can give insight into features of the network dynamics that may not be apparent in numerical results using more detailed models. We therefore turn now to some such simplified models.

1.2.2   Integrate-and-fire neurons

The dynamics of the membrane potential in the subthreshold and suprathreshold regions are essentially independent. This suggests a simplified model in which only the subthreshold dynamics leading up to the generation of a spike are treated explicitly and a spike is put in "by hand" when the threshold is reached. Generally, one can simplify Eqn. (1.1) further in the subthreshold regime by linearizing it around an equilibrium value V_0. This gives the subthreshold dynamics exactly for small disturbances around this equilibrium. Since these dynamics are linear, they are equivalent to a simple parallel combination of a capacitance and some effective conductance G_{eff}. For the (leaky) integrate-and-fire model, one adopts this simple description for all subthreshold V. The only parameter characterizing the neuron is the membrane time constant $\tau = RC = C/G_{\mathrm{eff}}$.

This description is generally simplified further by replacing the factor $V - V_{\mathrm{s}}$ in the synaptic current $G_{\mathrm{s}}(V - V_{\mathrm{s}})$ by $V_0 - V_{\mathrm{s}}$, a constant for each type of synapse. Then the current produced by a presynaptic spike is (for a given type of synapse) always the same, independent of the membrane potential. Thus, below threshold, the model is specified by the simple equations of motion

$$\tau \dot{V}_i + V_i = \hat{I}_i(t). \tag{1.10}$$

Here i labels the neurons in a network, and $\hat{I}_i(t) \equiv R(V_{\mathrm{s}} - V_0)G_{\mathrm{s}i}(t)$. I have ignored the external injected current in (1.1) and moved the synaptic term to the right-hand side so that the excitatory input now appears as a positive $\hat{I}_i(t)$. I have also set the zero of the membrane potential as the equilibrium value when there is no synaptic input.

Finally, one frequently studies the case where $g_{\mathrm{s}}(t - t')$ in (1.2) is taken to be a delta function. Then the synaptic current (for a given kind of channel) is given by

$$\hat{I}_i(t) = \sum_j J_{ij} \sum_k \delta(t - t_{ij}^{\mathrm{d}} - t_j^k), \tag{1.11}$$

with J_{ij} measuring the net strength of the input from neuron j to neuron i, t_j^k the time of the kth spike of neuron j, and t_{ij}^{d} a transmission delay time. This simplification is not necessary for many calculations, because the model is already linear without it, but it is often useful when one is only interested on timescales longer than the 1 ms or so it takes for a single synaptic event.

With this simplification, the dynamics are simple to describe in words: Each presynaptic spike produces a discontinuous jump in the membrane potential of strength J_{ij}/τ (upward for excitatory input, downward for inhibitory). Between such events, V_i just decays exponentially toward zero with time constant τ.

All of this was for the subthreshold dynamics. It has to be supplemented by a condition for when a spike should occur and what one does after it. Having chosen it, one integrates the (linear) equation of motion until the threshold is reached. Then the membrane potential is reset to a subthreshold value and the integration of the linear equation begins again. Both the threshold and the reset value are generally chosen phenomenologically. Sometimes a refractory time is added: The integration of the linear equation does not start again until a certain time after the spike. Thus the model neuron has four parameters: the membrane time constant, the threshold, the reset value, and the refractory time. For a network, one must specify in addition the synaptic and transmission delay matrices J_{ij} and t_{ij}^{d}.

The feature that the dynamics between spikes is linear obviously lends itself to efficient simulations of networks. One doesn't have to integrate the equations step by step numerically when the analytic solution of a simple linear model is available. To a good approximation the two things a neuron does – integration of synaptic input and spiking – are separate and independent. The present model is a reasonable qualitative description of the former, suitable when we don't care about details of the spike generation dynamics.

Considering for the moment only the DC part $\hat{I}_i^{\mathrm{DC}}$ of the synaptic input, we can see from (1.10) that V will reach an equilibrium level $V_{0i} = \hat{I}_i^{\mathrm{DC}}$, provided that this is less than the threshold. The fluctuations in the synaptic input will then produce a kind of correlated random walk in $V_i(t)$, with correlation time τ. When it happens to reach the threshold, a spike occurs (leading to synaptic input to other neurons), followed by a reset. After a time $\approx \tau$, the potential will have recovered to about V_{0i}, and the random walk starts again.

This description is consistent if the magnitude of the DC and fluctuating parts of the synaptic input are such as to put the equilibrium level somewhat below threshold. However, the resulting spike rate is extremely sensitive to the magnitudes of both the DC and fluctuating parts.

When the DC input is large, the neuron will tend to fire rather regularly, as the membrane potential rises repeatedly from its reset value to threshold. In the other limit, where the equilibrium membrane potential is well below threshold, the firing is random, with an extremely low rate. The rate is exponentially sensitive to external drive. Both these limiting cases are in disagreement with the observed firing characteristics of cortical neurons (irregular firing, with the rate smoothly input-dependent). To achieve the observed behavior, the driving strength has to be very finely tuned.

Shadlen and Newsome [11] observed that in a model with both excitatory and inhibitory input one could manipulate the mean and the standard deviation of the input current separately. One could thus have a net DC input, the difference between the mean excitatory and mean inhibitory inputs, which was not necessarily large compared to the typical fluctuations in the input. However, placing the equilibrium membrane potential the right distance from the threshold still required some fine tuning. We will study a model below in which this balance can be achieved automatically through the dynamics of the network.

1.2.3 Binary (Ising) neurons

While networks of integrate-and-fire neurons lend themselves to easy simulations, it is still very hard to do consistent theoretical analysis on them without making uncontrolled approximations. Therefore for analytic modeling we make a further (and more drastic) idealization [9] which will permit us to construct a simple, exactly soluble network model.

We will now deal with two-state neurons. They are either in a firing ($S_i = 1$) or a nonfiring ($S_i = 0$) state. At any time they receive synaptic input

$$h_i(t) = \sum_j J_{ij} S_j(t) \tag{1.12}$$

dependent on the states of all the other neurons. Each neuron updates its state stochastically, on average once per time interval τ: if $h_i(t) \geq \theta$, S_i becomes 1; if $h_i(t) < \theta$, S_i becomes 0. The τ is identified with the membrane time constant of more microscopic descriptions. Thus, all presynaptic neurons which have fired no longer than τ ago contribute to the "membrane potential" $h_i(t)$. (But $h_i(t)$ is not exactly a membrane potential, since it could be above the threshold θ for a while before the neuron updates itself.) The "on average once per τ" can be made more precise: The updating times constitute a Poisson process with a rate $1/\tau$.

For this model, networks of N randomly connected neurons can be solved exactly in the large-N limit. It clearly constitutes an extreme simplification of the real system. It is even a significant idealization of a network of integrate-and-fire neurons. Nevertheless, it does catch the essential basic feature of neural network dynamics: Neurons change their state in response to the synaptic input, with a memory extending roughly τ into the past. We will test the reliability of the model by comparing its predictions with results of simulations on integrate-and-fire and Hodgkin-Huxley neurons.

1.3   Local Cortical Network Dynamics

Almost everything in the preceding section is classical knowledge. The Hodgkin-Huxley theory is from the early '50s, and the integrate-and-fire model from around 1900. The analysis we are going to make here is rather new, even though the basic experimental finding that it deals with – the irregularity of cortical neuronal firing – is many decades old. It has taken some time to get the right perspective on the problem and apply appropriate tools to its solution.

The model we study was introduced by van Vreeswijk and Sompolinsky [9]. It contains N_E excitatory and N_I inhibitory neurons (Fig. 1.4). Thus there are four synaptic coupling parameters: excitatory-to-excitatory (J_{EE}), excitatory-to-inhibitory (J_{IE}), inhibitory-to-excitatory (J_{EI}), and inhibitory-to-inhibitory (J_{EE}). Each cell in each population receives input with fixed probabilities K/N_E, K/N_I from every cell in the two populations. We consider the dilute limit $K/N_E, K/N_I \ll 1$, but with mean connectivities $K \gg 1$. Thus each neuron receives many inputs, but their number is still a small fraction of the total number of neurons in the network. These inequalities are consistent with what is known about connectivities within a cortical column ($K \sim 5000$, $K/N \sim 5 - 10\%$). I will follow van Vreeswijk and Sompolinsky and use the index $k = 1, 2$ for excitatory and inhibitory populations, respectively. Thus the inverse updating rates, thresholds, neuron activity variables, and synaptic inputs acquire this index (τ_k, θ_k, S_k^i and h_k^i) and the synapse from neuron j in population l to neuron i in population κ is called J_{kl}^{ij}.

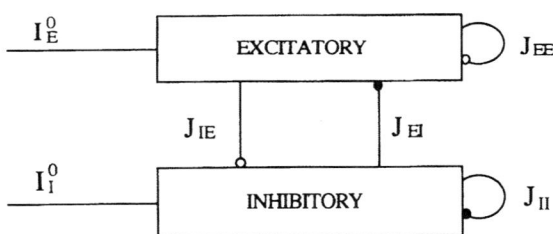

FIGURE 1.4. A schematic diagram of the van Vreeswijk-Sompolinsky network. Excitatory connections are shown terminating at open circles, inhibitory connections at closed circles. I_E and I_I are currents injected into all cells in the respective populations. (From ref. 13)

Actually, to make the mean field theory we are about to do rigorous, we need the more restrictive condition $K \ll \log N_k$, which is not satisfied by numbers from cortical anatomy. However, we will go ahead and do the theory anyway, hoping it at least has some qualitative relevance to real cortical dynamics. Eventually, someone should calculate corrections to the mean field results, but this has not been done yet.

1.3.1   Mean field theory

It is common to scale parameters so that the interesting quantities – here, the fluctuations of the synaptic input h_k^i – are of order 1 in the mean field limit ($K \to \infty$). Thus, we scale the interactions so that the typical size of any nonzero J_{kl}^{ij} in the network is of order $1/\sqrt{K}$:

$$J_{kl}^{ij} = \frac{J_{kl}}{\sqrt{K}} \qquad (1.13)$$

with probability K/N_l; otherwise $J_{kl}^{ij} = 0$. The quantities J_{kl} are of order 1. For any sum of random independent terms like our h_k^i, the fluctuations will be smaller than the mean by a factor of $O(\sqrt{K})$. (It is here that we need the condition $K \ll \log N_k$, to guarantee independence of the different terms in the sum.) Thus, the average values of h_k^i will be of order $\sqrt{K}$ and the fluctuations will be of order 1.

In order to have all inputs to a neuron of the same order, we also have to scale the external inputs (representing input from outside the single cortical column we are modeling) like $\sqrt{K}$:

$$\hat{I}_k^{\text{ext}} = \sqrt{K} I_k, \qquad (1.14)$$

with I_k of order unity.

It may seem disturbing that the average h_k^i diverges for $K \to \infty$. However, it is an inescapable feature if we want the fluctuations of the h_k^i to be of order 1. And we need the latter if we are to have firing rates between 0 and 1: If, say, we scaled parameters so that the average h_k^i were finite as $K \to \infty$, their fluctuations would go to zero. h_k^i would become a sharply defined quantity, and it would be either above or below its threshold θ_k, so all neurons would fire either all the time or never. The way we will get out of this problem will be to have the *total* input, $h_k^i + \hat{I}_k^{\text{ext}}$, be of order 1, even though the two terms in it are individually of order $\sqrt{K}$.

As we will be looking for asynchronous, randomly firing states, the quantities we want to calculate are the average firing rates (per τ)

$$m_k = \langle S_k^i \rangle \equiv \frac{1}{N_k} \sum_{i=1}^{N_k} S_k^i \qquad (1.15)$$

for the two populations. That is, m_k is just probability that $S_k^i = 1$, i.e., the probability that $h_k^i + \hat{I}_k^{\text{ext}} > \theta_k$. This means we have to examine

$$h_k^i = \sum_{l=1}^{2} \sum_{j=1}^{N_l} J_{kl}^{ij} S_l^j. \qquad (1.16)$$

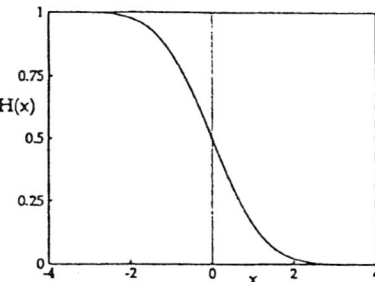

FIGURE 1.5. The complementary error function H(x).

In a random firing state, this is a sum of a large number of independent random numbers, so by the central limit theorem it is Gaussian. The total mean input (including the external part) is thus

$$\tilde{h}_k = \sqrt{K} \left(\sum_{l=1}^{2} J_{kl} m_l + I_k \right), \tag{1.17}$$

and its variance is

$$\sigma_k^2 = \sum_{l=1}^{2} J_{kl}^2 m_l. \tag{1.18}$$

Hence m_k, i.e., the probability that $h_k^i + \hat{I}_k^{\text{ext}} > \theta_k$, is just

$$m_k = \mathrm{H}\left(\frac{\theta_k - \tilde{h}_k}{\sigma_k} \right), \tag{1.19}$$

where H is the complementary error function (Fig. 1.5)

$$\mathrm{H}(x) = \int_x^\infty \frac{\mathrm{d}x}{\sqrt{2\pi}} \exp(-\tfrac{1}{2}x^2). \tag{1.20}$$

H(x) is a decreasing sigmoidal function of x. For $x \to -\infty$, H(x) $\to 1$, and for $x \to \infty$, H(x) $\to 0$. Eqn. (1.19) is the basic mean field equation for this network. It is exact for the model with synaptic strengths and external inputs defined by (1.13) and (1.14) in the limit $K \to \infty$. The crucial tool in deriving it was the use of the central limit theorem, which depended on the large connectivity K and the strong dilution.

Conditions for balanced asynchronous firing

This net does not always exhibit irregular, asynchronous firing at rates below saturation. Since, as we noted above, $\tilde{h}_k$ (1.17) scales like $\sqrt{K}$ while σ_k is of order 1, we need the leading term in the total input to vanish, with a correction of order 1. Thus, taking for simplicity the excitatory couplings $J_{EE} = J_{IE} = 1$ and defining

$$J_E \equiv -J_{EI}; \qquad J_I \equiv -J_{II} \tag{1.21}$$

for the inhibitory ones, we require

$$m_E - J_E m_I + I_E = 0 \tag{1.22}$$
$$m_E - J_I m_I + I_I = 0. \tag{1.23}$$

We can solve (1.22) and (1.23) for the rates m_E and m_I:

$$m_E = \frac{J_I I_E - J_E I_I}{J_E - J_I} \tag{1.24}$$

$$m_I = \frac{I_E - I_I}{J_E - J_I} \tag{1.25}$$

A little analysis shows that in order to have $0 < m_k < 1$ for both populations, we need

$$J_E > 1, \tag{1.26}$$

i.e., the inhibition acting on the excitatory neurons must be stronger than the self-excitation,

$$J_E > J_I, \tag{1.27}$$

i.e., the inhibitory neurons should inhibit the excitatory ones more than they inhibit themselves, and

$$\frac{I_E}{I_I} > \frac{J_E}{J_I}, \tag{1.28}$$

the external drive on the excitatory neurons can not be too weak (otherwise we could not get a positive m_E).

The state of the network when these conditions are fulfilled is one with random asynchronous firing at rates which are below (and possibly low compared with) saturation (1 spike per τ). The condition that the $O(\sqrt{K})$ term in the average input vanish does not require fine tuning of the model parameters: The firing rates of the populations depend continuously on the J_{kl} and the I_k, as shown by Eqns. (1.24) and (1.25). There is a dynamical balance – if we increase, say, the external input I_E by a little (but still satisfy the inequality (1.28), the inhibition will adjust itself self-consistently with the excitation so that the rates only change a little. We do not find the extreme sensitivity to input strength that we noticed in the single-neuron model. Rather, the firing rates for the populations depend linearly on the external drives I_k (within the region satisfying the inequality (1.28)), qualitatively consistent with neurophysiology. Thus, this model exhibits a network mechanism to enforce the Newsome-Shadlen model of balanced excitation and inhibition dynamically.

As previously observed, this solution is for a model with stochastic dynamics. However, one can get the same (mean-field) solution for a deterministic model in which each neuron is updated periodically, with a fixed interval Δt_k^i between updatings, and these intervals vary randomly from neuron to neuron, with a distribution $P(\Delta t) \propto \Delta t \exp(-\Delta t/\tau)$. This model is somewhat artificial, but it illustrates an important point: One can get asynchronous, irregularly firing states in a completely deterministic network. Of course, deterministic chaos is not a new story, but it is significant that it emerges so easily in a model that takes some account of neurophysiological reality.

Going in the opposite direction, it is also worth remarking that the same mean field theory applies (again for $K \ll \log N_k$) to a model in which all the synaptic strengths are chosen from their distribution independently at every updating step. In statistical-physics jargon, we would call this "annealed" rather than "quenched" randomness. Indeed, in real life this kind of stochasticity is present – synapses are not deterministic, but can fail (apparently) randomly.

On the other hand, the question of whether the dynamics of such cortical networks is intrinsically chaotic is nevertheless a critical point of principle. In the following subsection we will study whether it holds in simulations of more realistic finite systems.

Stability

It is possible to extend the steady-state mean field analysis to time-varying rates; the result is simply that Eqn. (1.19) is replaced by

$$\tau_k \frac{dm_k}{dt} + m_k = H\left(\frac{\theta_k - \tilde{h}_k}{\sigma_k}\right). \tag{1.29}$$

These can be linearized around the stationary solution, leading to equations of the form

$$\frac{d\delta m_k}{dt} = \sum_l M_{kl} \delta m_l(t). \tag{1.30}$$

The stability condition is just that the eigenvalues of the 2×2 matrix $\mathbf{M}$ both be negative. This is a somewhat complicated problem to analyze in detail, but the general nature of the result is that the ratio τ_E/τ_I exceed a critical value. This critical ratio depends on the parameters J_{kl} and I_k, but it is a number of order unity, typically a bit less than 1. The eigenvalues are both in general of order $\sqrt{K}$, so when the asynchronous state is stable it is very stable, and when it is unstable it is very unstable.

The above analysis is for the van Vreeswijk-Sompolinsky model [9]. In a corresponding treatment for integrate-and-fire neurons, valid for low firing rates [12], the stability equations take the same form (1.30).

This analysis applies to small perturbations, or order $1/\sqrt{K}$. One can also study effects of large perturbations (where the H( ) in (1.29) effectively becomes a step function). Details can be found in the paper by van Vreeswijk and Sompolinsky [9].

1.3.2 Simulations with spiking neurons

In our group we have simulated networks of up to several thousand spiking neurons, both integrate-and-fire and Hodgkin-Huxley ones, to test this theory [12, 13, 14]. We get similar results for the two kinds of models: Stable asynchronous states for a wide range are found for parameter values approximately satisfying the inequalities (1.26)-(1.28), although we also see states with oscillations (of varying degrees of regularity) in the firing rates.

Fig. 1.6 shows the typical course of the postsynaptic potential of a single neuron in an integrate-and-fire network.. One can see a couple of spikes and, between them, the fluctuations of the membrane potential. These fluctuations are driven by the synaptic input (1.11) from other neurons in the network; thus they are essentially random. Consequently, so are the times they reach threshold and produce a spike. Hence the picture is self-consistent: random presynaptic firing leads to random postsynaptic firing.

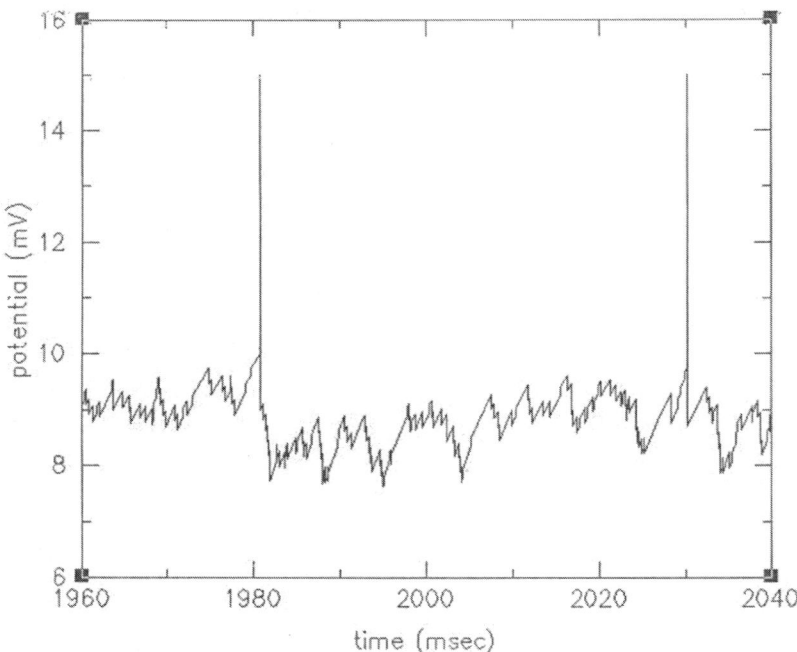

FIGURE 1.6. Membrane potential of an integrate-and-fire neuron in a network of 1000 excitatory and 1000 inhibitory neurons in an asynchronous firing state.

To be more precise, we can say that between spikes the membrane potential executes Brownian motion of a particle in a parabolic potential driven by white noise and subject to friction. This process is Gaussian with a standard deviation proportional to the rms effective input noise and has a correlation time equal to the membrane time constant τ. Thus, roughly speaking, once per τ the membrane potential makes independent samplings of its Gaussian distribution. The upper tail of this distribution extends above the threshold, and the weight in this tail gives the firing rate (relative to $1/\tau$).

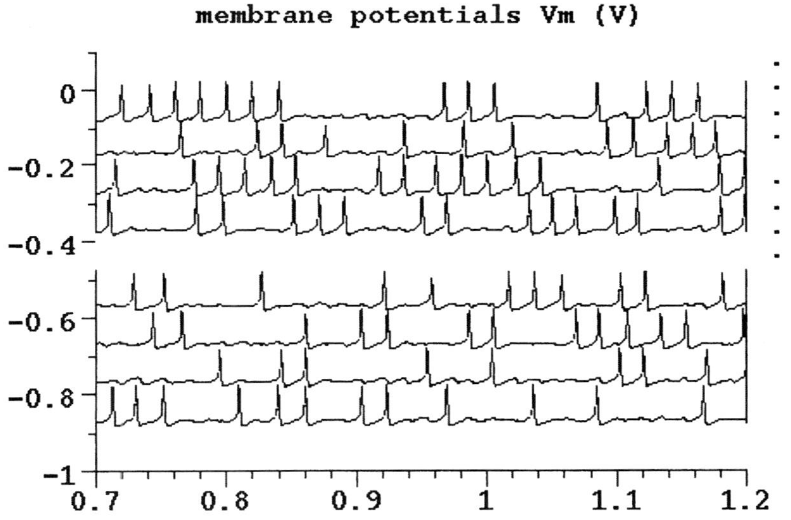

membrane potentials Vm (V)

FIGURE 1.7. HH neurons: membrane potential of 4 excitatory cells (top) and 4 inhibitory cells (bottom) in a 500-cell net. (Time is in sec., and the membrane potential is in V.) (From ref. 13)

We find the same kind of behavior for networks of Hodgkin-Huxley neurons [13]. Fig. 1.7 shows membrane potential traces for 8 neurons in such a net, and Fig. 1.8 shows the membrane potential, as well as the excitatory and inhibitory synaptic currents that drive it, in an 80-ms period just before the firing of one of the neurons.

If we increase the external drive, the resulting new firing rate increases smoothly, but not dramatically (Fig. 1.9). This shows the action of the dynamical balance between excitation and inhibition: If we were dealing with an isolated neuron, driven by external noise, with a low firing rate (a small tail of the membrane potential distribution above threshold), raising the drive strength would lead to huge relative rate changes (because of the form of the Gaussian tail). Here this is averted by the self-consistent dynamical balance between populations in the network.

The net breaks down if we try to make the rates very low (the neurons stop firing). This happens when there are not enough synaptic inputs within the neuron's memory time (τ) that the central limit theorem description of the membrane potential fluctuations is valid. It is amusing that the condition for being able to do an easy theory are essentially the same with the condition for the stability of the net. (I say "essentially" because, as can be seen in Fig. 1.9, there is a transition region where the rates, while still nonzero, fall below the linear dependence on external drive that they have at large values of the drive.)

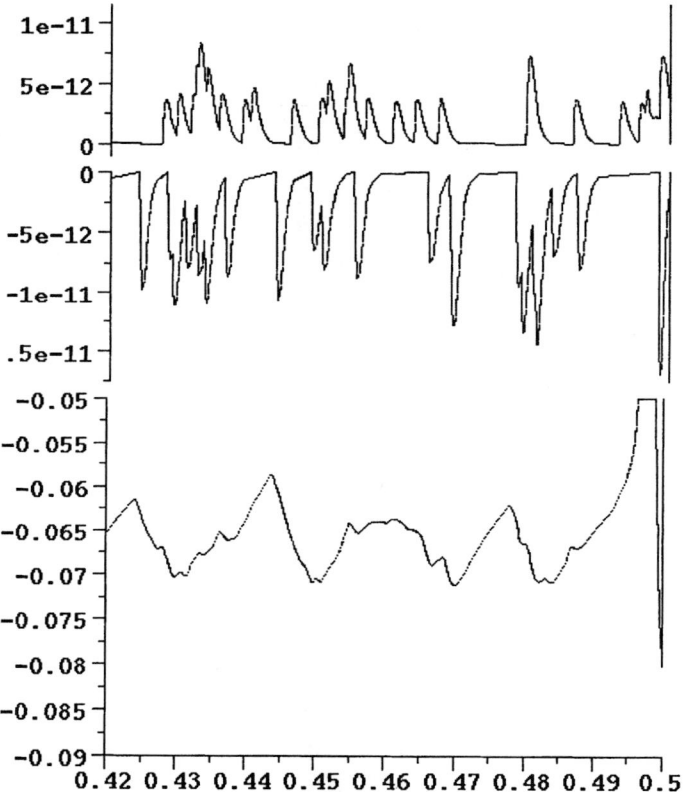

FIGURE 1.8. Close-up of the membrane potential in an interval 10τ preceding an action potential. Top: excitatory currents. Middle: inhibitory currents. Bottom: membrane potential. (From ref. 13)

The fact that the neurons are firing at constant rates is shown in Fig. 1.10, where the cumulative number of firings is plotted for the two populations as a function of time. Constant rates are reflected as straight lines.

The independence of the inputs to different neurons is essential to the asynchronous state. If we increase the concentration of synapses too much, different neurons will feel correlated input and will tend to fire in a corre-

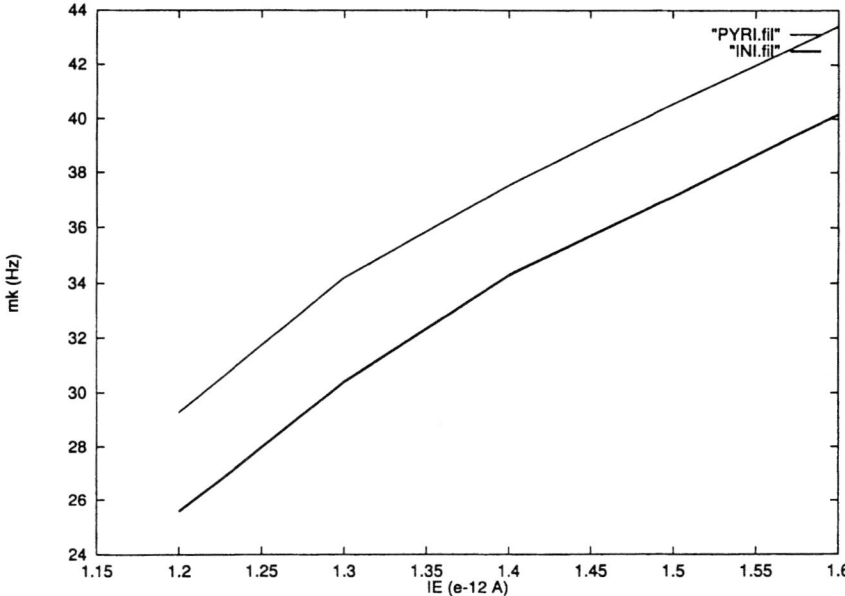

FIGURE 1.9. Firing rates of the two populations for as a function of I_E. The ratio I_E/I_I is kept constant. (From ref. 13)

lated fashion, leading to globally synchronous firing patterns. These state may exhibit either periodic or chaotic behaviour.

Such behavior (Fig. 1.11) is reflected in wiggles in the cumulative firing plots. The details of this behavior, as well as whether we find it at all, depend on details such as (in these cases), the distribution of transmission delays, in addition to the degree of dilution.

We can demonstrate explicitly the chaos in this (deterministic) model, comparing the evolution of two copies of the network. In one, we applied an extra input current pulse to one neuron at $t = 0$, making it fire prematurely. We then compared the states of the two systems in the following way. We divided the time into intervals of width 2 ms. For each neuron in time interval t we defined $S_i^c(t)$ equal to 1 if the neuron fired in that interval and 0 if it didn't. We distinguished the copies by the superscript c; $c = 0$ for the unperturbed copy and $c = 1$ for the perturbed one. We then computed the normalized Hamming distance between the copies:

$$D(t) = \frac{1}{N} \sum_i [S_i^0(t) - S_i^1(t)]^2. \tag{1.31}$$

Fig. 1.12 shows the strongly chaotic behaviour of the network. The almost discontinuous sharp rise is consistent with the mean field prediction [9].

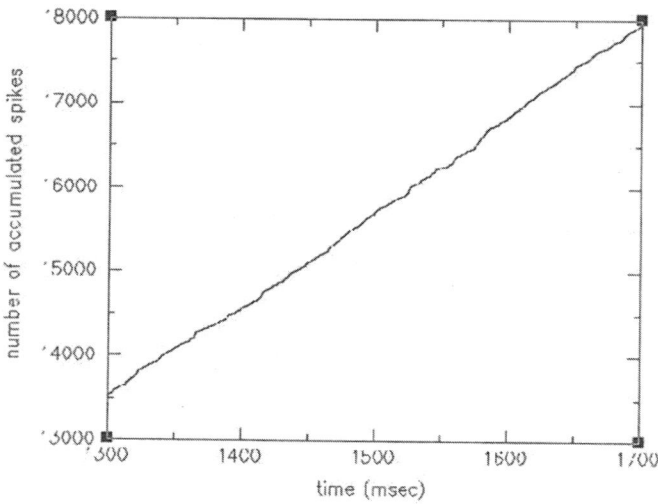

FIGURE 1.10. Cumulative firings as a function of time for a net in an asynchronous state.

1.4   Collective Computation: Associative Memory

We turn now to a problem at a more macroscopic level: associative memory. It is relevant here as an example of modeling at the level of cognitive phenomena and function. It is a particularly simple, yet essential, cognitive computation, so it has been popular as a subject for theoretical modelling. The models discussed here can be traced back to work in the '60s and '70s by Grossberg, Willshaw, Kohonen and others. The particular version I will discuss first is due to Hopfield [15]. My treatment is lifted straight out of my book with Anders Krogh and Richard Palmer [16].

We will restrict our attention to networks of binary units S_i. Sometimes we will use $S_i = \pm 1$ and sometimes $S_i = 0, 1$. Unless otherwise stated we assume random sequential updating.

The basic task in associative memory is to store a set of p patterns ξ_i^μ in such a way that, when presented with a new pattern ζ_i, the network responds by producing whichever one of the stored patterns most closely resembles ζ_i. Thus, such a memory should correct errors if ζ_i is a corrupted version of one of the ξ_i^μ, as well as fill in possibly missing parts of such a stored pattern. (Here the patterns are labeled by an index μ.)

1.4.1   Hopfield model

We consider first the symmetric case studied by Hopfield, with $S_i = \pm 1$. The patterns are assumed to be random and independently chosen with

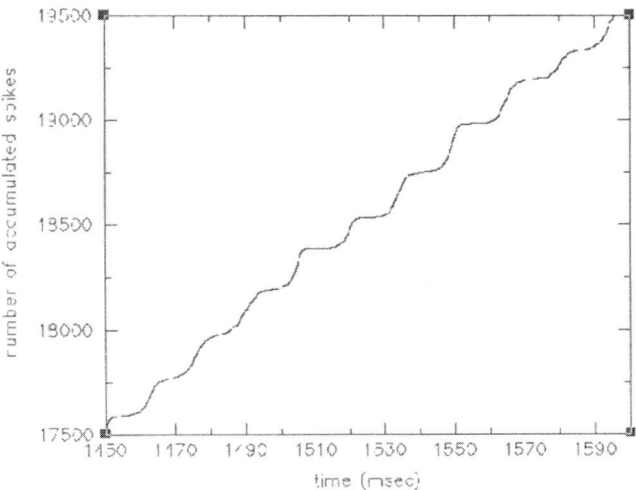

FIGURE 1.11. Cumulative firings as a function of time for a net exhibiting oscillations in its firing rates.

$\xi_i^\mu = \pm 1$ with equal probability. And we take the dynamics to be intrinsically noisy, with

$$\text{Prob}(S_i = \pm 1) = f_\beta(\pm h_i) \equiv \frac{1}{1 + \exp(\mp 2\beta h_i)} \qquad (1.32)$$

with h_i the net input to unit i. The parameter β is an inverse temperature. In the limit of low temperature ($\beta \to \infty$), the unit just follows the sign of its net input h_i exactly. At finite β it makes errors. As $\beta \to 0$, it just takes the "right" value sgn h_i with a probability slightly greater than $\frac{1}{2}$ (by something proportional to h_i). Thus, for a given h_i,

$$\langle S_i \rangle = \tanh(\beta h_i). \qquad (1.33)$$

The parallel with the behavior of a spin in a magnetic field is evident.

For the Hopfield model, one takes the connections in the network according to the prescription

$$J_{ij} = \frac{1}{N} \sum_\mu \xi_i^\mu \xi_j^\mu. \qquad (1.34)$$

This is inspired by Hebb's idea that simultaneous pre- and postsynaptic activity leads to synaptic strengthening [4]. It isn't the optimal formula, but it gives a simple case that can be solved exactly.

This leads to mean field equations for the average activities

$$\langle S_i \rangle = \tanh\left(\frac{\beta}{N} \sum_{j,\mu} \xi_i^\mu \xi_j^\mu \langle S_j \rangle \right). \qquad (1.35)$$

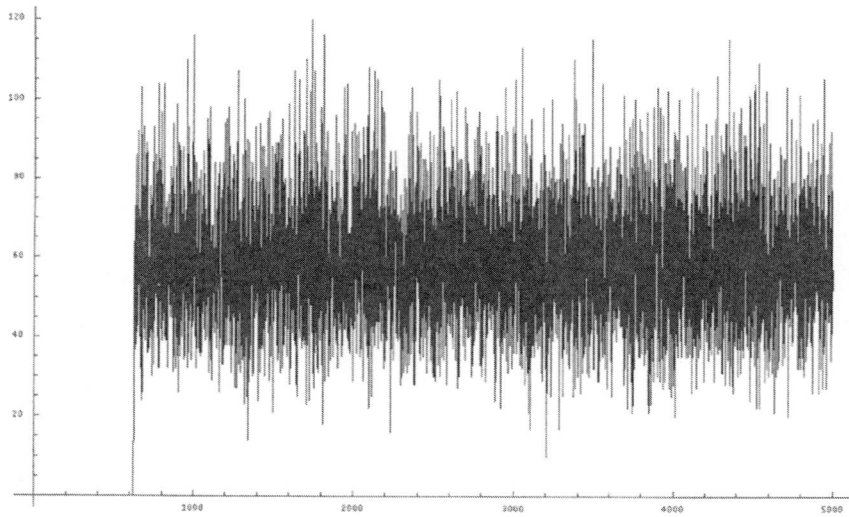

FIGURE 1.12. Hamming distance between perturbed and unperturbed copies of an integrate-and-fire network as a function of time. The correlation between the two nets almost immediately reaches the totally random level.

These have *memory solutions* with $\langle S_i \rangle \propto \xi_i^\mu$ for any μ. To find the constant of proportionality, we just plug $\langle S_i \rangle = m\xi_i^\nu$ into (1.35) and we get

$$m = \tanh\left(\frac{\beta m}{N} \sum_{j,\mu} \xi_i^\nu \xi_i^\mu \xi_j^\mu \xi_j^\nu\right). \tag{1.36}$$

The sum can be split up into a term with $\mu = \nu$ and the rest. The rest is $1/N$ times a sum of $N(p-1)$ random independent terms of unit magnitude, so it is of order $\sqrt{p/N}$.

Thus if the number of stored patterns $p \ll N$, the problem is quite trivial; we just get states correlated with the stored patterns, with a strength m given by the solutions of

$$m = \tanh(\beta m). \tag{1.37}$$

This is just the Weiss mean field equation for a ferromagnet. It has a Curie temperature $T_c = \beta_c^{-1} = 1$. Above T_c the system is disordered and exhibits no correlation with the stored patterns. Below T_c, we find collectively ordered states correlated with them. Just below T_c, m rises from zero $\propto \sqrt{T_c - T}$. At low T, the errors in retrieval are exponentially rare. If we put the system in an initial state ζ_i, it will fall naturally into a region of configurations near the nearest stored pattern. This is the basic physics of associative memory retrieval in the Hopfield model. When p becomes of order N, things become more complicated, but the basic physics of the collective retrieval remains the same.

Let us now look at this more complicated situation, which has some interesting physics. It is useful to define the *load parameter*

$$\alpha = \frac{p}{N}. \tag{1.38}$$

To analyze the mean field equation in the finite-α case, we will need to consider the overlap of $\langle S_i \rangle$ with all the patterns:

$$m_\nu = \frac{1}{N} \sum_i \xi_i^\nu \langle S_i \rangle. \tag{1.39}$$

For the one being retrieved, this is of order 1, while the rest are of order $1/\sqrt{N}$. They will have random sign, so it is their mean square value, normalized to $1/N$,

$$r = \frac{1}{\alpha} \sum_{\nu \neq 1} m_\nu^2 \tag{1.40}$$

that will come into the calculation. (We will suppose that it is pattern number 1 whose retrieval we are studying.)

Capacity calculation

The following heuristic calculation [17] is not rigorous, the correct treatment requires the use of the replica method [18] and has been explained in extenso elsewhere [19]. In terms of the overlaps m_ν, the mean field equations (1.35) take the form

$$m_\nu = \frac{1}{N} \sum_i \xi_i^\nu \tanh \left(\beta \sum_\mu \xi_i^\mu m_\mu \right), \tag{1.41}$$

and for $\nu \neq 1$ we separate out explicitly the terms with $\mu = 1$ and $\mu = \nu$:

$$m_\nu = \frac{1}{N} \sum_i \xi_i^\nu \xi_i^1 \tanh \left[\beta \left(m_1 + \xi_i^\nu \xi_i^1 m_\nu + \sum_{\mu \neq 1, \nu} \xi_i^\mu \xi_i^1 m_\mu \right) \right]. \tag{1.42}$$

Here I have taken advantage of the freedom to move factors such as ξ_i^ν in and out of the tanh, because they are ± 1 and the tanh is odd.

The first term in the argument of the tanh is large (order 1) by hypothesis, because it is pattern 1 that we are retrieving, and the last term is large because there are about p terms in it. But the second term is small, order $1/\sqrt{N}$, so we can expand:

$$
\begin{aligned}
m_\nu &= \frac{1}{N} \sum_i \xi_i^\nu \xi_i^1 \tanh \left[\beta \left(m_1 + \sum_{\mu \neq 1, \nu} \xi_i^\mu \xi_i^1 m_\mu \right) \right] \\
&+ \frac{\beta}{N} \sum_i \left\{ 1 - \tanh^2 \left[\beta \left(m_1 + \sum_{\mu \neq 1, \nu} \xi_i^\mu \xi_i^1 m_\mu \right) \right] \right\} m_\nu.
\end{aligned} \tag{1.43}
$$

We now assume that the small overlaps m_μ, $\mu \neq 1$, are independent random variables with mean zero and variance $\alpha r/p$, as suggested by (1.40). In the second line of (1.43), $\xi_i^\mu \xi_i^1$ is random and independent of m_μ, so by the central limit theorem the site average $N^{-1} \sum_i$ is effectively an average over a Gaussian "noise" $\sum_{\mu \neq 1, \nu} \xi_i^\mu \xi_i^1 m_\mu$ of variance αr. Thus (1.43) reduces to

$$m_\nu = \frac{1}{N} \sum_i \xi_i^\nu \xi_i^1 \tanh\left[\beta\left(m_1 + \sum_{\mu \neq 1, \nu} \xi_i^\mu \xi_i^1 m_\mu\right)\right] + \beta m_\nu - \beta q m_\nu, \quad (1.44)$$

or

$$m_\nu = \frac{N^{-1} \sum_i \xi_i^\nu \xi_i^1 \tanh[\beta(m_1 + \sum_{\mu \neq 1, \nu} \xi_i^\mu \xi_i^1 m_\mu)]}{1 - \beta(1 - q)}, \quad (1.45)$$

where

$$q = \int \frac{dz}{\sqrt{2\pi}} e^{-\frac{1}{2}z^2} \tanh^2[\beta(m_1 + \sqrt{\alpha r}z)]. \quad (1.46)$$

Now we can calculate r. We just square (1.45):

$$m_\nu^2 = \left[\frac{1}{1 - \beta(1 - q)}\right]^2 \frac{1}{N^2} \sum_{ij} \xi_i^\nu \xi_i^1 \xi_j^\nu \xi_j^1 \tanh\left[\beta\left(m_1 + \sum_{\mu \neq 1, \nu} \xi_i^\mu \xi_i^1 m_\mu\right)\right]$$

$$\times \quad \tanh\left[\beta\left(m_1 + \sum_{\mu \neq 1, \nu} \xi_j^\mu \xi_j^1 m_\mu\right)\right] \quad (1.47)$$

and average the result over patterns. Since pattern ν does not occur inside the tanh's, the pattern factors $\xi_i^\nu \xi_i^1 \xi_j^\nu \xi_j^1$ outside the tanh's can be averages separately, and only the $i = j$ term survives. Then, the remaining average of the tanh's just gives a factor of q as in (1.43). The result is independent of ν, so from (1.40),

$$r = \frac{q}{[1 - \beta(1 - q)]^2}. \quad (1.48)$$

We also need an equation for m_1. Using the same approach, starting again from (1.41) with $\nu = 1$, it is easy to obtain

$$m_1 = \int \frac{dz}{\sqrt{2\pi}} e^{-\frac{1}{2}z^2} \tanh[\beta(m_1 + \sqrt{\alpha r}z)]. \quad (1.49)$$

The three equations (1.46), (1.48), and (1.49) can now be solved simultaneously for m_1, q, and r. In general, this must be done numerically. We examine only the $T \to 0$ ($\beta \to \infty$) limit. In this limit it is clear that $q \to 1$ but the quantity $C \equiv \beta(1 - q)$ remains finite. We can use

$$\int \frac{dz}{\sqrt{2\pi}} e^{-\frac{1}{2}z^2} (1 - \tanh^2 \beta[az + b])$$

$$\approx \frac{1}{\sqrt{2\pi}} e^{-\frac{1}{2}z^2} |_{\tanh^2 \beta[az+b]=0} \times \int dz (1 - \tanh^2 \beta[ax+b])$$

$$= \frac{1}{\sqrt{2\pi}} e^{-b^2/2a^2} \frac{1}{a\beta} \int dz \frac{\partial}{\partial z} \tanh \beta[az+b]$$

$$= \sqrt{\frac{2}{\pi}} \frac{1}{a\beta} e^{-b^2/2a^2} \tag{1.50}$$

and

$$\int \frac{dz}{\sqrt{2\pi}} e^{-\frac{1}{2}z^2} \tanh \beta[ax+b]$$

$$\rightarrow \int \frac{dz}{\sqrt{2\pi}} e^{-\frac{1}{2}z^2} \operatorname{sgn}[az+b]$$

$$= 2H\left(\frac{-b}{a}\right) - 1 = \operatorname{erf}\left(\frac{b}{\sqrt{2}a}\right) \tag{1.51}$$

Our three equations thus become

$$C \equiv (1-q) = \sqrt{\frac{2}{\pi \alpha r}} \exp\left(-\frac{m^2}{2\alpha r}\right)$$

$$r = \frac{1}{(1-C)^2}$$

$$m = \operatorname{erf}\left(\frac{m}{\sqrt{2\alpha r}}\right), \tag{1.52}$$

where we have written m for m_1. Setting $y = m/\sqrt{2\alpha r}$, we obtain a single equation,

$$y\left(\sqrt{2\alpha} + \frac{2}{\sqrt{\pi}} e^{-y^2}\right) = \operatorname{erf} y, \tag{1.53}$$

to solve numerically (a graphical analysis is instructive). One finds a limiting capacity

$$\alpha_c \approx 0.138. \tag{1.54}$$

Doing (one-step) replica symmetry breaking in the context of the replica method [19] correctly, which is a hairy business, one finds [20] $\alpha_c = 0.138186$. Thus the approximation (1.54) is pretty close. At higher T, $\alpha_c(T)$ shrinks, until for $T \geq 1$ no patterns can be stored. The phase diagram in T, α space has interesting features but won't be discussed here. The main point has been to show how one can use statistical mechanical methods to find a nontrivial property of a simple model associative memory.

1.4.2 Sparse-pattern model

If we are thinking of modeling real neural networks, the $\pm$ symmetry of the Hopfield model (half the units active in any pattern) is not realistic. Much closer to reality is the limit of sparse patterns (almost all units off in any pattern) and sparse activity. It was this kind of limit that was first studied by Willshaw and collaborators [21]. Here we follow a more recent treatment by Tsodyks and Feigel'man [22]. This model is easier to solve than the symmetric Hopfeld model, since in the sparse limit, the counterpart of the self-consistent calculation of the quantity r above becomes trivial.

Here we take units $S_i = 1, 0$ for firing and not firing, respectively. The p patterns are chosen randomly as $\xi_i^\mu = 1$ with probability $f \equiv n/N \ll 1$ and $\xi_i^\mu = 0$ with probability $1 - f$. Thus f is the fraction of neurons active in a given pattern, and n is the size of the cell assembly that encodes a pattern. Because of the sparsity, a given pair of patterns have a very small overlap. However, we will deal with a case where there are so many patterns that each neuron typically participates in many patterns.

In the deterministic ($T = 0$) limit the dynamics of the model are defined by

$$S_i(t) \to \Theta[\sum_j J_{ij} S_j(t) - \theta], \tag{1.55}$$

with $\Theta(\ )$ the unit step function, θ a threshold, and random sequential updating, as before. The Hebbian model for the synaptic strengths is

$$J_{ij} = \frac{1}{n} \sum_\mu \xi_i^\mu \xi_j^\mu - \frac{pf}{N}. \tag{1.56}$$

In this model we do not have separate excitatory and inhibitory populations, as in our earlier modeling. Just as in the symmetric Hopfield model described previously, we assume any neuron can act in both ways. The global inhibitory term is for stability.

Let us do the mean field theory and capacity calculation for this model. We will assume we are retrieving pattern 1, and the retrieval order parameter is

$$m = \frac{1}{n} \sum_i \xi_i^1 S_i. \tag{1.57}$$

We also need the total average activity

$$fQ = \frac{1}{N} \sum_i S_i. \tag{1.58}$$

Now the net input to neuron i is

$$
\begin{aligned}
h_i &= \sum_j J_{ij} S_j \\
&= \frac{1}{N} \sum_{\mu,j} \xi_i^\mu \xi_j^\mu S_j - \frac{pf}{N} \sum_j S_j
\end{aligned}
$$

$$
= \frac{1}{N}\sum_j \xi_i^1 \xi_j^1 S_j + \frac{1}{N}\sum_{\mu \neq 1, j} \xi_i^\mu \xi_j^\mu S_j - \frac{pf}{N}\sum_j S_j
$$

$$
= m\xi_i^1 + \delta h_i. \tag{1.59}
$$

Using familiar arguments, the noise term δh_i is Gaussian with zero mean and variance

$$
\langle \delta h_i^2 \rangle = \alpha f Q (1 + n f Q). \tag{1.60}
$$

Since we need $nf = n^2/N$ of order 1 at most, we get a sensible model for $n = O(\sqrt{N})$, i.e., $f = O(1/\sqrt{N})$.

Thus, for neurons in assembly 1, the net input is $b + m +$ a Gaussian noise, while for the rest of the neurons it is just $b +$ the Gaussian noise. This is enough to calculate m and Q self-consistently. We get

$$
\begin{aligned}
m &= \int_{\theta - m}^{\infty} \frac{dh}{\sqrt{2\pi \langle \delta h^2 \rangle}} e^{-\frac{1}{2} h^2 / \langle \delta h^2 \rangle} \\
&= \mathrm{H}\left(\frac{\theta - m}{\sqrt{\langle \delta h^2 \rangle}} \right) \tag{1.61}
\end{aligned}
$$

and

$$
fQ = fm + (1 - f)\mathrm{H}\left(\frac{\theta}{\sqrt{\langle \delta h^2 \rangle}} \right), \tag{1.62}
$$

with $\langle \delta h^2 \rangle$ given by (1.60). These are the mean-field equations for this model. A treatment away from the sparse limit would also have an equation for a quantity analogous to r (1.48), but here r and Q are essentially identical.

The threshold θ is still a free parameter. It can be chosen so that, insofar as is possible, most of the distribution for the neurons in the active assembly are above threshold, while most of the others lie below it.

The capacity can be estimated by a simple argument. We have the picture of Gaussian distributions of membrane potentials for the two classes of neurons – in and not in assembly (pattern) number 1. They have the same width, $\sqrt{\alpha f Q(1 + n f Q)}$, but their means are displaced by m relative to each other. This difference is maintained self-consistently. This picture cannot be self-consistent if the tail of the distribution of membrane potentials for the neurons not in the assembly contains more neurons than there are in the assembly. That is, the second term in (1.62) has to be smaller than the first. Thus, setting $Q \approx 1$, $m \approx 1$, we have

$$
\int_1^{\infty} \frac{dh}{\sqrt{2\pi \alpha f(1 + nf)}} e^{-\frac{1}{2} h^2 / \alpha f(1 + nf)} \ll f, \tag{1.63}
$$

i.e.,

$$
\mathrm{H}\left(\frac{1}{\sqrt{\alpha f(1 + nf)}} \right) \ll 1 \tag{1.64}
$$

Using the asymptotic form of H,

$$H(x) \approx \frac{1}{\sqrt{2\pi}x} e^{-\frac{1}{2}x^2} \qquad (1.65)$$

we get, to logarithmic accuracy, the capacity

$$\alpha < \frac{1}{f(1+nf)\log(1/f^2)}. \qquad (1.66)$$

This criterion can be derived more systematically. Note that if it were not for the log, it would just be the statement that the separation between the means of the two Gaussians should be greater than their standard deviations. The log factor appears because there are many more neurons outside the assembly than within it. Effectively, the means have to be separated by $\log(1/f^2)$ standard deviations instead of just 1.

If we try to apply this to a cortical column of 10^5 neurons and assume a sparseness of $f \sim 10^{-2}$ (i.e., 1000 neurons per assembly) we get a capacity of something like $\alpha N \approx 10^6$ patterns. This is subject to a lot of uncertainties, of course, but it shows that a cortical network could have a large capacity.

A weakness of any such model, however, is that the picture implies that almost all the cells in an active assembly would be firing all the time. While there is some evidence for stationary memory attractors [23], they are never characterized by such high rates. A possible resolution is to incorporate features of the balanced-network picture of Sect. 1.3 into the present model. In the next section we will consider another possible mechanism for removing this flaw.

1.4.3   Memory with time-dependent patterns

We now consider an associative memory model in which the remembered items are encoded in a time-dependent fashion, i.e., as sequences of patterns. This does not imply that the items encoded are themselves necessarily time-dependent entities, only that they are represented by time-dependent patterns.

Hitherto we have stressed the irregularity of neuronal firing. But is it possible that there is some hidden structure in it – that the timing of spikes actually carry information, but we are blind to this structure because we don't know what it is encoding? Temporal encoding in neural processing has been speculated on for a long time [24], since neurons could obviously be more efficient if they exploited this degree of freedom. Recently, there is accumulating evidence that the timing of spikes relative to others emitted by the same or other neurons may be reproducible on significantly shorter timescales [25, 26, 27, 28, 29, 30, 31, 32].

Here I will review briefly one set of experiments [26, 27] which suggest that some temporal structure in cortical neuronal spike trains may be reproducible from trial to trial with a precision at the 1-ms level. I will then describe a class of models, inspired by these observations, for how a cortical column might function. As the name suggests, these feature the synchronized firing of successive pools of neurons – so-called "synfire chains." I will address two basic questions:

- How this synfire activity can maintain its coherence with millisecond-level timing precision, despite the order-of-magnitude-larger membrane time constant and the presence of noise and inhomogeneities.

- How such a network might function as a kind of associative memory for encoding afferent stimuli.

Experimental background

The experiments of the Jerusalem group [26, 27] were done in the frontal cortex of macaque monkeys performing a simple behavioral task. In response to a visual cue, the animal was supposed to make a hand movement to one of two possible keys which had been illuminated briefly earlier in the trial. The range of delay periods during which the animal had to remember which key had been illuminated was 1–32 seconds, varying randomly from trial to trial. Blocks of trials of this task ("GO" trials) were alternated with blocks of "NO-GO" trials, which had the same format except that the animal was to refrain from making the hand movement. Spike trains were isolated from 8-10 neurons at a time and measured throughout each trial. (Variations on this experimental paradigm were used in some of the experiments, but the present description is sufficient for our purposes here.)

At first glance, the measured spike trains do not exhibit obvious reproducible temporal structure. The overall firing rate of some neurons rises significantly at some points in the task (e.g., around the time the hand movement is planned or executed), but the individual spike times in different trials do not appear to coincide.

However, such an analysis will only be sensitive to temporal structure which is reliably locked to the laboratory clock (i.e., to external stimuli). Suppose the neurons are sending the same signals in each trial, but starting at different times. Then one should look for *internal* locking of temporal structure in the neuronal signals, rather than locking to external events. One therefore has to try varying the temporal reference point for different trials and testing in what degree spike timing is then reproducible from trial to trial.

Carrying this program out reveals that such structure does exist. In particular, one finds triplets of spikes with interspike intervals up to 100 ms or so occuring reliably in most trials. The second and third spikes in these triplets may come from the same neuron as the first or from different ones. In any case, their relative timing appears precise to within a millisecond or two. In between these spikes are many others whose timing appears random. Thus, it looks as if a small part of the spike train (e.g. 3 spikes out of 30 or so in the period of enhanced activity just before the hand movement) has surprisingly precise temporal structure, but the rest still looks random.

But there is more to the story. One can find *different* shifts of the internal reference times for different trials, for which different triplets appear to repeat. That is, it is as if the neuronal signals contains many of these triplets, each with precise internal timing, but initiated at times which are random with respect to each other. In some cases, it proves possible to account for almost all the activity in a 500-ms long portion of the data in terms of such triplets. Furthermore, the data can be described in this way for both the "GO" and "NO-GO" blocks of trials, but the particular triplets found for the two conditions are different, suggesting that these signals are functionally significant, not just some side effect of the cortical dynamics.

Mathematically, this kind of temporal structure is revealed in singular structure in the 3-time correlation function

$$C_{ijk}(t_1, t_2, t_3) = \langle S_i(t_1) S_j(t_2) S_k(t_3) \rangle. \tag{1.67}$$

Here i, j, and k label neurons; these indices can in general all be different, but they need not be. In a period with temporal homogeneity (in practice, 500 ms or so during which the overall firing rate is fairly constant), $C_{ijk}(t_1, t_2, t_3)$ is a function only of the time differences, say, $t_2 - t_1$ and $t_3 - t_2$. A perfectly reproducible triplet will appear as a delta-function spike in C at particular values of these time differences.

To estimate $C_{ijk}(t_1 - t_2, t_2 - t_3)$, one simply makes a two-dimensional histogram of spike time differences from the data. Because of the finite sample size, there will be some random variation in the population of the bins, and some large peaks might appear by chance. However, if there are sufficient data to get a good estimate of the probability of such random fluctuations, it is a fairly simple matter to pick out (by eye) the peaks that represent real repeating triplets.

Fig. 1.13 shows an example of such a histogram, with a "skyscraper" popping up among the low buildings, indicating an anomalously frequent occurence of triplets with these particular interspike intervals.

Why is the analysis concentrated on triplets of spikes, rather than pairs, quartets, etc? Pairs are uninteresting for the simple reason that the triplets found appear to contain a broad range of first and second interspike intervals. Quartets and higher-order multiplets would be interesting, but they are are hard to observe for two reasons. First, if succesive spikes in a multiplet are separated by 100 ms or so, we cannot hope to find many large multiplets if the data we analyze only cover 500 ms or so. Second, if we want to look for, say, quartets, we have to construct our histogram with a three-dimensional array of bins rather than a two-dimensional one. The correspondingly larger number of bins makes the statistical problem of distinguishing real peaks from chance fluctuations more difficult, and this barrier is reached already for quartets (for the present data set sizes).

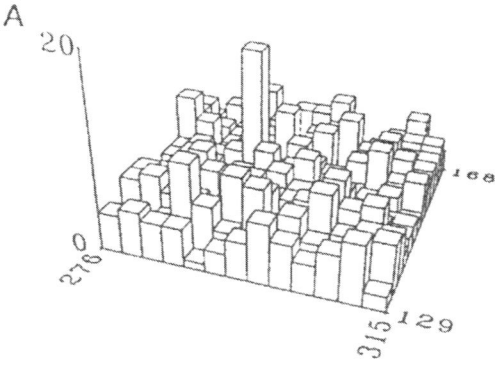

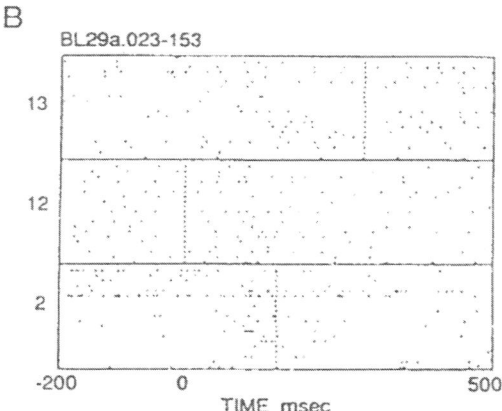

FIGURE 1.13. An example showing a highly synchronized triplet of three correlated firings. In (B), each dot represents a firing of cell 13, 12, or 2 respectively. Each line represent a recording. All spikes are aligned around the first spike of the pattern at the internal clock set to zero. Note that the pattern is composed of isolated spikes and not as a part of bursting neurons. In (A), spike times in windows relative to each other are shown. A single triplet combination of firing times is repeated significantly more often than others. (From ref. 26)

The question of when a peak such as that shown in Fig. 13 is significant is a subtle one. The answer is sensitive to general assumptions about the statistics of the spike trains, such as whether they are Poisson processes. Recent work by the group of Barry Richmond [33] shows examples of repeated patterns which appear to be significant if one assumes Poisson statistics, but which are revealed as not so when the true (empirical) firing statistics are employed in the test.

Perhaps the best test is to jitter the spikes in the data (say, by ± a few ms) and see what happens to the peaks in the histogram for $C_{ijk}(t_1, t_2, t_3)$. If the effect is real, they should disappear when the jittering exceeds the precision of their timing. If the effect is a statistical accident, then new peaks should appear (by similar accidents), as in the data studied by Oram et al [33].

The statistical questions are nontrivial and remain open. However, if these apparent effects are real, they are very interesting from a biophysical and network dynamical point of view.

Dynamical stability

Abeles proposed one way to achieve reproducible temporal structure in a network [2]. Start with a pool of neurons that all fire simultaneously (i.e., within a millisecond or so of each other). Suppose that these are all (or at least mostly) connected by strong excitatory synapses to a second pool, those are connected similarly to a third pool, and so on (Fig. 1.14).

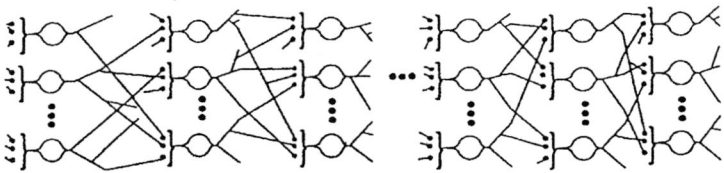

FIGURE 1.14. A synfire chain with incomplete feedforward connectivity. The neurons in each pool are plotted in successive columns. The structure is similar to a multilayered feedforward network. (From ref. 26)

The synchronized firing of each pool will lead to synchronized firing of the succeeding one, and thus to a well-defined wave moving through the system. This is the picture called a "synfire chain." In this form it is essentially a feedforward layered neural network. If the dynamics of this network followed an external clock, the synchronization of the firing in successive pools would be trivial. However, in real life there is no such clock, and the stability of this synchronization becomes an interesting problem. The first analysis of this problem was given by Abeles [2]; the discussion here is based on that.

We make the analysis using leaky integrate-and-fire neurons. The membrane or postsynaptic potential (PSP) $V_i(t)$ of neuron i obeys

$$\frac{dV_i}{dt} + \frac{V_i(t)}{\tau} = \sum_j K_{ij} \sum_m \delta(t - t_{ij}^d - t_j^m). \tag{1.68}$$

Here τ is the membrane leakage time constant, $K_{ij} \equiv J_{ij}/\tau$ rescaled synaptic strengths, t_{ij}^d the transmission delay time between neuron j and neuron i, and t_j^m the time of the mth spike emitted by neuron j. Thus, each spike of neuron j produces a jump in V_i of size K_{ij} (positive for excitatory synapses, negative for inhibitory ones), and in between these jumps V_i decays toward its resting value (defined to be zero) at a rate $1/\tau$. When V_i reaches the neuron's threshold θ_i, the cell fires a spike, the potential is reset to zero, and eqn. (1.68) takes over again.

As we noted in Section 1.2 the PSP does not change discontinuously, since it takes a finite time (of the order of a ms) for the synaptic current to flow (see Eqn. (1.2)). However, the rise and fall time τ_s for the synaptic current is an order or magnitude or so smaller than the membrane time constant τ, so, viewed on the timescale of τ, the jumps in $V_i(t)$ are quite sudden.

If a neuron is a member of one of the pools (call it pool 1) in a synfire chain, as previously described, it receives an unusually strong positive input (i.e., it gets a sudden net increase in PSP) just after the cells in the previous pool (call it pool 0) in the chain fire. If their firing was nearly synchronous and the transmission delay times for all the synaptic connections between cells in the two pools are nearly the same, the total rise in PSP will just be $\sum_j K_{ij}$, where the sum is over the neurons that fired in the previous pool. That is, there is no time for the potential to decay significantly while the incoming volley raises it toward threshold. The condition for this to be true is simply that the spreads in presynaptic firing times and transmission delays, as well as the synaptic current time τ_0, must be small compared to the membrane time constant, which is about 10-20 ms.

We suppose that $\sum_j K_{ij}$ is sufficient to make the cell i fire. Now consider how $V_i(t)$ rises toward threshold in *different* cells i in pool 1, considering first the limit where all the neurons in this pool are identical, all delay times and synaptic strengths from pool 0 to pool 1 are equally strong, and there is no input to pool 1 other than that from pool 0. In that case, $V_i(t)$ is exactly the same for all i, so all the pool 1 cells will fire simultaneously, irrespective of how spread-out the firing times of the pool 0 are, provided only that the leakage decay of the PSP does not weaken its rise so much that it does not reach threshold. Thus, under these conditions, firing synchronization in pool 0 at the 10-ms level would be sufficient to produce perfect synchronization in pool 1.

Now let us relax some of the idealizations made in this argument. First, the synaptic strengths are not uniform (indeed, some of them may be zero), and pool 1 neurons can have different thresholds and receive part of their input from different sources than pool 0. These inhomogeneities produce a variation across the population of pool 1 cells in the value (relative to threshold) of V_i, with some characteristic size ΔV. It is not hard to estimate ΔV from simple models of the inhomogeneity in these parameters; the details are not important here. What is important is that they lead to a spread of firing times in pool 1, equal approximately to

$$(\Delta t)_1 = \frac{\Delta V}{\dot{V}}, \tag{1.69}$$

where $\dot{V}$ is the rate of rise of the PSP, which we estimate as

$$\dot{V} = \frac{\bar{\theta}}{(\Delta t)_0}, \tag{1.70}$$

with $\bar{\theta}$ the average threshold and $(\Delta t)_0$ the spread of firing times in pool 0. Combining Eqns. (1.69) and (1.70), we find

$$(\Delta t)_1 = \frac{\Delta V}{\bar{\theta}} (\Delta t)_0. \tag{1.71}$$

That is (Fig. 1.15), the synchronization is improved from one pool to the next by what is essentially a signal-to-noise factor: the ratio of the PSP needed to reach threshold to the rms PSP fluctuations.

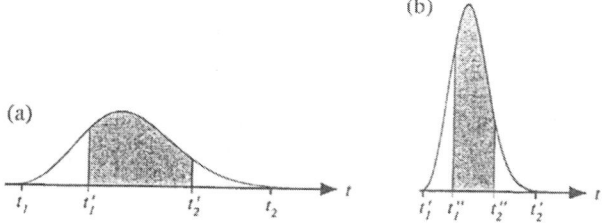

FIGURE 1.15. A schematic representation of the distribution of firing times in consecutive pools of of a synfire chain. (a) pool p, (b) pool $p + 1$. The shaded area show the time interval in which all the neurons in the next pool fire. (From ref. 34)

If this were the whole story, the network would approach perfect synchronization exponentially. What limits this precision is the variability of the transmission delays t_{ij}^{d} and the times τ_{s}^{ij} for synaptic current flow. Calling the rms fluctuations of these Δt^{d} and $\Delta \tau_{\mathrm{s}}$, respectively, putting them into Eq. (1.69) and using Eq. (1.70) again leads to

$$(\Delta t)_{p+1} = \frac{\Delta V}{\bar{\theta}} [(\Delta t)_p + \Delta t^{\mathrm{d}} + \Delta \tau_{\mathrm{s}}], \tag{1.72}$$

where p labels the successive pools in the chain. The extra spreading from Δt^{d} and $\Delta \tau_{\mathrm{s}}$ counteracts the tendency toward better synchronization that we saw in Eq. (1.71). For $p \to \infty$, the resulting precision is

$$(\Delta t)_{\mathrm{eq}} = \frac{\Delta t^{\mathrm{d}} + \Delta \tau_{\mathrm{s}}}{\bar{\theta}/\Delta V - 1}, \tag{1.73}$$

provided the signal-to-noise ratio $\bar{\theta}/\Delta V$ is bigger than 1. We see that the firing can be synchronized much better than the spreads in transmission times or synaptic current flow times if the signal-to-noise ratio is large. This level of timing precision is compatible with known values of transmission delays, synaptic current durations, and membrane potential fluctuations [2, 26, 34]. If the signal-to-noise ratio is less than one, on the other hand, $(\Delta t)_p$ increases from pool to pool and synchronization is lost.

We can now see how ms-level spike timing precision can be achieved despite the initial naive expectation based on the argument that neurons are sloppy temporal integrators. The dynamics in the synfire regime are governed by a different set of characteristic times, notably synaptic current flow times and transmission delays, all in the ms-level range. The membrane time constant τ does not come into play at all, provided the synchrony is maintained at a precision much shorter than τ itself. This only requires that τ be large enough that leakage not significantly weaken the PSP as it rises during the input volley from the previously-firing pool. A larger τ actually helps synchronization, rather than weakening it, as the naive argument would have held. Closely related arguments for a different model have been made by Gerstner et al. [35, 36].

Large signal-to-noise ratios can be achieved if the pool size n is large. If $\bar{K}$ is a typical synaptic strength, the critical signal magnitude $\bar{\theta} = \mathcal{O}(\bar{K}n) \propto n$, while the contribution to ΔV from inhomogeneity in synaptic strength and dilution scales with n like $\sqrt{n}$. As this chapter will show, a part of the contribution to ΔV from noisy inputs also scales like $\sqrt{n}$, while the rest of it is n-independent, as is the part coming from threshold variability. So the signal-to-noise ratio can always be improved by taking larger pools.

Thus, in the large signal-to-noise ratio limit, the system effectively functions as if it had a clock. In the following we will investigate how a synfire network can function as a memory device for encoding external signals.

Recurrent synfire networks: associative memory and capacity

The foregoing description of a synfire chain was essentially as a feedforward network. This is unrealistic; cortical processing is done in networks with extensive interconnection and feedback. Here we examine a simple model that incorporates feedback but is still easily soluble [34, 37, 38].

This model encodes stimuli as synfire chains. That is, one imagines there is a set of stimuli, each of which activates a certain subset of neurons in the network. From these initial conditions, the network evolves through a particular synfire sequence of states. The basic idea is that these spatiotemporal activity patterns are the encodings of the stimuli, to be read by later processing stages. We do not concern ourselves here with how this reading is to be carried out. Our only concerns are that

- The different stimuli should be encoded by distinguishable synfire sequences. This is necessary if the rest of the brain is to be able to distinguish between the different stimuli.

- Small changes in a stimulus should not lead to different encodings, i.e., the network should perform error correction on its inputs. To put it another way, synfire chains should be *attractors* of the dynamics.

Of course, these two demands are complementary: the better error correction we insist on, the fewer distinct stimuli the network can encode.

The model we study here is related to the previously-discused Tsodyks-Feigel'man sparse associative memory model. We have a set of N neurons $S_i(t) = 0, 1$ obeying the equations of motion

$$S_i(t + 1) = \Theta[\sum_j J_{ij} S_j(t) - \theta], \tag{1.74}$$

with synaptic strengths J_{ij}, threshold θ, and $\Theta(\ )$ the unit step function. We are now taking time to be digital; that is, we are assuming the synchronization mechanism described in the preceding section provides us with an effective clock. Note that these neurons are not integrate-and-fire ones like those described by Eq. (1.68); they have no memory of the previous time step at all. We will see later how to take account of the possible buildup of PSP from previous steps (at least approximately); for now we stick to the simple memoryless model.

The synaptic weights J_{ij} are constructed in such a way as to make particular synfire sequences $\{\xi_i^\mu(t)\}$, $\mu = 1, \ldots, P$ attractors of the dynamics. The sequences are each taken to be T time steps long ($t = 0, \ldots, T$). We take each $\xi_i^\mu(t)$ to be randomly 1 or 0 with probabilities f and $1 - f$, respectively, independently for different t or μ. As before, the sparseness f is taken of order $1/\sqrt{N}$. (For application to real cortical tissue f should be $\mathcal{O}(10^{-3})$, so the pool size for a single step in the sequence n is of order 100). We take the total number PT of pools to be of order N and define the load parameter as

$$\alpha = \frac{PT}{N}. \tag{1.75}$$

The formula for the synaptic weights is

$$J_{ij} = \frac{J}{n} \sum_{\mu=1}^{P} \sum_{t=0}^{T} \xi_i^\mu(t+1)\xi_j^\mu(t). \tag{1.76}$$

That is, there are excitatory connections of strength J between successive pools (labeled by the index t) for each sequence μ. As previously described, this tends to make the different pools active in the specified sequences. Formula (1.76) includes only one-step synaptic delays. A more complete model with independently-fixed synapses for all delays [39] can store correspondingly more patterns.

The balancing of these excitatory couplings could be done with a constant global inhibition term as in (1.56), but here we will do something even simpler. (The results we will be interested in will be the same.) We employ what we can call an n-winner-take all (n-WTA) model. In this model, there is no explicit inhibition. Instead, at each time step, the threshold is adjusted so that exactly n neurons fire. This essentially enforces a constraint $Q = 1$ (or, for states with s chains simultaneously active, $Q = s$), simplifying the mean field theory a bit.

To derive the mean field theory, let us suppose that a fraction m of the neurons in pool 1 in sequence 1 are active. We compute the PSP in Eqn. (1.74):

$$h_i = \sum_j J_{ij} S_j = \frac{J}{n} \sum_{\mu,\tau,j} \xi_i^\mu(\tau+1)\xi_j^\mu(\tau) S_j$$

$$= \frac{J}{n} \left[\sum_j \xi_j^1(1) S_j \right] \xi_i^1(2) + \frac{J}{n} \sum_{\{\mu,\tau\}\neq\{1,1\},j} \xi_i^\mu(\tau+1)\xi_j^\mu(\tau) S_j$$

$$= Jm\xi_i^1(2) + \delta h_i. \tag{1.77}$$

As in our previous problem, in the limit of small f the quantity δh_i is a Gaussian random variable with variance

$$\langle \delta h^2 \rangle = J^2 \alpha f(1 + nf) \tag{1.78}$$

The meaning of Eqn. (1.77) is this: The neurons in pool 2 of the active sequence (1) have a strong excitatory PSP because they get input from the currently active pool 1. In addition, every neuron, whether in this pool or not, has an additional random PSP which originates from the random overlaps between different pools in the second line of Eqn. (1.77). Because the standard deviation of this effective noise is of $\mathcal{O}(1)$, some neurons in the pool that should fire next have subthreshold PSPs, while others not in that pool have high enough PSPs to fire. For a given threshold θ, this reduces the average pool activity m at the next time step while increasing that of neurons outside the active pool. As time goes on, m changes and settles down to a fixed point value. If at some time m is reduced (say, by some extrinsic noise) to a value slightly below its fixed-point one, it will increase on the next step and approach the fixed-point value again.

Perfect propagation of a prescribed synfire chain $\{S_i^\mu(t)\}$ is not possible, since the Gaussian PSP distributions have tails extending out to $\pm\infty$. However, the fixed point value of m may not be too much less than 1.

We can now write the mean-field equation for m:

$$m = \mathrm{H}\left(\frac{\theta - Jm}{\sqrt{J^2 \alpha f(1 + nf)}}\right). \tag{1.79}$$

This just says that m is the fraction of the pool which should be active that actually is above threshold. The threshold is adjusted so that the total activity (normalized to the pool size),

$$Q = m + \frac{1}{f}\mathrm{H}\left(\frac{\theta}{\sqrt{J^2 \alpha f(1 + nf)}}\right) = 1. \tag{1.80}$$

These equations are exact in the limit of a large network ($N \to \infty$, $f = \mathcal{O}(1/\sqrt{N})$, $n = \mathcal{O}(\sqrt{N})$.) They are the same as for for static patterns; here we simply have to measure the overlap m at time t with respect to the pool $\{\xi_i^\mu(t)\}$ that should be active at that time (working in the "moving frame").

It is apparent that if we try to make the network capable of running through either too many or too long sequences, the effective noise variance (1.78) will grow, and its performance (as measured by the order parameter m) will be degraded. There is a capacity, a maximum α_c of the load parameter (1.75) beyond which the network will break down, just as in the static-pattern model. The estimation of α_c is the same as before. However, this is a capacity for the total number of steps in all the pattern sequences, not the total number of sequences. Thus, one pays a price in reduced number of possible stored sequences proportional to their length.

This calculation can be extended simply in several directions. First, it is simple to add extrinsic noise to the intrinsic term δh_i in Eqn. (1.77) (representing inputs to our local cortical network from the rest of the brain). Let this extra noise have variance B^2. Then the modified signal-to-noise criterion reads

$$J^2 \alpha_c f(1 + nf) + B^2 = \frac{J^2}{\log(1/f^2)}. \tag{1.81}$$

leading to a reduced capacity

$$\alpha_c = \frac{1 - (B^2/J^2)\log(1/f^2)}{f(1 + nf)\log(1/f^2)}. \tag{1.82}$$

A little noise goes a long way: to give a finite α_c for $f \to 0$ requires that B^2 scale like $1/\log(1/f^2)$.

It is also possible to estimate qualitatively how the capacity would be changed if we replaced the memoryless neurons used in this analysis with integrate-and-fire ones, which accumulate changes in their PSP over time. Returning to the case of zero extrinsic noise, an effective noise with variance given by Eqn. (1.78) acts for the entire time between firings, and in this interval the PSP obeys

$$h_i(t+1) = [h_i(t) + \delta h_i(t)]e^{-1/\tau}, \tag{1.83}$$

so the accumulating variance $A(t)$ evolves according to

$$A(t+1) = [A(t) + A_0]e^{-2/\tau}, \tag{1.84}$$

with $A_0 = J^2 \alpha f(1 + nf)$. In the limit $t \gg \tau \gg 1$, $A(t) \longrightarrow A_0 \tau/2$, i.e., the effective accumulated noise variance saturates at a value $\tau/2$ times greater than the one-step noise variance. Thus, if the interval between firings is much greater than τ (as it will certainly be for small enough f), the capacity is simply reduced from the previous estimate by this factor (about 5 in real cortex if we take the basic time step as 1 ms and τ as 10 ms).

Hebbian learning of synfire chains

In real life, synaptic weights are determined by a self-organizing process that depends on the experience of the animal. Hebb [4] proposed that a general principle of this process might be that synapses that got "used" (i.e., for which presynaptic firing was followed by postsynaptic firing) would get strengthened at the expense of those that were not. This temporal asymmetry makes learning sequences, rather than stationary attractors, natural [19, 40, 41, 42], and a statistical analysis of this dynamical process is possible [43].

Consider a network described by our n-WTA model, with initially random synapses. Suppose it is put into some initial state by an external stimulus. (We assume for the sake of simplicity here that the stimulus is short-lived, and that its only relevant effect on our network is fixing its initial condition. This condition will have to be relaxed in future, more realistic treatments.) The network will then evolve through a sequence of states $S_i^1(t)$. A second stimulus will produce a different sequence, $S_i^2(t)$, and so on. These sequences will be almost completely different, because the dynamics of a network with completely random synapses is chaotic (cf. Sect. 1.3). Thus, even stimuli which are quite similar (i.e., which put the network into initial conditions which are quite similar) will lead to sequences which are quite different after a few time steps. These sequences are not suitable as synfire chains for just this reason.

However, with Hebbian learning, the synapses from the sets of neurons active at one time step to those active at the next will be strengthened, and repetitions of the stimulus will strengthen them further. This can make these sequences into synfire attractors that encode the different stimuli.

Of course, the net learning strength should not be too small (the chains will not be stable attractors in the presence of noise), or too big (the basins of attraction of different chains will overlap, with the result that the corresponding stimuli cannot be distinguished).

In his thesis [13], Thomas Lauritzen studied synfire chain propagation in networks of Hodgkin-Huxley neurons. He started with balanced networks of the type discussed in Sect. 1.3 (500 neurons, 12 randomly chosen excitatory and 12 randomly chosen inhibitory inputs to every neuron). After dividing the 250 excitatory neurons arbitrarily into 25 pools of 10 neurons each, he then ordered the pools arbitrarily and replaced 7 of the 12 excitatory inputs each cell received by ones from neurons in the preceding pool. Thus a structure similar to Fig.. 1.14 was built up, but retaining a good deal of the randomness of the original net.

He then repeated the measurements of the Hamming distance between the network and a copy in which one neuron is made to fire prematurely. Fig. 1.16 shows that the two copies remain in nearly the same state for about 80 ms before they are driven apart. Hence, hand-constructed synfire chains can fire reliably, also in more realistic model networks such as this one.

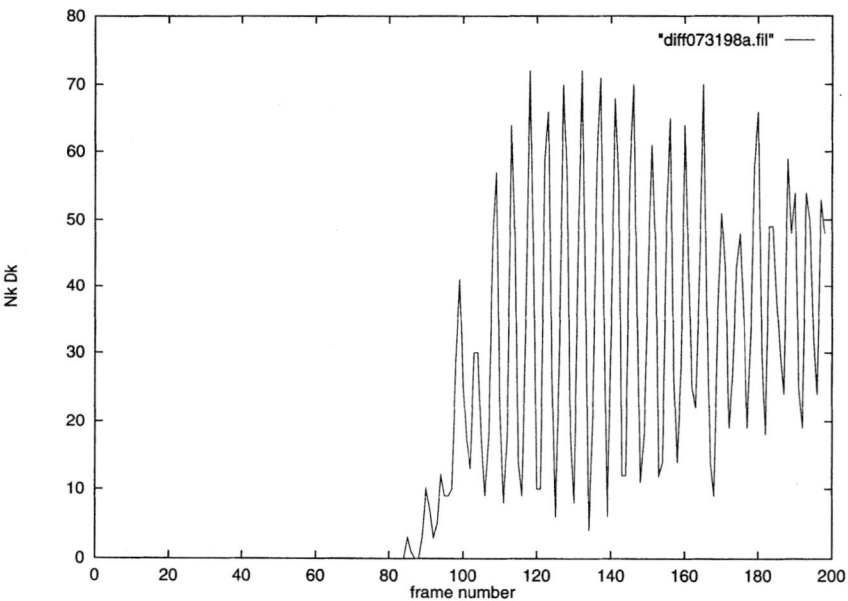

FIGURE 1.16. $N_k D_k$ for the synfire net. The time frames are 2ms wide. Perturbation of the net occurs at t = 0.1 s, corresponding to time frame number 50. (From ref. 13)

Finally, he studied whether synfire chains might begin to form through unsupervised Hebbian learning. To do this, he started again with a random balanced net and identified the sets of excitatory neurons that fired in the first 2 ms (pool 1), the next 2 ms (pool 2), and so on. He then made or strengthened the synapses between neurons in successive pools defined this way. Increases of 25% and 50% relative to the synaptic strengths in the original net were tried. While this did not lead to stable long chains,

Hamming-distance measurements like those in Fig. 16 did show a slower growth of $D(t)$ than in the completely random net. Fig. 1.17 shows how the neurons within a pool became better synchronized after learning, indicative of an incipient synfire chain formation.

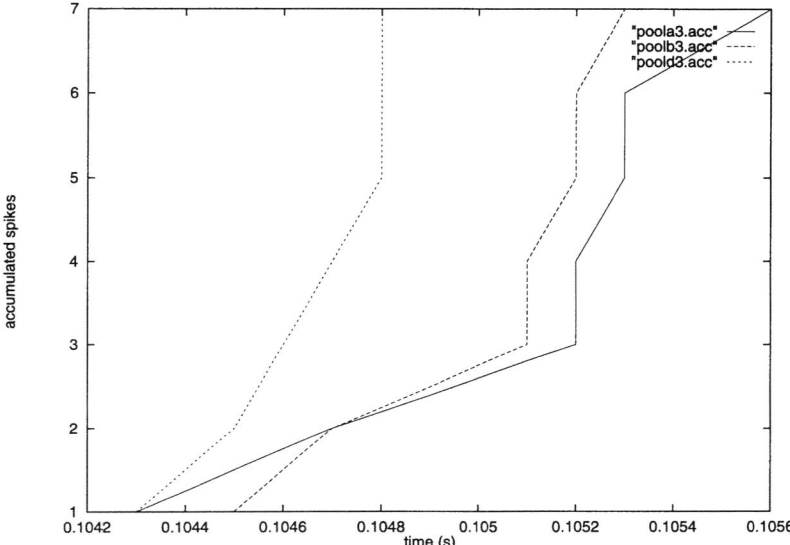

FIGURE 1.17. Accumulated spikes in the third firing pool. The firing is plotted for three simulations: No Hebb connections, solid line; Hebb connections with 20% of regular synaptic strength, long dashed; and Hebb connections with 50% of regular synaptic strength, short dashed. (From ref. 13)

In his thesis, Andrea Fazzini [14] took this line of investigation a step closer to biology. Using a more realistic learning rule (including synaptic adaptation and reduction of synaptic effectiveness when the postsynaptic firing precedes the presynaptic one [44]), he studied the learning of spatiotemporal patterns induced by an external stimulus to an originally-chaotic network. Repeated synchronized stimulation of an initial pool of neurons of the right size turns out to lead to highly reproducible firing patterns.

1.5 Concluding Remarks

The problems we have studied here, from the single-neuron level to that of cognitive computation, illustrate a number of the fundamental issues in neural modeling and neural computation. In particular, they show the need for different kinds of models and methods, both theoretical and computational, at different levels of description and for connecting our understanding at different levels.

In such a wide field the choice of specific problems to address in an introductory article is to some extent arbitrary. It is influenced to some degree by the author's current research interests. In the present case, this article originated as lecture notes at a summer school, so the choice of topics was also partially determined by what other lecturers were covering. Nevertheless, I hope that the treatment here may serve as a useful introduction to neural modeling problems at several levels and as background for readers' own further investigations.

1.6   Acknowledgments

It is a pleasure to thank a long list of collaborators: Andrius Bernotas, Andrea Fazzini, Michael Herrmann, Thomas Lauritzen, Adam Prügel-Bennett, and Sergio Solinas, whose work was responsible for a large part of what I could present here.

This article was originally prepared as lecture notes for the 1999 School on Neural Information Processing at the Abdus Salam International Centre for Theoretical Physics in Trieste. I would like to thank the co-organizers Sara Solla and Riccardo Zecchina, Miguel Virasoro (who had the original idea to hold the school), and all the students for interesting discussions and feedback.

1.7   REFERENCES

[1] E. Kandel, J. H . Schwartz, and T. M. Jessel, *Principles of Neural Science* (Elsevier, New York, 1991).

[2] M. Abeles, *Corticonics: Neuronal Circuits of the Cerebral Cortex* (Cambridge University Press, Cambridge, 1991).

[3] A. W. Roe, S. L. Pallas, J-O. Hahm and M. Sur, *Science* **250** 818-820 (1990).

[4] D. O. Hebb, *The Organization of Behavior* (Wiley, New York, 1949).

[5] L. Abbott and P. Dayan, *Theoretical Neuroscience*, MIT Press, to be published (2000).

[6] T. M. Cover and J. A. Thomas, *Elements of Information Theory* (Wiley, New York, 1992).

[7] R. Rieke, D. Warland, R de Ruyter van Steveninck and W. Bialek, *Spikes: Exploring the Neural Code* (MIT Press, Cambridge, MA, 1997).

[8] D. J. Amit and N. Brunel, *Cerebral Cortex* **7** 237 (1997); *Network* **8** 373 (1997).

[9] C. van Vreeswijk and H. Sompolinsky, *Science* **274** 1724 (1996); *Neural Computation* **10** 1321 (1998).

[10] J. Keener and J. Sneyd, *Mathematical Physiology* (Springer, New York, 1998) Chapter 4.

[11] M. N. Shadlen and W. T. Newsome, *Curr Opin Neurobiol* **4** 569-579 (1994); *ibid* **5** 248-250 (1995).

[12] S. Solinas, Laurea thesis, Univ. Bologna (1999).

[13] T. Z. Lauritzen, MS thesis, Copenhagen Univ. (1998).

[14] A. Fazzini, Laurea thesis, Univ. Bologna (1999).

[15] J. Hopfield, *Proc Nat Acad Sci USA* **79** 2554-2558 (1982), **81** 3088-3092 (1984).

[16] J. Hertz, A. Krogh and R. G. Palmer, *Introduction to the Theory of Neural Computation* (Addison-Wesley, Redwood City, CA, 1991).

[17] T. Geszti, *Physical Models of Neural Networks* (World Scientific, Singapore, 1991).

[18] D. J. Amit, H. Gutfreund and H. Sompolinsky, *Phys Rev Lett* **55** 1530-1533 (1985); *Ann Phys (NY)* **173** 30-67 (1987).

[19] J. L. van Hemmen and R. Kühn in: *Models of Neural Networks I*, E Domany, J L van Hemmen, and K Schulten, eds (Springer, Berlin, 1995) Chapters 1 & 7.

[20] H. Steffan and R. Kühn, *Z Phys* **95** 249-260 (1994).

[21] D. J. Willshaw, O. P. Buneman and H. C. Longuet-Higgins, *Nature* **222** 960 (1969).

[22] M. V. Tsodyks and M. V. Feigel'man, *Europhys Lett* **6** 101 (1988); M V Tsodyks, *ibid* **7** 203; *Mod Phys Lett B* **3** 555 (1989).

[23] Y. Miyashita and H. S. Chang, *Nature* **331** 68 (1988); Y. Miyashita, *ibid* **335** 817 (1988); D. J. Amit, N. Brunel and M. Tsodyks, *J Neurosci* **14** 6435-6445 (1994).

[24] D. M. MacKay and W. S. McCulloch, *Bull Math Biophys* **14** 127-135 (1952).

[25] W. Kruse and R. Eckhorn, *Proc Nat Acad Sci USA* **93** 6112-6117 (1996).

[26] M. Abeles, E. Vaadia, H. Bergman, Y. Prut, I. Haalman and H. Slovin, *Concepts in Neuroscience* **4** 131-158 (1993).

[27] M. Abeles, H. Bergmann, E. Margalit and E. Vaadia, *J Neurophysiol* **70** 1629-1638 (1993).

[28] A. Riehle, S. Grün, M. Diesmann and A. Aertsen, *Science* **278** 1950-1953 (1997).

[29] Z. Mainen and T. Sejnowski, *Science* **268** 1503-1506 (1995) .

[30] M. Meister, L. Lagnado and D. Baylor, *Science* **270** 1207-1210 (1995).

[31] R. C. deCharms and M. M. Merzenich, *Nature* **381** 610-613 (1996).

[32] J.-M. Alonso, W. Usrey and R. Reid, *Nature* **383** 815-819 (1996).

[33] M. W. Oram, M. C. Weiner, R. Lestienne and B. J. Richmond, *J Neurophysiol* **81** 3021-3033 (1999).

[34] M. Herrmann, J. Hertz and A. Prügel-Bennett, *Network: Computation in Neural Systems* **6** 403-414 (1995).

[35] W. Gerstner, *Phys Rev E* **51** 738-758 (1995).

[36] W. Gerstner, J. L. van Hemmen and J. D. Cowan, *Neural Comp.* **8** 1653-1676 (1996).

[37] J. Hertz and A. Prügel-Bennett, *Int J Neural Systems* **7** 445-450 (1996).

[38] J. Hertz, pp 135-144 in *Theoretical Aspects of Neural Computation*, K-Y. M. Wong, I. King and D-Y. Yeung, eds (Springer, Berlin, 1998).

[39] A. V. M. Herz, Z. Li and J. L. van Hemmen, *Phys Rev Lett* **66** 1370-1373 (1991).

[40] A. V. M. Herz, B. Sulzer, R. Kühn and J. L. van Hemmen, *Europhys Lett* **7** 663-669 (1988); *Biol Cybern* **60** 457-467 (1989).

[41] W. Gerstner, R. Ritz, and J. L. van Hemmen, *Biol Cybern* **69** 503-515 (1993).

[42] E. Bienenstock, *Network: Computation in Neural Systems* **6** 179-224 (1995).

[43] R. Kempter, W. Gerstner and J. L. van Hemmen, *Phys Rev E* **59** 4498-4514 (1999).

[44] H. Markram and M. Tsodyks, *Nature* **382** 807-810 (1996).

2

The Gaze Control System

John van Opstal

ABSTRACT The geometrical arrangement of the extra-ocular muscles
enables rotations of the eyes in three dimensions (3D). Helmholtz [29] has
noted that the oculomotor system should therefore account for the noncommutative properties of 3D rotations. This principle entails that, in a
series of rotations, the final orientation of the eye depends on the order in
which they are generated. As a consequence, this property could jeopardize
the stability of the perceived visual world and greatly complicate visuomotor control. However, for head-fixed saccadic eye movements the problem is
elegantly circumvented by Listing's law, which imposes precise geometric
constraints on the ocular rotation axis. Recent findings have shown that
somewhat different constraints, described by Donders' law, apply to conditions where both the eye and head are free to move [11], [79], [71]. A
current controversy is whether these constraints are due to a neural control strategy [42], [51], [57], [75], [71], [72], or whether they result mainly
from passive mechanical properties of the motor plants in combination with
commutative neural controllers [60], [61], [66]. In this chapter, a review is
given of the 3D kinematic principles of rotations that underlie the control
of saccadic gaze shifts. It is argued that the experimental evidence strongly
supports the notion that both Donders' and Listing's law have a major
neural component.

2.1 Introduction

Due to the highly inhomogeneous distribution of photo receptors on the
retina, optimal visual resolution is only achieved for targets projected on the
central retina (the fovea). In primates, the fovea has an effective diameter of
less than a degree. Therefore, whenever the primate visual system decides
to further explore the details of a peripheral visual target, the current gaze
line must be precisely redirected at the object. Such a gaze shift is generated
by the saccadic gaze control system.
The saccadic orienting response is extremely rapid (peak velocities in monkey may exceed 1000 deg/s) and flexible, since in general, goal-directed
gaze shifts may be achieved by an eye movement only, or by a myriad of
different combinations of eye-, head-, and body rotations. Because of its
clear function and apparent simplicity, the gaze control system has been
studied extensively in many laboratories over the past two decades, and

experimental data have been accounted for by an increasing number of detailed quantitative models.

Well into the seventies and early eighties, most studies were confined to gaze shifts in one dimension (1D; typically horizontal). Initially, recordings were made of head-fixed ocular saccades only, but later also horizontal head-movements were incorporated in these early studies [1], [25].

Additional new insights into the saccadic system emerged when movements were recorded with two degrees of freedom (2D; horizontal and vertical) [21] [52] [23]. In recent years, however, the study of the gaze control system has been further extended to incorporate all three dimensions simultaneously (3D; horizontal, vertical, and torsional), and allowing for both the eye(s) (e.g. [15], [30], [47], [59], [75]) and head [26], [70], [79] to move freely. This paper will be specifically concerned with the concepts and results from these 3D studies.[1]

2.2   The Gaze Control System in One and Two Dimensions

The two important concepts dominate the visuomotor literature, and are important for the present discussion, are the notions of *internal feedback* and *velocity integration*.

Internal feedback in saccade programming:

When a visual stimulus is presented on the peripheral retina, the saccadic system might program an accurate eye movement to foveate the target, solely on the basis of the retinal error signal (defined as the difference between the current gaze direction and the desired gaze direction). Based on the following considerations, however, the retinal error, although sufficient at first glance, is not the only source of information on which the saccadic programmer relies:

- Saccades accurately compensate for any intervening eye movement that may occur between target presentation and saccade initiation. For example, in a flashed double-step paradigm, the subject has to generate saccades to two briefly flashed targets, both extinguished well before the first movement onset. Despite the absence of visual feedback, the two saccades accurately land on both targets. A second saccade is even accurate when the second target is flashed *during* the first saccade (e.g. [35]).

---

[1]An excellent and nontechnical state-of-the-art review of the topics covered in this paper, and more, is provided in the proceedings of a recent workshop on 3D eye-, head-, and limb movements, held in Tübingen, Germany, 1995 [16].

- Monkeys can generate an accurate eye movement to an extinguished flashed target in darkness, even after the eyes have been driven away from their initial position, within the reaction time period, by electrical stimulation of the midbrain superior colliculus (SC; [49])

- Accurate saccades can be generated to auditory targets in darkness, although the acoustic frame of reference is head- centered, rather than eye-centered (e.g. [17]). For this, the saccadic system needs to know the absolute position of the eyes in the orbit.

Internal feedback in saccade generation

It was further noted by Robinson and colleagues [24], [64] that visual feedback is far too slow to be of any use for the accurate control of rapid eye movements (visual delays are in the same order of magnitude as typical saccade durations, roughly 50-60 ms). Since normal saccades have quite stereotyped kinematics with relatively little scatter, the possibility exists that they are generated by a preprogrammed pattern generator in the brainstem. However, strong additional evidence supports the idea that also the brainstem saccade generator is controlled by internal feedback:

- Saccades remain accurate despite a considerable, e.g., drug-induced or fatigue-related, variability in their kinematics (see [41]).
- Saccades that have been interrupted in midflight (by brief electrical microstimulation of the saccade gating system, embodied by the brainstem omnipause neurons (OPNs, [43]) and the rostral SC [50]) accurately re-acquire the target in complete darkness.

Velocity integration

The motoneuron signal that innervates the horizontal extraocular muscles is well characterized by a pulse-step activity pattern. The pulse is an intensive, phasic burst of action potentials, and is derived from short-lead burst neurons in the paramedian pontine reticular formation [24]. The output of these latter neurons, $b(t)$, correlates well with a linear combination of instantaneous eye velocity, $\dot{e}(t)$, and acceleration, $\ddot{e}(t)$, and is needed to rapidly accelerate the plant against the viscous drag of surrounding tissues [24]. These premotor neurons are considered to embody the horizontal saccade generator, because lesions in this area completely and specifically abolish the occurrence of horizontal saccades [36].

The static elastic forces of the muscles are overcome by an eye-position related tonic innervation (the step). Robinson proposed that such a signal could be obtained from the burst neurons by time-integrating the phasic burst [24], [64]:

$$n(t) = \int_{t_0}^{t_1} b(t) \cdot dt \;\; + \;\; n(t_o) \cdot \exp(-t/T) \qquad (2.1)$$

The existence of such a neural integrator stage has been convincingly confirmed by lesion studies of the nucleus prepositus hypoglossi (NPH) and the medial vestibular nuclei (MVN). Under normal conditions, the neural integrator (NI) has a time constant of about $T=20$ s; lesions in this area, however, cause the time constant for horizontal fixations to drop well below 1 s [7].

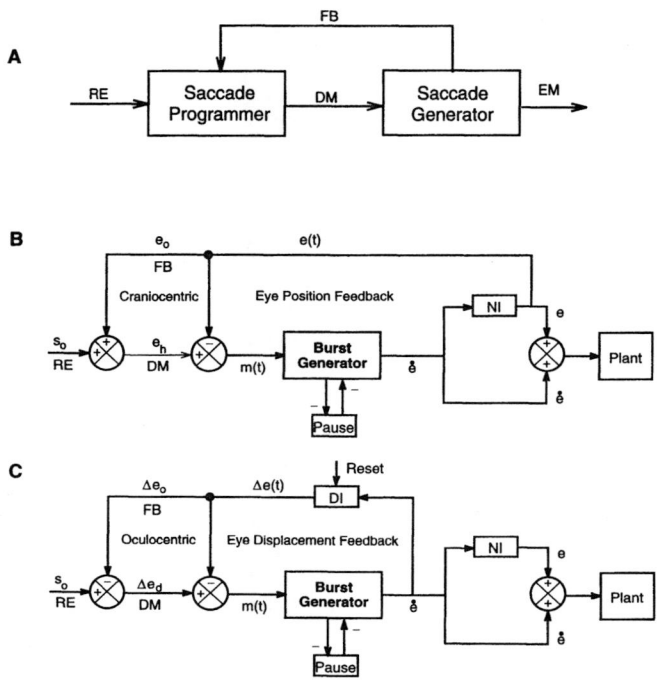

FIGURE 2.1: (**A**) Overall organization of the saccadic system. The saccade programmer issues a desired movement command (DM) to the brainstem saccade generator. The programmer may update DM on the basis of feedback (FB) about intervening eye movements (EM). Also the generator is driven by local feedback (not shown). RE: retinal error. (**B**) In Robinson's eye position feedback model [64], the desired movement command is eye position in the head (e_h). Feedback from the neural integrator (NI) about inital (e_o) and current absolute eye position $(e(t))$ is provided to the programmer and the generator. The timing of the latter is controlled by an inhibitory gating mechanism embodied by pause neurons. s_o: retinal error; $m(t)$: current motor error; $\dot{e}$: eye velocity. (**C**) Eye displacement feedback model of Jürgens et al. [41]. The eye displacement signal (initial: Δe_o, current: $\Delta e(t)$) is computed by a resettable displacement integrator (DI).

The 1D oculomotor plant.

In models of the saccadic system (Fig. 2.1B,C), the brainstem pulse-step signal matches the dynamics of the horizontal plant. This is achieved by letting the net innervation of the oculomotor neurons, given by a linearly weighted sum of the NI ($n(t)$) and burst-cell outputs ($b(t)$), precisely counteract the muscle stiffness, K, tissue viscosity, V, and plant inertia, J, respectively. That is, if the pulse-step motoneuron innervation, MN(t), is given by:

$$MN(t) = C \cdot n(t) + D \cdot b(t) \tag{2.2}$$

with C and D representing an internal model (as neural weights) of the plant's dynamics, then the actual eye velocity of the system is given by:

$$\dot{e}(t) = \frac{1}{V} \cdot [MN(t) - K \cdot e(t) - J \cdot \ddot{e}(t)] \tag{2.3}$$

For the system to have matched dynamics, it follows that $n=(K/C)e$, and $b=(1/D)(V\dot{e} + J\ddot{e})$, respectively. When the inertia of the globe may be ignored ($J \approx 0$), the burst neurons simply encode eye velocity according to $b=(V/D)\dot{e}$.

One-dimensional models

Since visual feedback plays no role in the experimental observations described above, it was hypothesized by Robinson that the saccade generator is driven by an internal feedback circuit in which a desired eye movement (issued by the saccade programmer) is continuously compared with an efference copy of the actual movement [24], [64].
In the current literature there is some controversy on the nature of the signals involved in the different feedback mechanisms. In Robinson's original local feedback model [64], the saccadic burst generator is supposed to be driven by a neural estimate of the dynamic motor error, $m(t)$, by subtracting the desired eye position in the head, e_h, from an efference copy of current eye position, $e(t)$:

$$m(t) = e_h - e(t) \tag{2.4}$$

The latter signal is derived from the position neural integrator (NI, see Eqn. (2.1); Fig. 2.1B). The former signal was obtained in the model by adding the retinal error of the target, s_o, with a neural estimate of the initial eye position, e_o (sampled from the initial state of the NI, $n(t_o)$):

$$e_h = s_o + e_o \tag{2.5}$$

A problem with the position feedback model, however, was the lack of evidence for a neural representation of e_h. In addition, the model did not provide a role for the SC, which was generally held to be part of a crucial stage in saccade visuomotor programming, because it provides direct input to the burst cells. Yet the neurophysiological evidence has indicated that the SC rather seemed to encode a desired static displacement signal for the eyes, Δe_d, irrespective of initial eye position [63].

Therefore, recent displacement feedback models proposed that the dynamic motor error is obtained by comparing the collicular signal with an efference copy of actual eye *displacement* (e.g. [41]; Fig. 2.1C):

$$m(t) = \Delta e_d - \Delta e(t) \qquad (2.6)$$

The latter signal is produced by a feedback resettable displacement integrator (DI), which integrates eye velocity, just like the position NI, but needs to be reset to zero after each saccade. To account for the double-step results, and the intersaccadic eye-displacement experiments (see above), the former signal was not directly taken as the initial retinal error of the target, s_o, but was assumed to take the displacement of the previous saccade, Δe_o, into account, by *remapping* the target into oculocentric coordinates:

$$\Delta e_d = s_o - \Delta e_o \qquad (2.7)$$

Although the revised model solved one problem (it accommodated the SC), it created a new one (lack of neurophysiological support for a DI).

Two-dimensional ocular control

The vertical saccadic pulse-step generator in the brainstem consists of short-lead burst cells in the rostral interstitial nucleus of the medial longitudinal fasciculus (riMLF) [40], and the neural integrator in the interstitial nucleus of Cajal (iC) [4]. Neurophysiology has revealed that cells in this burst generator are actually tuned to movements in the vertical/torsional plane ([34] for review, and [4], [10]). Both saccade generators receive a common input command from the SC, which represents a desired displacement of the eye (Δe_d) irrespective of the initial eye position (see also above). The SC signal is spatially encoded in a topographic motor map, in the sense that neighbouring recruited regions encode similar saccade vectors by their *location* within the map, rather than by the intensity of the neural activity (e.g., [37], [54], [63]). The transformation of this oculocentric *vectorial* signal into the appropriate innervation patterns of the extraocular muscles, has become known as the *spatial-temporal* transformation stage, and involves two different processes:

(a)   Vector decomposition (VD) into the horizontal and vertical saccade components.

(b)   Pulse generation (PG) of motor error, $m(t)$, into an eye velocity command, $\dot{e}(t)$.

Because the latter stage is assumed to be nonlinear (described by a saturating function that accounts for the observed saccade kinematics:

$\dot{e} = a[1 - \exp(-b \cdot m)]$; [23], [24], [64]), the order in which these two processes are implemented matters (i.e., it is a noncommutative operation: VD·PG $\neq$ PG·VD).

An interesting question therefore is: are the horizontal and vertical components of an oblique saccade generated by independent pulse generators, or are they somehow coupled?

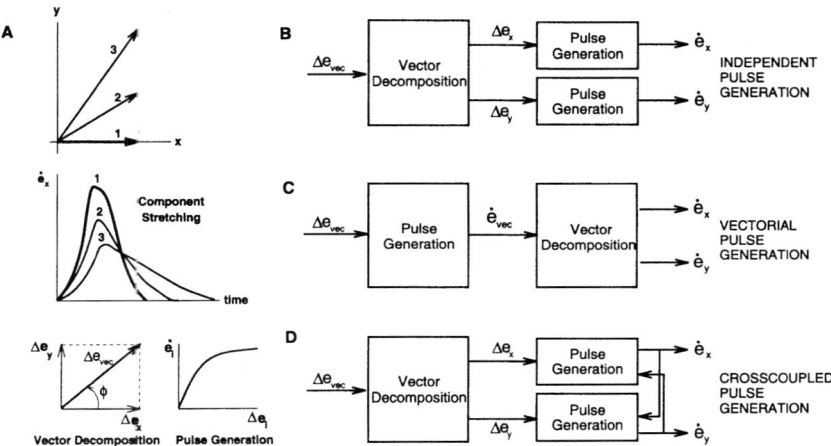

FIGURE 2.2: Two-dimensional extensions of the saccadic system should account for the fact that oblique saccades are approximately straight, and that the peak velocity and duration of horizontal and vertical saccade components systematically depend on saccade direction, ϕ ('stretching') (A). Note that the order of the vector decomposition (linear) and pulse generation (nonlinear) processes matters. (B) In the Independent model, vector decomposition precedes pulse generation. This model therefore predicts that component velocity is independent of saccade direction (no stretching). (C) In the Common Source scheme, a vectorial pulse generator precedes the decomposition of the vectorial eye velocity signal. (D) The Crosscoupling scheme resembles the Independent model, but the pulse generators are mutually coupled. The latter two models produce both stretching and straight saccades.

A detailed behavioral and model study of monkey oblique saccades revealed that both saccade components are tightly synchronized and coupled such that approximately straight saccades are elicited in all directions [23]. For example, the kinematics of a fixed horizontal component, when compared for a set of oblique saccades in different directions (having different vectorial amplitudes and vertical components), systematically depend on the direction of the saccade vector. The peak velocities of the horizontal and vertical components of the saccade vector depend on the saccade direction, such that they have matched durations (*component stretching*; Fig. 2.2A). In this way, a given component always reaches its highest peak velocity when executed in isolation (either purely horizontal (nr. 1 in Fig. 2.2A), or purely vertical movements) [23], [52]. It is generally believed that this phenomenon reflects a property of the neural organization of the saccade generator.

Although this finding excludes an independent control of the horizontal and vertical burst generators (Fig. 2.2B), two conceptually different schemes coexist that both account for these findings (see Fig. 2.2C,D). In the *cross-coupling scheme* (Fig. 2.2D), the horizontal and vertical burst-generators receive a decomposed horizontal/vertical motor error input from the SC, like in the independent scheme, but mutually couple their velocity outputs at the brainstem level [52]. In contrast, in the so-called *common source scheme* (Fig. 2.2C), a vectorial burst generator issues a vectorially encoded pulse command, that is subsequently decomposed into the respective horizontal and vertical velocity components [23]. Note, that straight saccades and stretching are emergent properties of this latter model, whereas they are a specific design feature in the crosscoupling model.

Eye-head movement control in 1D and 2D

When the head is also allowed to move, the position of the eye in space (gaze in space, g_s) now comprises both the position of the eye in the head, e_h, and the position of the head in space, h_s, according to:

$$g_s = e_h + h_s \qquad (2.8)$$

The nature of combined horizontal eye-head movements has been studied extensively in human, cat, and monkey ([1], [25], [62]). Initially, Bizzi and colleagues proposed that head-free gaze saccades are, like head-fixed gaze saccades, programmed as an ocular saccade, independent of the occurrence and size of a concomitant head movement. According to this so-called *oculocentric hypothesis*, the vestibulo-ocular reflex (VOR) would cancel any contribution of the head to the gaze shift by causing the eyes to counter-rotate by the same amount [1]. Despite its simplicity, a serious flaw of this model is that the system would not enable accurate gaze shifts beyond the oculomotor range. In addition, several experiments have shown that the action of the VOR is actually suppressed during gaze saccades. These and other observations have led to the conclusion that the oculocentric hypothesis is strictly valid only for gaze shifts smaller than ∼10 deg.

As an alternative for the oculocentric hypothesis, the conceptual oculo-motor model (see above) was extended to a gaze control model in the head-free condition [25]. According to this *gaze feedback hypothesis*, an internally created, instantaneous gaze motor-error (i.e., desired gaze∼ minus current gaze displacement) is used to drive the oculomotor system. In this way, the accuracy of gaze saccades can be maintained, regardless of head movements, even if the VOR is suppressed during the movement.

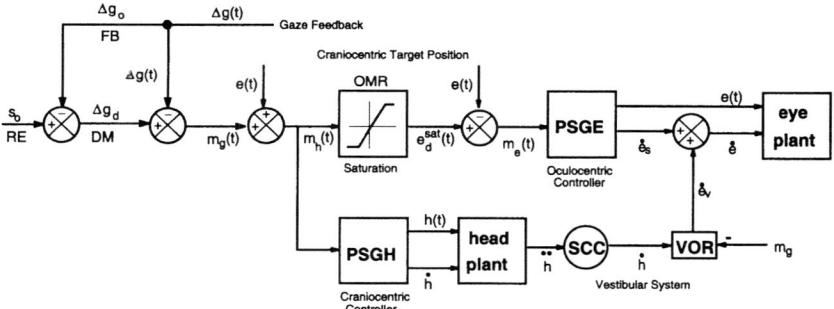

FIGURE 2.3: Two-dimensional eye-head coordination model, adapted from [21]. Eye and head are driven by a desired gaze displacement signal, Δg_d (emanating from the SC) that is updated by feedback about current gaze displacement, $\Delta g(t)$. Note, that this part of the model could alternatively be formulated in terms of a desired gaze-position signal in space (see e.g. Fig. 2.13). The head is driven by a current head motor-error signal, $m_h(t)$. The same signal is also fed into a 2D neural representation of the (elliptical) oculomotor range (OMR) to determine the desired position of the eye in the head, e_d^{sat}. The eye pulse-step generator is then driven by instantaneous eye motor-error, $m_e(t)$ (cf. with Robinson's model; for clarity, local feedback is omitted). The VOR is inhibited by the current gaze motor error signal, $m_g(t)$. As this signal falls to zero during the gaze shift (i.e. eye is on target), the VOR is reactivated by the ongoing head movement.

Note, that the concept of gaze feedback by itself does not specify the head motor command. It was initially proposed, on the basis of gaze control studies in the cat, that both the oculomotor system and the head motor system are controlled by the *same* internally created gaze motor-error signal [20]. Although several behavioral and neurophysiological studies seemed to provide support for this *common gaze model*, a recent analysis of the eye-, head- and gaze trajectories in two dimensions suggested that the two motor systems are driven by signals expressed in their own frames of reference (oculocentric and craniocentric, respectively [21]). For example, it was found that when eyes and head start from unaligned initial positions, both motor systems make a goal-directed saccade toward the target. Since eye and head then simultaneously move in quite different directions, they must be driven by different inputs. Fig. 2.3 presents a simplified scheme that summarizes the ideas of gaze feedback in combination with different eye- and head controllers. The scheme is qualitatively supported by recent data obtained from SC stimulation in the head-free monkey [19]. For specific details, the reader is referred to [21].

Although not detailed in Fig. 2.3, a complete model of the gaze control system should also incorporate the dynamics of the head motor system. In its simplest 1D form, the head is modeled by a linear, overdamped, second-order differential equation (e.g. [20], compare to Eqn. (2.3)). In contrast to the eye, the head inertia is substantial and should not be ignored. In addition, for head movements in nonhorizontal directions, there may also be a considerable translational component to the movement due to the flexibility of the neck vertebrae that cause the rotation axes of the head-neck system not to intersect in a single point (e.g. [46]). We will, however, not deal with the latter problem in this chapter.

2.3   New Aspects for Eye Rotations in 3D

That the eye should be considered as a rigid sphere rotating about its head-fixed center [34] can be ignored as long as movements are confined to the horizontal plane (only rotation about a fixed, vertical axis, so that parametrization by a single scalar, the horizontal angle, suffices). As soon as movements are allowed in more dimensions, however, the ocular rotation axes may differ from saccade to saccade, and a full 3D approach is called for.

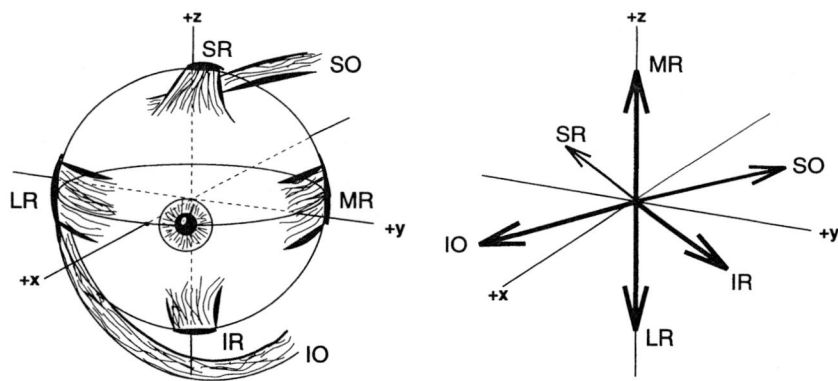

FIGURE 2.4: Left: frontal view of the right eye's six extraocular muscles, shown with the eye rotated about 20 degrees leftward from the primary direction (+x). Right: Schematic of the muscle rotation axes, represented in the head-fixed, primary coordinate system, (x, y, z). Note that the antagonist pairs (LR-MR, SR-IR, and SO-IO) pull in three, approximately orthogonal directions. The torque vectors are also approximately aligned with the on-directions of the semicircular canals.

The three antagonistic pairs of extraocular muscles pull in approximately orthogonal directions, so that movements of the eye are in principle allowed three independent degrees of freedom [3]: The horizontal recti (LR and MR) rotate the eye about a vertical axis ($\mathbf{z}$) to generate horizontal gaze shifts, whereas the vertical recti (SR, IR) and oblique muscles (SO, IO) define two orthogonal rotation axes in the vertical-torsional ($\mathbf{y}$-$\mathbf{x}$) plane (Fig. 2.4). Despite the apparent simplicity of the mechanical organization of the oculomotor system, it also introduces an interesting problem, that was noted more than a century ago by Helmholtz [29] but has received renewed attention in more recent studies [13], [28], [30], [59], [66], [73], [81].

The problem arises because 3D rotations of rigid bodies are *noncommutative*. As a consequence, the order in which two consecutive rotations about different axes are performed determines the final orientation of the object [22]. The problem is absent as long as movements are constrained to one dimension only, and because of this, it has been ignored in oculomotor studies for a long time. Yet, it has been recognized that this property could have important consequences for the way in which eye (and head) movements are programmed and generated [34], [75], [76].

If gaze shifts would be controlled without accounting for the noncommutativity of 3D rotations, problems could arise for the visual perceptual system, for visuomotor control, as well as for the motor system proper.

First, the orientation on the retina of fixed objects in the visual world would depend on the history of previous eye rotations. Since typically about three saccades per second are generated in all directions and from all possible initial eye positions, the ability to perceive a stable visual world, and to precisely control binocular alignment for adequate depth perception, would be seriously challenged.

Second, the visual information on the retina is, by itself, not sufficient to produce accurate eye movements (see following text). The absolute position of the eye in the head should be incorporated to prevent systematic mislocalizations of the stimulus, especially for targets presented in the retinal periphery with eccentric initial eye positions, or with considerable torsion [6], [9], [42], [75]. Finally, also from the motor point of view, problems could arise if the motoneuron signals to the plant would not account for the non-abelian property. However, as will be discussed, a precise assessment of this problem requires a detailed model of the plant dynamics.

Fig. 2.5 illustrates the problem for the vestibuloocular reflex (VOR) by letting the eye maintain fixation of a stationary point in space after two different sequences of head rotations. Obviously, a properly functioning VOR requires that the amount of head rotation is exactly canceled by an equal and opposite rotation of the eyes in the head. Like for the saccadic system, it is generally accepted that the vestibular input, which at the level of the vestibular afferents is proportional to head angular velocity, is neurally integrated (by the *same* NI as in Fig. 2.1B) to generate the tonic eye position signal. However, Tweed and Vilis [75] pointed out that in 3D, eye position is *not* obtained by taking the time-integral of eye angular velocity.

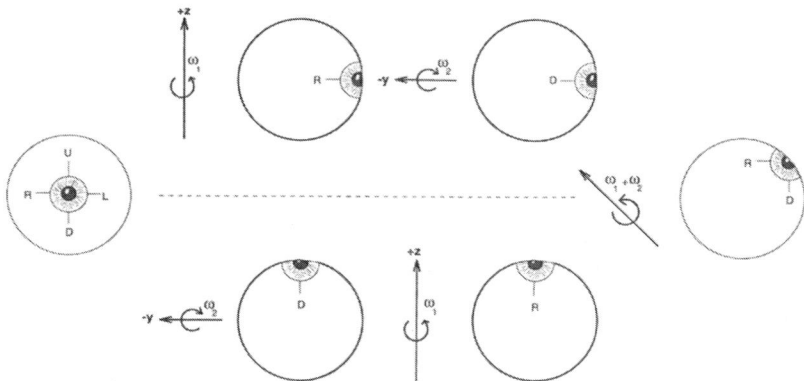

FIGURE 2.5: Noncommutativity of rotations illustrated for a perfectly functioning VOR that compensates for two consecutive 90 degrees head rotations about
the head-fixed -z (rightward head movement) and +y (downward) axes, respectively (top). The orientation of the eye in the head is shown (as if viewing onto
the subject's nose). Reversing the order of the two head-rotations (bottom) yields
different final eye positions, although the total amount of rotation is equal for the
two cases. This property results because eye position is not the integral of eye
angular velocity. Rotation about an axis that is the vectorial sum of the (simultaneously presented) angular velocity signals (center, right) yields yet a different
eye position without torsion. After [78].

The two exaggerated examples in Fig. 2.5 clearly illustrate that the VOR
brings the eyes in very different positions in the head (to the extreme left
and with 90 degrees counterclockwise torsion for the first series of head rotations, but upward and 90 degrees clockwise torsion for the second example).
Yet, it is also obvious that the total time integral of the two consecutive
angular velocity signals for both cases (cf. $\int (\omega_1 + \omega_2)\, dt$ vs. $\int (\omega_2 + \omega_1)\, dt$),
is identical $(\pi/2(\mathbf{e_y} - \mathbf{e_z}))$. Thus, if the tonic oculomotoneuron firing would
be produced by a neural integrator that receives eye angular velocity as an
input, the eyes would have ended up in identical, but wrong, positions [78],
[75]. Recently, a realistic version of the vestibular thought experiment, car-
tooned in Fig. 2.5, was performed which clearly demonstrated the capacity
of the brain to deal with the noncommutativity problem [72].
It is important to stress at this point, that the logic of this example does not
depend on any mechanical plant model. It is only assumed that identical
oculomotoneuron firing patterns (regardless of their origin) lead to identical
eye positions in the head. The principles of noncommutative rotational
kinematics therefore require that the central nervous system somehow takes
the integrator problem of Fig. 2.5 into account. Since the neural integrator
is thought to be shared by all oculomotor subsystems (not only the VOR,
but also by the pursuit, the saccadic, and vergence systems), this problem
affects the cornerstone of gaze control theories [31], [75], [76].

To successfully deal with the underlying concepts in a quantitative way, however, one has to apply the mathematical framework of 3D rotational kinematics. The main principles, deemed relevant for the eye-head gaze control system, will therefore be briefly introduced. Some of the computational details, when not immediately needed for comprehending the line of thought, have been delegated to the Appendix.

2.4   Mathematics of 3D Rotational Kinematics

Any finite rotation of a rigid body can be fully described by a real, orthogonal, proper 3×3 matrix (i.e., with determinant $+1$, and transpose $R^T=R^{-1}$), and all 3D rotations make up the special, nonabelian (i.e. noncommutative), orthogonal group $SO(3)$ [74].

There are many ways to describe the orientation of a rigid body. Here, we will not review all the different methods (such as rotation matrices, and the twelve different sets of Euler angles, see e.g. Goldstein [22] for a detailed treatment). Rather, we will present those parametrizations that have been particularly useful in their applications to the oculomotor and eye-head control systems. We start out with the basic equation describing the rotation of an arbitrary vector, and then proceed with the framework of quaternions and the Euler-Rodrigues coordinates.

2.4.1   Finite rotations

According to Euler's theorem, the general displacement of a rigid body, when one point of the body remains fixed in space, is uniquely described by a single-axis rotation, with the rotation axis passing through the fixed point [22].

To describe eye orientations in 3D, we adopt a right-handed, head-fixed Cartesian coordinate system (as in Fig. 2.4), with the origin in the center of the eye. The $\hat{e}_x$-axis points forward (clockwise torsional rotations positive), the $\hat{e}_y$-axis lies along the interocular line connecting the two centers (downward rotations positive), and the $\hat{e}_z$-axis points upward (leftward rotations positive).

The fixed-axis rotation of an arbitrary vector, $\mathbf{u}$, over some finite angle, ρ (in radians), around axis $\hat{\mathbf{n}}$ (a unit vector, see Fig. 2.6), is then given by:

$$R(\hat{\mathbf{n}}, \rho)\,\mathbf{u} = (\mathbf{u} \bullet \hat{\mathbf{n}})\hat{\mathbf{n}} + \sin(\rho)(\hat{\mathbf{n}} \times \mathbf{u}) - \cos(\rho)\hat{\mathbf{n}} \times (\hat{\mathbf{n}} \times \mathbf{u}) \qquad (2.9)$$

in which $\bullet$ is the scalar (dot) product and $\times$ the vector (cross) product.

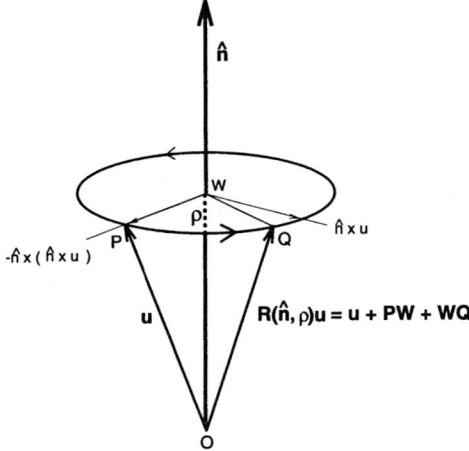

FIGURE 2.6: Rotation of vector **u** over finite angle ρ about axis $\hat{\mathbf{n}}$.

Note that $\mathbf{u} = (\mathbf{u} \bullet \hat{\mathbf{n}})\hat{\mathbf{n}} - \hat{\mathbf{n}} \times (\hat{\mathbf{n}} \times \mathbf{u})$, from which a Taylor expansion yields in good approximation:

$$R(\hat{\mathbf{n}}, \rho)\, \mathbf{u} = \mathbf{u} + \rho \hat{\mathbf{n}} \times \mathbf{u} + \frac{1}{2}\rho^2 \hat{\mathbf{n}} \times (\hat{\mathbf{n}} \times \mathbf{u}) + \mathcal{O}(\rho^3) \qquad (2.10)$$

For fixed-axis rotations, only the angle ρ is a function of time, and thus $R = R(\hat{\mathbf{n}}, \rho(t)) = R(t)$. The angular velocity, $\omega(t)$, with which **u** spins around the rotation axis $\hat{\mathbf{n}}$ is a vector defined by

$$\omega(t) \equiv \dot{\rho}(t)\hat{\mathbf{n}} \qquad (2.11)$$

($\dot{\rho}$ is the time derivative of ρ). The angular velocity is related to the rotation $R(t)\mathbf{u}$ by the well-known kinematic relation:

$$\begin{aligned} \dot{R}(t)\mathbf{u} &= \omega(t) \times R(t)\mathbf{u} \\ &\approx \dot{\rho}(t)\left[\hat{\mathbf{n}} \times \mathbf{u} + \rho\left((\hat{\mathbf{n}} \bullet \mathbf{u})\hat{\mathbf{n}} - \mathbf{u}\right)\right] \end{aligned} \qquad (2.12)$$

where we have used Eqn. (2.10) and Eqn. (2.11). Here, $\dot{R}(t)\mathbf{u}$ will be designated the *coordinate velocity* vector. It is the velocity with which the tip of **u** rotates around the axis. Note from Fig. 2.6 and Eqn. (2.12), that for a given angular velocity, the coordinate velocity vector depends on both the orientation of **u** with respect to the rotation axis, and on its length. An important question for understanding 3D gaze control, briefly touched upon in the discussion of Fig. 2.5, is whether the saccadic short-lead burst neurons in the brainstem encode a coordinate velocity or an angular velocity signal. As will be explained in the next section, the distinction has important implications for the neural organization of the saccadic system,

and for its interactions with other oculomotor subsystems (like the vestibular system, the smooth pursuit system, and the vergence system). Before
elaborating further on these matters, we will first provide an outline of
the mathematical framework within which these concepts can be elegantly
represented. For a more detailed treatment, the reader is referred to [2],
[74].

2.4.2 Quaternions

W.R. Hamilton (1843) was the first to propose a closed algebraic framework
for the multiplication of 3D vectors. The result of his work was that such
an algebra had to be described by hypercomplex, four-dimensional entities,
which he named quaternions [2]. Quaternions have a particularly useful and
natural relationship with fixed-axis rotations, and it is for that reason that
their properties will be reviewed in somewhat more detail in this section.
A quaternion is written as a four-component complex quantity defined by:

$$q \equiv q_o + \mathbf{I} \bullet \mathbf{q} \tag{2.13}$$

with $\{q_o, q_x, q_y, q_z\} \in \mathcal{R}$, the four quaternion components (q_o is called the
scalar part of the quaternion, $S(q)$, and $\mathbf{q} = (q_x, q_y, q_z)$ is its vector part,
$V(q)$). The adjunct (or complex conjugate) of the quaternion is given by
$q^\star = q_o - \mathbf{q} \bullet \mathbf{I}$. The components of the complex vector, $\mathbf{I} = (i, j, k) \in \mathcal{C}$
obey the following noncommutative, cyclic relations[2]:

$$i^2 = j^2 = k^2 = i \cdot j \cdot k = -1$$
$$i \cdot j = -j \cdot i = k \qquad j \cdot k = -k \cdot j = i \qquad k \cdot i = -i \cdot k = j \tag{2.14}$$

Main properties of quaternions

Armed with these definitions, it is now possible to multiply two arbitrary
quaternions, say $p = p_o + \mathbf{p} \bullet \mathbf{I}$ and $q = q_o + \mathbf{q} \bullet \mathbf{I}$. Applying Eqn. (2.13)
and Eqn. (2.14), the following important rule for the quaternion product
is obtained:

$$pq = p_o q_o - \mathbf{p} \bullet \mathbf{q} + (p_o \mathbf{q} + q_o \mathbf{p} + \mathbf{p} \times \mathbf{q}) \bullet \mathbf{I} \tag{2.15}$$

---

[2]It can be readily verified that the properties of the components of the complex
vector $\mathbf{I}$ can be identified with the Pauli spin matrices $[\sigma_i]$ which are the generators of
the SU(2) group (the special group of complex, hermitean 2×2 unitary matrices; see e.g.
[22], [74]):

$$[\sigma]_o = \begin{pmatrix} 1 & 0 \\ 0 & 1 \end{pmatrix}, \quad [\sigma]_1 = \begin{pmatrix} 0 & 1 \\ 1 & 0 \end{pmatrix}, \quad [\sigma]_2 = \begin{pmatrix} 0 & -i \\ i & 0 \end{pmatrix}, \quad [\sigma]_3 = \begin{pmatrix} 1 & 0 \\ 0 & -1 \end{pmatrix},$$

which establishes the local isomorphism between SO(3) and SU(2) ($1 \Leftrightarrow [\sigma_o]$, $i \Leftrightarrow [\sigma_1]$,
$j \Leftrightarrow [\sigma_2]$, $k \Leftrightarrow [\sigma_3]$).

Thus, multiplication of two quaternions results in a new quaternion (hence, they form a closed algebraic system). Note that, in general, the vector part of the product quaternion, $V(pq)$, does not lie in the plane spanned by **p** and **q**. More important, however, $pq \neq qp$ whenever **p** and **q** are not parallel (or zero). This latter property, the noncommutativity of quaternion multiplication, is deeply connected (see as follows) to the noncommutativity of rotational kinematics briefly outlined above.

The length (or norm) of a quaternion is given by

$$|q| \equiv \sqrt{qq^\star} = \sqrt{q_o^2 + \mathbf{q} \bullet \mathbf{q}} \qquad (2.16)$$

Taken together, Eqn. (2.13) can be parametrized, for reasons that will become clear below, by

$$q = |q| \cdot (\cos(\theta) + \sin(\theta) \cdot \hat{\mathbf{e}} \bullet \mathbf{I}) \qquad (2.17)$$

Here, $\hat{\mathbf{e}}$ denotes the unit vector (or axis) of the quaternion, with $e_i = q_i / \sqrt{q_x^2 + q_y^2 + q_z^2}$, and θ is the quaternion angle ($\cos(\theta) = q_o/|q|$).

Note that in this framework, a 3D vector can be regarded as a quaternion with scalar part zero; any real number is a quaternion with vector part zero; and any complex number is a quaterion with vector components $(q_y, q_z) = 0$.

The *inverse* of a quaternion is obtained by the demand $qq^{-1} = q^{-1}q = 1$, and it can be readily verified from Eqn. (2.15), Eqn. (2.16), and Eqn. (2.17) that it is determined by[3]:

$$q^{-1} = \frac{1}{|q|} \cdot (\cos(\theta) - \sin(\theta) \cdot \hat{\mathbf{e}} \bullet \mathbf{I}) = \frac{q^\star}{|q|^2} \qquad (2.18)$$

Relation with fixed-axis rotations

Now what is the relation between a quaternion, q, and the rotation of an arbitrary vector, **u**, about an axis, $\hat{\mathbf{n}}$, over angle ρ (see above)? An elegant theorem that follows from the rules of quaternion calculus, described above, says that if one parametrizes a quaternion by Eqn. (2.17), then the new vector **u**′:

$$\mathbf{u}' = q\mathbf{u}q^{-1} = R(\hat{\mathbf{n}}, 2\theta)\, \mathbf{u} \qquad (2.19)$$

In other words, the unit vector, $\hat{\mathbf{e}}$, of the quaternion is directed along the rotation axis $\hat{\mathbf{n}}$, and vector **u** is rotated by twice the quaternion angle

---

[3]One may now also define the *quotient* of two quaternions, p and q, as the solution, r, of an equation of the following type: $p = qr$. However, rather than writing $r = p/q$ for the solution, one should distinguish the two different possibilities: $r_1 = pq^{-1}$, and $r_2 = q^{-1}p$, which are generally unequal due to the noncommutativity of quaternion multiplication. In this case, r_2 is the correct solution (left quotient), which was obtained by left-multiplying both sides with q^{-1} (the right quotient r_1 is the solution for $p = rq$).

around this axis. This important property may be verified in Appendix A-1.

Without loss of generality, it is customary to constrain the norm $|q| = 1$, so that the following unit quaternion fully parametrizes the rotation of an arbitrary vector about $\hat{\mathbf{n}}$ over angle ρ:

$$q = \cos(\rho/2) + \sin(\rho/2)\ (\hat{\mathbf{n}} \bullet \mathbf{I}) = \exp(\frac{\rho}{2} \cdot \hat{\mathbf{n}} \bullet \mathbf{I}) \tag{2.20}$$

Geometrically, there is an interesting correspondence between the space of unit quaternions, and the unit sphere in 3D Cartesian space. In particular, the real quaternion $q = 1$ (i.e., $\theta = 0$) corresponds to any single point on the unit sphere, whereas $q = -1$ (i.e., $\theta = \pi$) describes any great circle on the unit sphere (in both cases, $\hat{\mathbf{e}}$ is arbitrary). All other non-real quaternions (with $\theta \in (0, \pi)$) correspond to arcs (of length θ radians) that are part of the great circle perpendicular to the quaternion (or polar) axis $\hat{\mathbf{e}}$.

Angular and coordinate velocities in q representation

As briefly mentioned in the introduction to this section, a proper understanding of 3D gaze control requires that a distinction be made between the angular and the coordinate velocity vectors. If a quaternion, q, may be identified with the rotation $R(\hat{\mathbf{n}}, \rho)\mathbf{u}$, then the time derivative of the quaternion, $\dot{q} = dq/dt$, corresponds to the coordinate velocity, Eqn. (2.12), of the rotating vector.

In Appendix A-2, a derivation is given for the relation between the coordinate velocity, $\dot{q}$, and the angular velocity, ω, in the quaternion parametrization. The result is given by the following simple equation [75]:

$$\dot{q} = \frac{\omega q}{2} \tag{2.21}$$

By applying the product rule Eqn. (2.15) for quaternions one finds immediately for the vector part of Eqn. (2.21):

$$\dot{\mathbf{q}} = \frac{1}{2}(q_o \omega + \omega \times \mathbf{q}) \tag{2.22}$$

(note that $\omega_o = 0$) and, similarly, the reverse relation is obtained by right-multiplication of both sides in Eqn. (2.21) with q^{-1}:

$$\omega = 2\dot{q}q^{-1} = 2(q_o\dot{\mathbf{q}} - \dot{q}_o\mathbf{q} + \mathbf{q} \times \dot{\mathbf{q}}) \tag{2.23}$$

It is now also possible to obtain the coordinate and angular acceleration of the body, by evaluating the time derivatives of Eqn. (2.21) and Eqn. (2.23). For example, for the coordinate acceleration one obtains:

$$\ddot{q} = \frac{1}{2}(\dot{\omega}q + \omega\dot{q}) \Rightarrow \ddot{\mathbf{q}} = \frac{1}{2}\left[q_o\dot{\omega} - \frac{\omega^2}{2}\mathbf{q} + \dot{\omega} \times \mathbf{q}\right] \tag{2.24}$$

2.4.3 Rotation vectors

A slightly different parametrization of a 3D rotation is provided by the Euler-Rodrigues coordinates [44], designated in the oculomotor literature as the *rotation vector* representation. The rotation vector is introduced here because it has some interesting geometric properties for modeling the saccadic system. One such property is that saccades in Listing's plane, when viewed as fixed-axis rotations, follow straight lines (i.e. shortest paths, or geodesics) in rotation vector space (see [30], and as follows). The rotation vector, and its inverse, is defined by:

$$\mathbf{r} \equiv \tan(\rho/2)\hat{\mathbf{n}} \quad \text{and} \quad \mathbf{r}^{-1} = -\mathbf{r} \tag{2.25}$$

The correspondence between quaternions and rotation vectors is therefore established by:

$$\mathbf{r} = \frac{V(q)}{S(q)} = \frac{\mathbf{q}}{q_o} \tag{2.26}$$

Obviously, a potential disadvantage of the rotation vector representation emerges for large angles: when $q_o \to 0$ (for $\rho \to 180$ deg), the length of the rotation vector rapidly approaches infinity and is therefore ill-defined. For head-fixed saccades, however, rotation vectors are particularly useful parametrizations.

By applying Eqn. (2.26) to the product of quaternions Eqn. (2.15), a simple rule for rotation vector multiplication can be derived that gives the combined result of rotation $\mathbf{r}_1$, followed by a second rotation, $\mathbf{r}_2$:

$$\mathbf{r}_2 \circ \mathbf{r}_1 = \frac{V(q_2 q_1)}{S(q_2 q_1)} = \frac{\mathbf{r}_1 + \mathbf{r}_2 + \mathbf{r}_2 \times \mathbf{r}_1}{1 - \mathbf{r}_1 \bullet \mathbf{r}_2} \approx \mathbf{r}_1 + \mathbf{r}_2 + \mathbf{r}_2 \times \mathbf{r}_1 \tag{2.27}$$

Here $\circ$ denotes the product of rotation vectors, and the approximation holds up to $\mathcal{O}(\rho^3)$. Using Eqn. (2.22) and Eqn. (2.23), one can also find the equivalent relations for the angular and coordinate velocities in the rotation vector representation. The exact results are (see [30] and Appendix A-2):

$$\omega = \frac{2(\dot{\mathbf{r}} + \mathbf{r} \times \dot{\mathbf{r}})}{1 + \mathbf{r} \bullet \mathbf{r}} \approx 2(\dot{\mathbf{r}} + \mathbf{r} \times \dot{\mathbf{r}}) \tag{2.28}$$

and, similarly, for the reverse relation:

$$\dot{\mathbf{r}} = \frac{1}{2}(\omega + \omega \times \mathbf{r} + (\omega \bullet \mathbf{r})\mathbf{r}) \tag{2.29}$$

which may also be compared to Eqn. (2.12).

2.5 Donders' Law and Listing's Law

Eye positions

In what follows, eye positions will be described by rotation vectors Eqn. (2.25), the coordinates of which are expressed in the head-fixed primary frame of reference (for definition, see as follows). Conceptually, this means that any 3D eye position is parametrized by the *virtual* rotation that brings the eye from the center of the head-fixed Cartesian coordinate system to the current position. The origin of the primary reference frame is, by definition, the zero position vector, $\mathbf{r} = \mathbf{0}$, and is called the *primary position* of the eye. It is important to note that the primary position is a geometrical concept that is not necessarily equal, or even close to, the center of the oculomotor range.

A *secondary* eye position is obtained by a rotation from the primary position about either the horizontal or vertical axis of the primary coordinate system. For example, a position 20 degrees to the left of primary position is described by the rotation vector $\mathbf{r} = (0, 0, 0.176)$ (positive rotation about the vertical axis), where the units are measured in half-radians. As a rule of thumb, every 0.01 half-radian roughly corresponds to $2 \cdot \arctan(0.01) \approx 1$ deg rotation about the axis.

A *tertiary* position is any eye position that obeys Listing's law (see as follows) and is not a primary or secondary position. Eye positions that do not follow Listing's law Eqn. (2.33) are called *quaternary* positions.

Donders' law

The fixation of a point target is a redundant motor task; the direction of the target (for each eye) is fully determined by two coordinates (azimuth and elevation re. fovea). The amount of torsion about the visual axis remains unspecified by the task.

Donders [29], [32] proposed, on the basis of his retinal after-image experiments, that for each target position in 3D space the gaze direction determines a unique orientation of the eye, independent of the trajectory followed to get to that position. In other words, ocular torsion is fully specified by the horizontal and vertical components of the gaze direction. This important rule is known as Donders' law, and restricts eye positions to a two-dimensional (2D) subspace:

$$\mathbf{r}_x = \text{function}(\mathbf{r}_y, \mathbf{r}_z) \qquad (2.30)$$

The precise geometry of this surface, however, is not specified by Donders' law and, as will be shown, appears to depend on the gaze orienting task. Since ocular torsion is constrained by a smooth function, Donders' law elegantly solves the problem of accumulation of torsion: Whatever the path followed by the eyes, torsion remains on the surface defined by Eqn. (2.30).

An illustrative example of a Donders' surface is provided by a gimbal system. In a gimbal system, a 3D orientation is described by three consecutive rotations about a nested set of axes. In physics, the Euler angles constitute such a gimbal system, and in fact, twelve different types of gimbals can be constructed to achieve the same objective [22].

In the so-called *Fick gimbal*, which will be relevant in the following discussion of 3D eye-head coordination, the orientation of the rotating body is described by making first a horizontal rotation about the *space-fixed* vertical axis ($\hat{\mathbf{e}}_z$, angle θ_F). Then, a second, vertical, rotation is performed about the new *body-fixed* horizontal axis ($\hat{\mathbf{e}}'_y$, angle φ_F), which has been rotated away from the space-fixed coordinate system by the first rotation. Finally, a third rotation is performed about the new, body-fixed frontal axis, ($\hat{\mathbf{e}}''_x$, angle ψ_F), inducing torsion. Rotations about body-fixed axes are known as *passive* rotations, since they involve a rotation of the coordinate system. In contrast, *active* rotations are about space-fixed axes, and generate the actual rotation of the body [22].

Suppose that gaze positions would somehow be generated by such a gimbal system. Then, if gaze would adhere to Donders' law Eqn. (2.30), its torsional component is determined by the horizontal (head-fixed) and vertical (eye-fixed) rotation angles only (θ_F and φ_F, respectively, i.e. a reduced, two-axis system; cf. Fig. 2.5). The resulting position of gaze is then computed as follows:

$$\begin{aligned}
\mathbf{r}_F &= \mathbf{r}'_{\text{eye}} \circ \mathbf{r}_{\text{head}} = \left(\mathbf{r}_{\text{head}} \circ \mathbf{r}_{\text{eye}} \circ \mathbf{r}_{\text{head}}^{-1}\right) \circ \mathbf{r}_{\text{head}} = \mathbf{r}_{\text{head}} \circ \mathbf{r}_{\text{eye}} \\
&= \left[-\tan\left(\theta_F/2\right) \cdot \tan\left(\varphi_F/2\right), \quad \tan\left(\varphi_F/2\right), \quad \tan\left(\theta_F/2\right)\right]
\end{aligned}$$
$$(2.31)$$

Note, that the passive rotation of the horizontal axis can be transformed into an active vertical rotation of the eye (in head-fixed coordinates), by reversing the order of the horizontal and vertical rotations (e.g. [22], [27], [74]).

Figure 2.7 illustrates the 3D "Fick" trajectories in rotation-vector space, when the eye first goes to target A from the primary position, $\mathbf{r} = \mathbf{0}$, in two steps (PQA), and then tracks a square array of targets in the clockwise direction (ABCDA). Note that the amount of torsion jumps between a^2 and $-a^2$ for every saccade, but that the 3D position of the eye in A is identical at the start and at the end of the trajectory. Therefore, no accumulation of torsion has occurred [26]. Note also that the Donders' surface for eye positions is strongly twisted in 3D rotation vector space. The torsional component of the Fick Donders' surface is taken from Eqn. (2.31) and is given by[4]:

$$\mathbf{r}_x = -\tan\left(\theta_F/2\right) \cdot \tan\left(\varphi_F/2\right) = -\mathbf{r}_y \cdot \mathbf{r}_z \qquad (2.32)$$

---

[4]In the (early) oculomotor literature this gimbal-induced torsion has been termed 'false torsion' [3]. To comply with Listing's law Eqn. (2.33), an additional rotation about the rotated visual axis, $\hat{\mathbf{e}}''_x$, is needed to null this torsion.

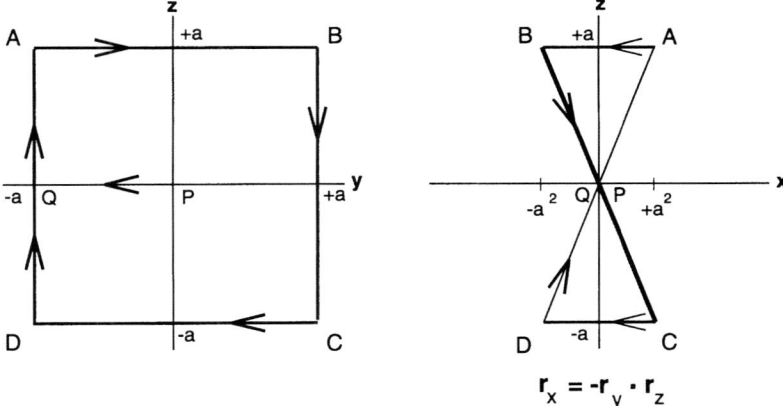

FIGURE 2.7: Donders' surface produced by a two-axis Fick gimbal system. Note that the resulting surface of gaze positions is markedly twisted in 3D rotation vector space. Thick line (BC) symbolizes positive y-components of gaze position, thin line (DA) negative y-components. As torsion jumps between $\pm a^2$, it does not accumulate during repeated clockwise tracking of the targets.

The nonlinear dependence of $\mathbf{r}_x$ on the gaze coordinates reflects the curvedness of the Donders' surface. In addition, the sign of ocular torsion changes, each time one of the gaze coordinates changes sign. The proportionality factor of -1 in Eqn. (2.32) is called the *twist score* of the Fick surface [26]. Different types of gimbal system give rise to different twist scores. For example, in the same way one may construct the Donders' surface for a Helmholtz gimbal, by reversing the order of the two rotations Eqn. (2.31) into a horizontal (head-fixed), followed by a vertical (eye-fixed) rotation. In this case, the twistscore will be $+1$.

Listing's law

Helmholtz put a specific constraint on the torsional component of eye position, by following a suggestion made earlier by the german physicist Listing [29]. According to this proposal, the oculomotor system reduces the Donders surface of eye positions Eqn. (2.30) to a plane (Listing's plane, LP) when the head is in an upright and stationary position, and the eyes fixate points at optical infinity (no vergence). Therefore, under these experimental conditions, the torsional component of eye position is a linear function of the horizontal and vertical gaze coordinates (Listing's law).
The unique direction perpendicular to LP is defined as the primary position. By a convenient choice of coordinates (which is the primary frame of reference, or the so-called Listing coordinates), all eye positions in Listing's plane have zero torsion (see e.g. Fig. 2.8 and below) [34], [59], [81]. Therefore, in the quaternion and rotation vector parametrizations, Listing's law

takes a very simple form:

$$\mathbf{r}_x = 0 \tag{2.33}$$

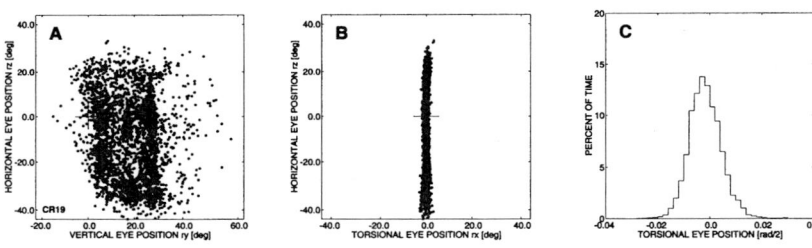

FIGURE 2.8: Listing's law for head-fixed conjugate eye movements of rhesus monkey CR. About 3500 positions of spontaneous eye movements in the light are plotted. Eye positions are plotted as components of rotation vectors (A) in the y,z-plane, and (B) in the x,z-plane, respectively (scale converted to degrees). Note that Listing's law is well-obeyed. Standard deviation of the width of Listing's plane (B) is only 0.6 degree. Panel (C) shows a histogram of the torsional components (in half-radians) of eye position for the same data set.

2.5.1   Listing's law for head-fixed saccades

Thanks to the recent development of accurate recording techniques [5], [33], 3D eye movements can be routinely measured with a high spatial and temporal resolution in both man and monkey. Although part of the available data suggests that Listing's law is only approximately valid during fixations [14] and during eye movements [15], [68], a majority of studies, performed in several different laboratories, has shown that it is obeyed with remarkable accuracy, both in human [39], [47], [76] and in monkey [10], [37], [38], [58] subjects (see Fig. 2.8). The law holds in good approximation for eye fixations, as well as dynamically during saccadic [37], [47], [59], [76] and smooth pursuit [39], [69] eye movements (but see also below). It is also equally-well obeyed for saccades toward auditory targets in complete darkness [17], which suggests that Listing's law does not exclusively serve a visual purpose.

Typically, the thickness of Listing's plane (given by the standard deviation of ocular torsion in the primary reference frame, Fig. 2.8B) is in the order of 0.5-0.7 deg for monkeys [37], [59], and about 1.0 deg for human subjects [47]. The plane's thickness is very similar for eye fixations, and for dynamic conditions such as smooth or rapid eye movements along a straight, or along a highly curved trajectory [47].

At this point, it is important to realize that neither Donders' law nor Listing's law simply follow from the infallible laws of physics applied to the plant mechanics; in principle, there are many different ways in which the oculomotor system could have reduced the number of degrees of freedom for gaze orientations.

For example, eye positions produced by the VOR do *not* obey either Donders' or Listing's law. During torsional vestibular stimulation, $\omega_h = (\dot{\rho}, 0, 0)$, the eye may look into the primary gaze direction ($\mathbf{r}_y = \mathbf{r}_z = 0$) with many different torsional components (violation of Donders' law, Eqn. (2.30)). In the absence of any additional mechanism, ocular torsion would accumulate until it would reach the physical limits of the oculomotor range (see Fig. 2.5). However, the vestibular system prevents such an accumulation by generating torsional "reset" quick-phases of nystagmus that repeatedly bring the eye back to a position close to zero torsion [8], [37].

A second example concerns the binocular viewing of targets in depth. A large body of experimental evidence from different laboratories shows that both eyes violate Listing's law when viewing near targets. The violation, however, follows a clear geometric pattern. Both eyes rotate temporally (excyclotorsion) when the elevation of the eyes is downward, whereas they rotate nasally (incyclotorsion) for upward elevations. Listing's law is preserved only when gaze elevation is zero (in the primary reference frame) [45], [48]. When plotting these eye positions in 3D space, the pattern resembles that of a Helmholtz two-axis gimbal system (see above), and therefore nicely adheres to Donders' law (a curved surface with a positive twist score close to one, i.e. a full-angle tilt of the angular velocity vector as function of eye elevation) [45], [80].

Yet, as will be argued, the exact properties of the oculomotor plant do have implications for the way in which the circuitry of the brainstem burst generators and neural integrators is organized, since it should compensate for any (nonlinear) peculiarities in the plant mechanics (see also above, for the 1D case). Regarding the modeling of the 3D saccadic system, however, it is reasonable to suppose that:

• Identical motoneuron firing patterns yield identical eye positions in the head, regardless of plant mechanics and task conditions.

• Ocular torsion may take any value within the oculomotor range (i.e., not restricted by plant mechanics), but it is constrained by the oculomotor task (i.e., the neural commands).

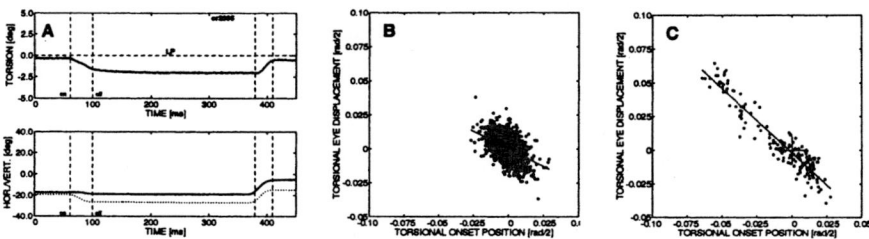

FIGURE 2.9: Spontaneous violations of Listing's law (A,B) by the saccadic system are reset by the next saccade. In panel (A) an example of such a violation (with about -2 degrees a particularly large one), as well as the subsequent reset saccade are both shown. Panel (B) shows, for the entire data set of Fig. 2.8, that there exists a highly significant relation (r=-0.49; N=3522) between torsional onset position of the eye (ordinate), and the saccadic torsional displacement of the following saccade (abscissa). Slope of the regression line is b=-0.52 (but note that violations and resets are pooled here). An accurate saccadic reset is also generated, when the violation is due to electrical microstimulation in the NRTP (C). The range of torsional onset positions is markedly expanded by this experiment (up to 10 deg). Correlation: r=-0.95, slope: b=-0.95; N=165 (i.e., no overshoots; after [57]). These data show that both the saccade burst generator and the neural position integrator of the ocular *saccadic* system are 3D.

2.5.2   Spontaneous violations of Listing's law

Although Listing's law is well obeyed by the monkey oculomotor system (at least for the head straight up and not moving, and the eyes looking at infinity), the data in Fig. 2.8C clearly indicate that it is not perfect. Occasionally, small (up to about 2.5 deg) spontaneous excursions from LP are made that are not due to blinks. Such torsional "errors" are relatively infrequent (in 23% of the saccades, averaged over six monkeys, the displacement exceeds 1.0 degrees) but, if present, the torsional component of the eye movement appears to be synchronized with the horizontal and vertical saccade components (see e.g., Fig. 2.9A). It should be noted, however, that torsional onset is usually better synchronized with the horizontal/vertical components, than torsional offset. As was observed for the horizontal and vertical components in earlier 2D studies (see above), component stretching seems to be a phenomenon that affects all three saccade components.

Thus, it appears that these torsional displacements are produced by the *saccadic* system. Closer inspection of the eye movement traces indicated that under these circumstances the eye does not drift back passively into the plane, but may stay at the new torsional level for several hundreds of msec (Fig. 2.9A). We observed that the system typically produces an active "reset" of ocular torsion by the next saccadic eye movement [31], [57], [58].

To further quantify this property, we analyzed to what extent the spontaneously occurring violations of Listing's law are corrected by the saccadic system. The results, valid for all monkeys studied so far, show a highly significant relation between the torsional onset position of the eye, r_x, and the subsequent torsional displacement, d_x, of the next spontaneous saccade (see Fig. 2.9B). Multiple linear regression on the data showed that the corrective torsional displacement is solely determined by the torsional onset position, and is neither related to the horizontal/vertical components of eye position, nor to the saccade direction [57].

The ability of the saccadic system to generate a torsional reset movement has also been shown by electrical microstimulation experiments in the monkey brainstem. Prolonged stimulation of the vertical/torsional burst generator in the riMLF produces a vertical/ipsi-torsional eye movement at constant velocity (torsion may reach levels exceeding 10 degrees; [40]). Microstimulation in the precerebellar nucleus reticularis tegmenti pontis (NRTP) yields staircases of horizontal/torsional saccades. Here, the direction of the torsional component depended on the location of the stimulation electrode within the NRTP [57]. Yet, also in these cases, the saccadic system generates a precise torsional reset at the next spontaneous saccade following the stimulation train (see Fig. 2.9C for an example of the NRTP stimulation results). It is important to realize that the large torsional components in these reset saccades were generated in the absence of any vestibular or optokinetic stimulation and were precisely aimed at Listing's plane.

These results therefore clearly indicate that both the saccadic burst generator as well as the eye position neural integrator carry a 3D eye-movement code, and that Listing's law is actively controlled by neural commands. As will be discussed below, however, the type of 3D signal used by these structures (i.e., coordinate velocity $\dot{r}$ vs. angular velocity, ω) cannot be decided on the basis of these experiments.

Interestingly, a localized reversible inactivation of the NRTP appeared to interfere with the capacity of the saccadic system to reset the torsional component into LP [57].

2.5.3   Parametrization of 3D saccades

From Listing's law Eqn. (2.33) it follows, that when the eye makes a movement from one Listing position, r_1, to a new Listing position, r_2, it can do so by spinning about the unique single-axis of rotation that is given by the quotient vector:

$$q = r_2 \circ r_1^{-1} \quad \text{because} \quad q \circ r_1 = r_2 \qquad (2.34)$$

Applying the product rule for rotation vectors Eqn. (2.27):

$$q = \frac{r_2 - r_1 + r_1 \times r_2}{1 + r_1 \bullet r_2} \approx d + r_1 \times d \qquad (2.35)$$

where the approximation holds up to $\mathcal{O}(\rho^3)$, and we have introduced the *difference vector* in LP:

$$d \equiv r_2 - r_1 \qquad (2.36)$$

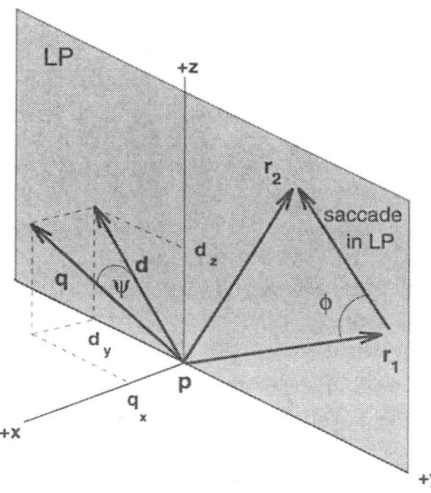

FIGURE 2.10: Eye positions in Listing's plane (LP, shaded) at the onset (r_1) and offset (r_2) of a head-fixed saccade. The difference vector, (d) lies in LP, but the quotient vector, (q) is tilted out of the plane by angle ψ. p: primary position.

Note that the quotient vector q is parallel to the angular velocity vector (cf. with Eqn. (2.28) and use the fact that $d \parallel \dot{r}$). Thus, despite that the initial and final eye positions belong to LP, the actual ocular rotation axis "tilts" out of LP (i.e., $q_x \neq 0$) by an amount that depends on the displacement amplitude, the inital eye position eccentricity, and the angle, ϕ, between r_1 and d:

$$q_x = r_{1,y} \cdot d_z - r_{1,z} \cdot d_y = |d| \cdot |r_1| \, \sin(\phi) = |d| \cdot r_1^{\perp} \qquad (2.37)$$

Although q can attain almost any angle with LP, this property has become known in the oculomotor literature as the *half-angle rule* (e.g., [16]; see Appendix A-3, for more details). The geometric relationship between the quotient and difference vectors and LP is further illustrated in Fig. 2.10.

A saccadic eye movement between two positions in LP can now be parametrized in two different ways [37], [54]. The first description considers a saccade as the unique *single-axis rotation* that brings the eye from initial position $\mathbf{r}_1$ into final position $\mathbf{r}_2$, through the quotient vector Eqn. (2.35). This vector is related to angular velocity by:

$$\mathbf{q} = \int_{t1}^{t2} \omega(t) dt \tag{2.38}$$

Note, that Eqn. (2.34) may be interpreted as a straightforward 3D extension of the concept of *eye motor error*. However, this vector depends not only on the difference between the initial and final eye positions (as in 1D and 2D saccade descriptions), but also on the absolute initial eye position (unlike the 1D and 2D case). Note also that in this parametrization the saccade vector has *three* degrees of freedom Eqn. (2.37), by following the half-angle rule, although the eye positions are constrained to a plane (Fig. 2.10).

Alternatively, a saccade may also be characterized as the *difference* vector between the two Listing positions Eqn. (2.36), which is related to the coordinate velocity by:

$$\mathbf{d} = \int_{t1}^{t2} \dot{\mathbf{r}}(t) dt \tag{2.39}$$

Since the difference vector is constrained to LP (at least for eye movements in LP), it has only *two* degrees of freedom.

A nontrivial property of the rotation vector description is that the trajectory of the eye for a single-axis rotation between two Listing positions, say $\mathbf{r}_1$ and $\mathbf{r}_2$, follows a straight line in LP. That is, for all times between saccade onset and saccade offset, the trajectory of the eye is determined by:[5]

$$\mathbf{r}(t) = \mathbf{r}_1 + \sigma(t) \cdot (\mathbf{r}_2 - \mathbf{r}_1) \tag{2.40}$$

where $\sigma(t)$ is the normalized saccade amplitude [30]. Such a simple geometric property of rotation vectors (i.e., saccades as geodesics in 3D space) could be advantageous for representing the spatial trajectory in the neural programming stage (e.g., in neural motor maps, such as the SC). As mentioned earlier, a fixed-axis rotation in quaternion space is represented by a

---

[5]To show that the trajectory of a single-axis rotation in Listing's plane is a straight line one considers the rotation Eqn. (2.35) as a function of time (i.e., the axis is fixed in space, but the rotation amplitude, $\tan(\rho_{12}/2) = \tan(\rho_{12}(t)/2) \equiv \sigma(t) \tan(\rho_{12}/2)$, with $\sigma(t)$ strictly increasing between 0 and 1). The movement from Listing's position $\mathbf{r}_1$ to $\mathbf{r}_2$ is then given by the time-dependent quotient vector Eqn. (2.34), $\mathbf{q}(t)$, applied to $\mathbf{r}_1$:

$$\mathbf{r}(t) = \mathbf{q}(t) \circ \mathbf{r}_1 \quad \text{with} \quad \mathbf{q}(t) = \sigma(t)(\mathbf{r}_2 \circ \mathbf{r}_1^{-1})$$

Substitution, and including only terms up to $\mathcal{O}(\rho^3)$, yields:

$$\mathbf{r}(t) \approx \sigma(t)\mathbf{q} + \mathbf{r}_1 + \sigma(t)\mathbf{q} \times \mathbf{r}_1$$

from which all cross-product terms vanish, and Eqn. (2.40) immediately follows. This result is also exact if the higher orders in rotation angle ρ are included [30].

great-circle arc on the unit sphere, which is not the shortest path between two points.

A further interesting and nontrivial interpretation of the difference vector Eqn. (2.36) is derived from its close correspondence to the *retinal error*, s_o, of a peripheral visual target (situated at Listing's position r_2), when the fovea is in the initial position, r_1. As further detailed in Appendix A-4 [30], [37], the horizontal and vertical coordinates of a visual target relative to the fovea are, in good approximation, linearly related to the components of the desired eye-displacement vector, d, and therefore *independent* of initial eye position:

$$
\begin{aligned}
s_{o,y} &\approx 2 \cdot d_z \\
s_{o,z} &\approx -2 \cdot d_y
\end{aligned}
\tag{2.41}
$$

where the approximation holds up to $\mathcal{O}(\rho^3)$ for the initial and final eye-position angles (relative to primary position). This approximation is very good in the periprimary oculomotor range (up to 15 degrees), but is still better than 12% for fixations up to 30 degrees away from primary position (which is about the range of normal saccade accuracy).

2.5.4 3D Models: The saccade programmer.

The two different parametrizations, d vs. q, suggest quite different possibilities for the neural organization of the saccadic system [9], [30], [37], [54], [66], [75], [77]. In principle, either description could represent the neural signal that is encoded by the saccade programmer as the desired position or displacement of the gaze line.

In Fig. 2.11A a visual double-step paradigm is illustrated for two different trials, in which the saccadic system has to program two different sequences of saccades to foveate the two targets: AB followed by BC, vs. AB' followed by B'C.

In the simplest 3D model (the *vector model*, [30], further detailed below), the programming stage of the saccadic system only involves *commutative*, vectorial computations, in which the coordinates of the next saccade vector are updated according to the approximation scheme suggested by Eqn. (2.41).

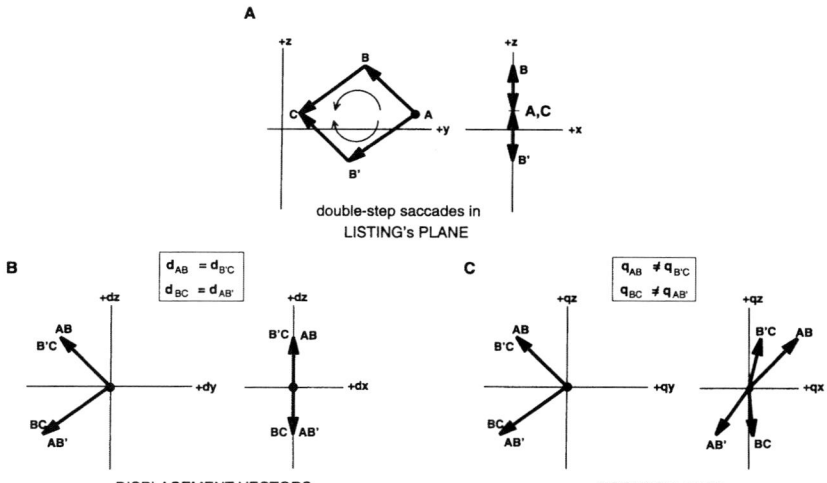

double-step saccades in
LISTING's PLANE

DISPLACEMENT VECTORS   ROTATION AXES

FIGURE 2.11: Illustration of the differences between the programming stages of the vector and quaternion models for two different double-step trials that move the eye between initial position A to final position C in LP (A) (see also Fig. 2.12). The vector model has a commutative saccade programmer, according to Eqn. (2.41). Since the resulting displacement vectors are independent of initial eye position, saccades in the double-step trials are identical (B). The quaternion model (C) has a noncommutative programmer that computes the required rotation axis coordinates through Eqn. (2.35). This results in different saccade codes for the different sequences.

With the eye initially in Listing position A, the two targets of the double-step jump, B (or B′) and C, define the retinal error vectors $\mathbf{s}_{AB}$ (or $\mathbf{s}_{AB'}$), and $\mathbf{s}_{AC}$, respectively.

Since both retinal error vectors correspond closely (up to $\mathcal{O}(\rho^3)$) to the respective eye-displacement vectors $\mathbf{d}_{AB}$ ($\mathbf{d}_{AB'}$) and $\mathbf{d}_{AC}$ in LP Eqn. (2.41), the vector model takes these approximative, but linear, estimates as the desired saccade program. For the different sequences, the two saccades are therefore computed as:

$$\mathbf{d}_{AB(\prime)} = \mathbf{s}_{AB(\prime)} \quad \text{and} \quad \mathbf{d}_{B(\prime)C} = \mathbf{s}_{AC} - \mathbf{d}_{AB(\prime)} \tag{2.42}$$

Note that in this model, the sequences A-B-C vs. A-B′-C involve the programming of two *identical* sets of saccade displacement vectors, that are to be executed in reversed order (commutative program, see Fig. 2.11B).

From an experimental point of view, this model therefore predicts that cell activity at sites encoding desired saccade displacement vectors in the double-step paradigm (as has been hypothesized e.g., for the cortical frontal eye fields (FEF), the posterior parietal cortex (PPC), and the midbrain SC) will show *no* difference for saccades into their movement field, regardless of the order in which they are elicited.

In the mathematically exact *quaternion model* (see below), the saccades are programmed as desired *rotations* from initial to final eye position Eqn. (2.35). Therefore, the retinal error vectors need to be transformed into the appropriate quotient vectors according to:

$$\text{A-B-C sequence:} \quad \mathbf{q}_{AB} = \mathbf{r}_B \circ \mathbf{r}_A^{-1} \quad \text{and} \quad \mathbf{q}_{BC} = \mathbf{r}_C \circ \mathbf{r}_B^{-1}$$
$$\text{A-B'-C sequence:} \quad \mathbf{q}_{AB'} = \mathbf{r}_{B'} \circ \mathbf{r}_A^{-1} \quad \text{and} \quad \mathbf{q}_{B'C} = \mathbf{r}_C \circ \mathbf{r}_{B'}^{-1} \tag{2.43}$$

In this model, the saccade codes of the programmer do differ when the sequence is reversed, because the rotation axis depends on the initial eye position (noncommutative quaternion multiplication, where $\mathbf{q}_x$ is of second order in the saccade amplitude, $\mathcal{O}(\rho^2)$; see Fig. 2.11C). Cell activity in a quaternion programmer is therefore expected to yield different firing patterns for the two double-step configurations. Moreover, electrical microstimulation at such a site (encoding the desired eye rotation $\mathbf{q}_{\text{site}}$) should lead to an eye-position dependent violation of Listing's law, since the stimulation would bypass the noncommutative programming stage [54], [55]:

$$\mathbf{r}_{\text{stim}} = \mathbf{q}_{\text{site}} \circ \mathbf{r}_o \quad \text{from which:} \quad \mathbf{r}_{\text{stim}}^x = |\mathbf{q}_{\text{site}}| \cdot |\mathbf{r}_o| \cdot \sin(\phi) \tag{2.44}$$

with ϕ the angle between $\mathbf{q}_{\text{site}}$ and initial eye position $\mathbf{r}_o$. Recent experimental data from the monkey SC have not provided support for the quaternion programming scheme, since both microstimulation and complete reversible collicular inactivation still resulted in saccades that obeyed Listing's law ($\mathbf{r}_{\text{stim}}^x = 0$) [54]. Moreover, movement fields appeared to be better described in terms of desired eye displacements in LP, $\mathbf{d}$, rather than in desired eye rotations, $\mathbf{q}$ [37], [54].

Still, conclusive evidence is lacking to either refute or prove either model of the saccade programmer for the following reasons: One complicating factor is the recent result that the movement fields of cells in the SC *do* seem to possess an eye-position related component [53]. However, changes in eye-position do not affect the optimal direction of the cell's movement field (i.e., the center of the movement field is unrelated to eye position) but, instead, seem to have a modulatory (multiplicative) influence on the cell's firing rate. A subsequent theoretical study showed that with a population of such so-called "gain-field" cells, the SC could provide an accurate estimate of the 3D desired rotation axis of the eye, despite the fact that the individual cells are tuned to relative eye displacements only [56].

Furthermore, the exact site of the saccade programming stage is still a matter of debate. So far, most studies assume that the deep layers of the SC issue a desired motor command to the burst generator. More recent experiments indicate, however, that the signal sent by the SC can still be modified substantially by downstream premotor structures without the SC being 'aware' of these changes. For example, Stanford and Sparks provided evidence that SC movement fields are better described in *visual* coordinates than in *motor* coordinates, because the systematic mislocalizations of saccades to remembered visual targets are not reflected in a cell's activity [67]. A similar conclusion was drawn by Frens and van Opstal on the basis of short-term saccadic gain adaptation experiments. In this paradigm, a consistent intrasaccadic target displacement causes the gain of saccade amplitudes to decrease accordingly. Yet the cell's movement-related activity does not change, but stays better tuned to the original (not displaced) retinal-error vector [18].

These experiments suggest the SC may actually send an (updated) *retinal error* signal to the brainstem, rather than a desired eye *motor* command. At the next level, this signal could be combined with signals about absolute eye- and head position and get transformed into the appropriate frame of reference for the control of orienting eye-, head-, and body movements. Such a suggestion has recently been made by Goossens and van Opstal [21] to explain the eye-head gaze-shift patterns of human subjects to auditory and visual targets (see Fig. 2.3), as well as by Tweed [79] on theoretical grounds (see Fig. 2.13).

2.5.5   3D Models: The saccade generator

The burst generator

In the subsequent stage of the saccadic system, the burst generator transforms a (current) motor error signal (constructed by internal feedback) into an eye velocity (and acceleration) signal (the pulse; see also above, Fig. 2.1B,C). The process of pulse-step generation also necessitates the integration of the eye-velocity pulse into a neural estimate of absolute eye position. Because the integral of eye angular velocity does not yield eye position (e.g., Fig. 2.5), an important question is whether the burst generator issues either a coordinate velocity, $\dot{r}$, or an angular velocity signal, ω.

In the two extreme versions of possible 3D saccade models presented here, the vector model (Fig. 2.12A) proposes that the burst generator provides a coordinate velocity output. In contrast, the quaternion model (Fig. 2.12B) assumes the generation of an angular velocity signal. There are several important consequences attached to these two conceptually different neural codes [9], [30], [75]:

- Because the "half-angle rule" requires the rotation axis to tilt with respect to LP as function of eye position (Eqn. 2.35), the activity of cells in the quaternion burst generator should depend on absolute eye position (i.e., the number of spikes in the burst is proportional to $|\mathbf{q}|$, Eqn. 2.38). By contrast the vectorial burst generator should be insensitive to changes in initial eye position (Eqn. 2.39).

- The input to the neural position integrator should be proportional to coordinate velocity (but may contain coordinate acceleration signals as well). In the quaternion model, an eye-position dependent, noncommutative transformation is needed to convert angular velocity into coordinate velocity (Eqn. 2.29). Note that the weeding out of angular acceleration signals is less trivial (Eqn. 2.24)..

- In the eye-displacement scheme (Fig. 2.1C), feedback of velocity should be transformed into current eye displacement, $\mathbf{d}(t)$. According to the quaternion model, this displacement is constructed by an eye-position dependent resettable integrator. The output of the burst generator in the vector model may be linearly integrated to provide eye displacement.

- In the implementation of feedback, a desired movement signal is compared to some internal neural estimate of the actual movement. In the vector model, these comparisons are achieved by commutative vectorial subtractions. For example, the input to the vectorial burst generator is a current motor error signal that is computed as a difference vector: $\mathbf{d}^\star(t) = \mathbf{d}_L - \mathbf{d}(t)$. In the quaternion model, however, the desired rotation is compared to the current movement by rotation vector *division* (Eqn. 2.35): $\mathbf{q}(t) = \mathbf{q}_d \circ \mathbf{d}^{-1}(t)$.

In the discussion of the 1D (and 2D) saccadic system, it became clear that the entire pulse-step generating circuit is needed to counteract the sluggishness (viscosity and elasticity) of the oculomotor plant. Therefore, the dynamics of the 3D oculomotor plant play an important role in the organization of the neural circuitry that is designed to control it, and may explain the need for either coordinate or angular velocity inputs as its drive.

3D oculomotor plant

To extend the 1D plant model to three dimensions, the motoneuron signal has to be treated as a 3D vector, $\mathbf{t}(t)$, which now represents the direction of the net torque acting on the plant. For example, if only the left lateral rectus muscle (LLR) is innervated, with the eye in the primary position, its action can be described by a torque acting in the positive vertical direction, i.e. $\mathbf{t}_{LLR}(t) \propto \mathrm{MN}(t) \cdot \hat{\mathbf{e}}_z$. As a result, the eye will rotate about the vertical axis, so that its angular velocity vector will be aligned with $\mathbf{t}$. The total torque acting on the plant is the vectorial sum of the individual muscle torques, and when the eye attains a fixed position, the net torque should be zero (because the eye does not translate, the sum of the forces on the globe is always zero). There is some controversy to what extent the direction of each muscle's torque vector depends on eye position, i.e., whether

$$\mathbf{t}_i \overset{?}{=} \mathbf{t}(\mathbf{r})_i \quad i = 1 \cdots 6 \tag{2.45}$$

This is not a trivial problem, since the total torque depends on the different forces acting on the globe. For every muscle, $\mathbf{t}_i = \mathbf{c}_i \times \mathbf{F}_i$, where

c_i is the position vector of the muscle pointing from the center of the eye to the point where it leaves the globe. This effective point of insertion may shift for different eye positions, thus changing the direction of that muscle's torque vector. The forces in turn are determined by each muscle's individual (eye-position dependent) and nonlinear length-tension and force-velocity relations (muscles behave as nonlinear springs, although part of the nonlinearity cancels when antagonists are paired).

So both the static elastic and dynamic viscous forces need to be known in order to formulate an adequate model of the 3D plant. Moreover, the elasticity and viscosity parameters of the plant are now described by 3×3 (possibly anisotropic) matrices, instead of by single scalars.

The following two, highly simplified 3D plant models capture the essentials of the problem, because each gives rise to a very different 3D saccade model (Fig. 2.12).

In the so-called *linear plant model*, [30], [73], it is assumed that the muscle torques *do* depend on eye position, but in such a way that when the net innervation of the motoneurons, $\mathbf{t}(t)$, is confined to Listing's plane, the eye will precisely follow the half-angle rule and move along a trajectory in Listing's plane. In this model, the burst signal, $\mathbf{b}(t)$ should be independent of eye position, and is taken to be proportional to the coordinate velocity vector, $\dot{\mathbf{r}}$. The neural integrator, $\mathbf{n}$ may linearly integrate the phasic burst signal (note that contributions of eye acceleration in this signal will not survive the integration, see Eqn. (2.1)). The linear plant model is therefore determined by the following dynamical system, which is a straightforward extension of the 1D Robinson model given in Eqn. (2.3) [30], [73]:

$$
\begin{aligned}
\dot{\mathbf{n}} &= \mathbf{b} \\
\mathbf{t} &= [C]\mathbf{n} + [D]\mathbf{b} \\
\dot{\mathbf{r}} &= [V]^{-1}(\mathbf{t} - [K]\mathbf{r})
\end{aligned}
\tag{2.46}
$$

Here, $[C]$ and $[D]$ are the 3×3 brainstem connection matrices to the three antagonist muscle pairs that embody a neural model of the plant, whereas $[K]$ and $[V]$ are the actual plant elasticity and viscosity tensors. Again, adequate pulse-step matching requires that $[C] = [K]$ and $[D] = [V]$ (In more realistic models, these matrices could incorporate eye position- and velocity-related nonlinearities, but since the brainstem-cerebellar circuitry may cope with any complexity of the actual plant, this wouldn't affect the basic argument.).

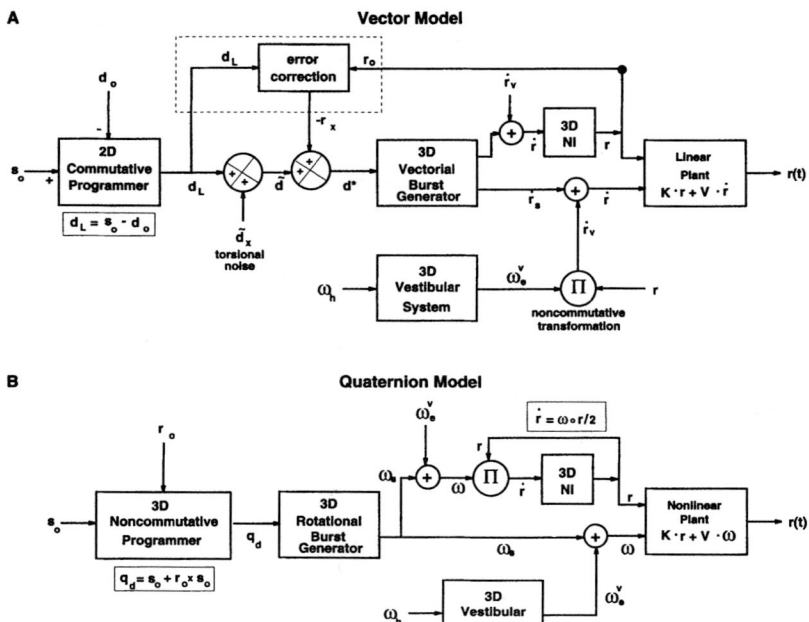

FIGURE 2.12: Two different proposals for the generation of head-fixed saccades (and VOR generation) in three dimensions. Both models are 3D extensions of the Jürgens et al. [41] displacement scheme (Fig. 2.1C).

In the vector model, the saccade programmer provides a commutative approximation of the saccade displacement vector in LP (see Eqn. (2.41) and Fig. 2.11B). The brainstem burst generator encodes the coordinate velocity of the saccade (which is 3D), and the linear plant lets angular velocity tilt, as required by Listing's law. As a result, head angular velocity from the vestibular system has to be transformed into coordinate velocity through a noncommutative transformation Eqn. (2.29). Occasional noise in the torsional channel brings the eye out of LP (by a 3D saccade), which is corrected for by the next (3D) saccade. This pathway presumably involves the NRTP and cerebellar vermis. Note that also the integrator has a 3D representation of eye position (after [30], [57], [58]).

In the quaternion model, the noncommutative programmer encodes the desired rotation about the angular velocity axis, q_d, which is obtained by multiplication of the retinal error vector (approximately equal to the desired displacement vector, see Eqn. 2.41) with the initial eye position: $s_o + r_o \times s_o$ (Eqn. 2.35 and Fig. 2.11C). To comply with the nonlinear plant, the burst generator issues saccadic eye angular velocity, ω_s. The neural integrator is therefore nonlinear, but the vestibular signal may now be transferred directly to the plant (after [9], [42], [75], [77]) (Local feedback loops have been omitted for clarity).

By contrast, in the *nonlinear plant model*, [75], [73], the muscle torques do *not* depend on eye position, so that the directions remain fixed in the head. As a consequence, the net innervation to the plant needs to account for the eye position-dependent tilt of the angular velocity vector (i.e., the burst generator should encode $\omega(t)$), and the input to the neural integrator, $\dot{\mathbf{n}}$, now depends nonlinearly on eye-position. This leads to the following set of model equations (where the noncommutative multiplication rule Eqn. (2.29) for rotation vectors has been applied):

$$
\begin{aligned}
\dot{\mathbf{n}} &= \mathbf{b} + \mathbf{b} \times \mathbf{n} + (\mathbf{b} \bullet \mathbf{n})\mathbf{n} \\
\mathbf{t} &= [C]\mathbf{n} + [D]\mathbf{b} \\
\omega &= [V]^{-1}(\mathbf{t} - [K]\mathbf{r})
\end{aligned}
\tag{2.47}
$$

Alternatively, pulling directions of eye muscles could even be assumed to be eye-fixed (i.e. the net-torque on the globe follows a "full-angle" rule), although this possibility seems unrealistic considering the current anatomical evidence [12]. Presumably, more realistic models of the plant will lie somewhere between the perfect half-angle and zero-angle extremes of Eqn. (2.46) and Eqn. (2.47, [9]). Recent theoretical studies have suggested that so-called muscle pulleys [12] may underlie an appropriate change in the effective point of insertion of the extraocular muscles [60], [61].

2.6   Head-free Saccadic Gaze Shifts in 3D

When the head is also free to move in 3D, the problem gets considerably more complicated. In this section we will restrict the description to the essential new points that arise. It should be first noted, that in this case the relevant variables are the head position and velocity in space, the eye position and velocity in space (or gaze), and the eye position and velocity in the head. Since the eye is moved by the head in the spatial frame of reference, the kinematics of gaze are described as the result of an active rotation of the head in space, followed by the *passive* rotation of the eye in the rotated head, i.e.

$$
\mathbf{g} = \mathbf{e}' \circ \mathbf{h} = \mathbf{h} \circ \mathbf{e}
\tag{2.48}
$$

where in the right-hand part of Eqn. (2.48) both vectors are expressed in spatial coordinates. The eye position in the head is therefore given by

$$
\mathbf{e} = \mathbf{h}^{-1} \circ \mathbf{g} = \frac{\mathbf{g} - \mathbf{h} + \mathbf{g} \times \mathbf{h}}{1 + \mathbf{g} \bullet \mathbf{h}}
\tag{2.49}
$$

and the ocular torsion in the head is therefore

$$
\mathbf{e}_x \approx \mathbf{g}_x - \mathbf{h}_x + \mathbf{g}_y \cdot \mathbf{h}_z - \mathbf{g}_z \cdot \mathbf{h}_y
\tag{2.50}
$$

Note that if gaze and head positions would adhere to Listing's law, the eye in the head in general will not. There are several complicating factors in the analysis and modeling of 3D head-free gaze shifts. First, the eye and the head may move at different speeds, in different directions, and usually do not start simultaneously [26], [21], [79], [70]. Second, the eye reaches the target well before the end of the (slower) head movement, which invokes the reactivation of the VOR. The vestibular system encodes head-angular velocity in 3D and is not concerned with Listing's law, or even Donders' law (see above). So how is the synergy between the oculomotor, the head motor and the vestibular systems organized?

Recent experiments in which large head-free gaze shifts were recorded have revealed the following principles [26], [70]:

- The head follows Donders' law, rather than Listing's law. Head position data can be reasonably well described by a surface produced by a two-axis Fick gimbal (with a twist score of about -0.5, see above).

- For oblique gaze shifts, the head moves predominantly in the horizontal direction, and much less in the vertical direction (i.e., against gravity). The reverse is true for the eye in the head. Because of this, gaze positions in space are also confined to a Fick-like Donders' surface resembling that of the head.

- *During* the gaze shift, the eye in space is *neither* constrained by Listing's law, nor by Donders' law. Instead, the gaze trajectories curve away from the Donders' surface determined by the static gaze positions. This finding can be understood when it is realized that the eye is driven toward a *saturated* position of the eye in space [11], [79], [71]. The 3D saturation prevents the eye from running into the borders of the oculomotor range (OMR; see also Fig. 2.3).

- *After* the gaze shift is over (i.e., eye is on target and the head at rest), gaze-in-space position again follows Donders' law, and eye position in the head is approximately constrained by Listing's law. The latter finding is approximate in the sense that LP is not exactly fixed in the head but shifts with head orientation re. gravity. Most notably, for head-roll positions, a static torsional counterroll of eye-in-head results that may reach about 5 deg.

In a recent 3D extension of current eye-head control models (such as discussed above, and shown in Fig. 2.3), Tweed [79] introduced several additional features that lead to a succesful reproduction of measured eye-head trajectories [71]. A schematic outline of this model is presented in Fig. 2.13. One important feature is that eye and head are controlled by their own oculocentric ($\mathbf{e}_d$) and craniocentric ($\mathbf{h}^*$) motor signals, respectively, driven in independent local feedback modes. The head motor system incorporates the experimental findings that the desired end position complies with Donders' law, and that the horizontal and vertical motor error components are different.

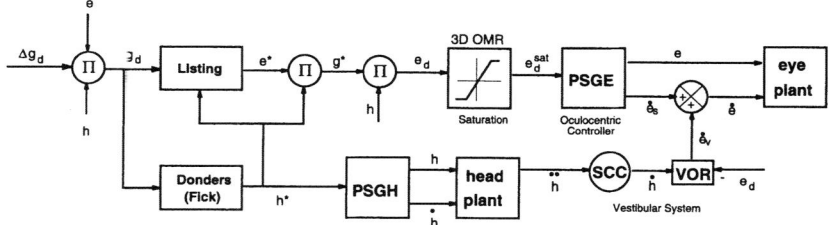

FIGURE 2.13: Three-dimensional proposal of eye-head control model by Tweed [79]. The model is driven by a desired eye position in space (gaze, g_d), which is transformed by the Listing operator into a desired eye position in the head ($e^\star$), such that the eye will obey Listing's law at the *end* of the eye- and head movement. Note, that due to the interaction with the VOR (at the moment of target acquisition), this signal should have three degrees of freedom (is *not* in LP!). The desired gaze shift is transformed by the Donders' operator (modeled by a Fick-like gimbal) into a desired head-in-space signal, $h^\star$, that drives the pulse-step generator of the head motor system. This signal is also transformed into desired gaze-in-space for the eye ($g^\star$). The ocular burst generator is driven by eye motor error (like in Robinson's model of Fig. 2.1), where the desired eye position in the head is clipped by a 3D saturation. The eye- and head plants are assumed to be linear. Π: noncommutative quaternion products. Local feedback loops are omitted for clarity. Compare also with Fig. 2.3. After [79].

A second novel property of the model concerns a proposal for the signal that drives the oculomotor system. In the model, the eye is controlled by a desired position in space ($\mathbf{g}^\star$), but in such a way that at the *end* of the gaze shift eye position in the head ($\mathbf{e}^\star$) will obey Listing's law. Note that this proposal includes a *neural* Listing's law operator interposed between the 2D representation of the SC output (which presumably encodes the updated initial gaze error signal, $\Delta\mathbf{g}_d$) and the oculomotor burst generator. Note also, that the desired eye position itself is *not* constrained by Listing's law (it may have a considerable torsional component), because it has to take the additional (Donders) head movement (and concomitant 3D VOR response) into account that may still proceed after the eye has acquired the target. This crucial prediction of the model has recently been confirmed by behavioural experiments [71].

The noncommutative computations embedded in the Listing operator incorporate both the position of the target in space (which is transformed from the 2D retinal error input, $\Delta\mathbf{g}_d$, into a spatial code, $\mathbf{g}_d$), and the desired position of the head as constrained by Donders' law ($\mathbf{h}^\star$). However, the desired eye position in space cannot be used directly by the oculomotor system, since the eccentricity of this signal often far exceeds the OMR. Therefore, before the signal is passed to the oculomotor burst generator, it is updated as a 3D clipped craniocentric eye position signal ($\mathbf{e}_d^{sat}$) that ensures that the eye motor error command incorporates the current head movement, and does not run into the physical limits imposed by the OMR (for specific details, see [79]).

Both the oculomotor and head pulse-step generators are matched to linear plant models, which ensures that the neural position integrators can operate in a linear mode, and that both burst generators issue coordinate velocity and acceleration signals (see also above). As in the model of Fig. 2.3, the VOR is kept inactivated by the current eye motor-error signal ($\mathbf{e}_d$).

2.7 Conclusion

The noncommutative kinematic principles of 3D rotations that underlie the control of orienting gaze shifts are most likely embedded to a large extent in the central nervous system and are not due to mechanical interactions at the level of the motor plant. Although linear mechanical models of the plant may in principle account for the "half-angle rule"-behavior of ocular saccades in Listing's plane (note that the behavior of the muscle torques under dynamic conditions is still unknown), they are unable to account for a myriad of other 3D kinematic aspects of gaze control:

- The existence of an active (saccadic) error correction mechanism that ensures that the eye stays very close to LP (within 0.6 degrees) despite substantial occasional errors. This mechanism prevents torsional accumulation (due to a random walk), and shows that the saccadic burst generator and eye position integrator represent eye movements in 3D, rather than in 2D retinal coordinates.
- The deficit in torsional error correction after local inactivation of the NRTP.
- The different movement strategies used by the eye and head control systems under head-fixed (Listing) and head-free (Donders) conditions.
- The necessity for a transformation of head (angular) velocity signals of the VOR into appropriate eye (coordinate) velocity signals that can be readily integrated by a position and displacement neural integrator.
- The Helmholtz gimbal-like behavior of the binocular disjunctive vergence system (full-angle elevation-dependence of ocular torsion).
- The systematic dependence of the orientation of LP on head position relative to gravity.
- The systematic shift of LP after unilateral inactivation of neurons in the vertical/torsional burst generator of the riMLF [65].
- The generation of a large predictive torsional saccade in a torsional eye-head gaze shift, that precedes the onset of the head movement [71].

Acknowledgments: This work was supported by the University of Nijmegen, The Netherlands, and the Human Frontiers Science Program (RG0174/1998-B). This chapter leans heavily on the outstanding work of many colleagues, most notably, Klaus Hepp, Tutis Vilis, Doug Crawford, and Doug Tweed. It is dedicated to the memory of our unforgettable assistant Vappu Furrer-Isoviita, and our dear colleague Volker Henn.

APPENDIX A-1: Proof that Quaternions Describe Rotations

Without loss of generality, we choose a right-handed coordinate system such that the rotation axis, $\hat{n}$, coincides with the $\hat{e}_z$ axis, and the vector $\mathbf{u}$ lies in the $(\hat{e}_x, \hat{e}_z)$-plane, making an angle ϕ with the rotation axis. Thus, $\hat{n} = (0, 0, 1)$, and the associated quaternions and vector are (from Eqn. (2.17) and Eqn. (2.18)):

$$
\begin{aligned}
q &= |q|(\cos(\theta) + \sin(\theta) \cdot k) \\
\mathbf{u} &= |u|(\cos(\phi) \cdot k + \sin(\phi) \cdot i) \\
q^{-1} &= |q|^{-1}(\cos(\theta) - \sin(\theta) \cdot k)
\end{aligned} \tag{A1}
$$

By applying the rules defined in Eqn. (2.14) it follows that

$$
\begin{aligned}
\mathbf{u}' = q\mathbf{u}q^{-1} &= |u|\cos(\phi)(qkq^{-1}) + |u|\sin(\phi)(qiq^{-1}) \\
&= |u| \cdot (\cos(2\theta)i + \sin(2\theta)j) \\
&= R(\hat{e}_z, 2\theta)\, \mathbf{u}
\end{aligned} \tag{A2}
$$

which is indeed the vector, $\mathbf{u}$, rotated around axis $\hat{e}_z$ (the invariant component under (A2), about an angle that is *twice* the quaternion angle θ. Note that the length of the quaternion does not play a role in describing the single-axis rotation. For this reason, it is customary to apply unit quaternions ($|q| = 1$).

APPENDIX A-2: Angular and Coordinate Velocity Vectors

First, note that the quaternion product Eqn. (2.15) obeys the following commutator relation:

$$
pq - qp = 2\,\mathbf{p} \times \mathbf{q} \tag{A3}
$$

which follows directly from Eqn. (2.15). Second, we use the result of Appendix A-1, that the rotation of an initial vector, say $\mathbf{s}_o$, can be described by the quaternion product Eqn. (2.19):

$$
\mathbf{s}(t) = q(t)\mathbf{s}_o q^{-1}(t) \tag{A4}
$$

Then, the coordinate velocity vector is obtained by time differentiation of (A4):

$$
\dot{\mathbf{s}}(t) = \dot{q}(t)\mathbf{s}_o q^{-1}(t) + q(t)\mathbf{s}_o \dot{q}^{-1}(t) \tag{A5}
$$

The right-hand factor of this equation may be evaluated as $\dot{q}^{-1} = -q^{-1}\dot{q}q^{-1}$, which follows from taking the time derivative of the definition $qq^{-1} = 1$.

Then, the identity operator $q^{-1}q$ is used once more by substitution in the left-hand part of (A5). By using (A3) and (A4), this yields

$$\dot{s}(t) = (\dot{q}q^{-1})s(t) - s(t)(\dot{q}q^{-1}) = 2V(\dot{q}q^{-1}) \times s(t) \tag{A6}$$

The scalar part of the quaternion $\dot{q}q^{-1}$, given by $S(\dot{q}q^{-1}) = \dot{q}_o q_o^{-1} + \dot{\mathbf{q}} \bullet \mathbf{q} = 0$. This follows from the fact that $S(\dot{q}q^{-1}) = d|q|^2/dt = 0$, because the quaternion length is fixed to 1.

So, to summarize

$$\dot{s}(t) = (2\dot{q}q^{-1}) \times s(t) \implies \omega(t) = 2\dot{q}q^{-1} \quad \text{and} \quad \dot{q} = \frac{\omega(t)q}{2} \tag{A7}$$

To obtain the equivalent relations for the rotation vector representation, we use the definition

$$\mathbf{q} = q_o\mathbf{r} \quad \text{and} \quad \dot{\mathbf{q}} = \dot{q}_o\mathbf{r} + q_o\dot{\mathbf{r}} \tag{A8}$$

and, for unit quaternions Eqn. (2.16):

$$q_o^2 = 1 - \mathbf{q} \bullet \mathbf{q} = \frac{1}{1 + \mathbf{r} \bullet \mathbf{r}} \tag{A9}$$

By substituting (A8) and (A9) into Eqn. (2.23), gives Eqn. (2.28) as a result.

Similarly, substitution of (A8) into Eqn. (2.22) yields:

$$\dot{q}_o\mathbf{r} + q_o\dot{\mathbf{r}} = \frac{q_o\omega + \omega \times q_o\mathbf{r}}{2} \tag{A10}$$

Noting that $\dot{q}_o = -(\omega \bullet \mathbf{q})/2$, and combining this with (A8) and (A10) yields Eqn. (2.29).

APPENDIX A-3: The Half-angle Rule.

The torsional component of the angular velocity vector of the eye, for saccades in Listing's plane, is given by Eqn. (2.28):

$$\omega_x = 2 \cdot (\mathbf{r}_y \cdot \dot{\mathbf{r}}_z - \mathbf{r}_z \cdot \dot{\mathbf{r}}_y) \tag{A11}$$

Suppose that the eye initially looks down (angle ρ_y) and makes a leftward saccade (say with average velocity $\bar{v}_z$):

$$\mathbf{r} = \tan(\rho_y/2)\hat{\mathbf{e}}_y \quad \text{and} \quad \dot{\mathbf{r}} = \bar{v}_z\hat{\mathbf{e}}_z \tag{A12}$$

It then follows that the angle, α, between ω and the primary direction, $\mathbf{p} = \mathbf{r} \times \dot{\mathbf{r}}$, is given by

$$\cos(\alpha) = \frac{(\omega \bullet \mathbf{p})}{|\omega| \cdot |\mathbf{p}|} = \frac{\tan(\rho_y/2)}{|\hat{\mathbf{e}}_z + \tan(\rho_y/2)\hat{\mathbf{e}}_x|} = \sin(\rho_y/2) \Rightarrow \psi = 90 - \alpha = \rho/2 \tag{A13}$$

Thus, when a saccade is made in a direction orthogonal to the eye position vector, the angle, ψ, between ω and LP is given by $\rho_y/2$, which is *half* the eccentricity of initial eye position. This property is known as the *"half-angle rule"*. Note, that in accordance with Listing's law, ω can in principle tilt by any angle between $\{-\rho_2, +\rho/2\}$. For example, when $\mathbf{r}$ and $\dot{\mathbf{r}}$ are parallel (e.g., the eye moves downward (or upward) from a downward fixation) the tilt angle ψ is zero. All eye movements that pass through the primary position have their angular velocity vector in LP.

APPENDIX A-4: Retinal Error and Motor Error Vectors in 3D
The finite rotation formula Eqn. (2.9) gives a straightforward recipe for calculating the orientation of a rotated body, once the rotation axis and angle are known. Suppose that the eye is in the initial eye position, $\mathbf{r}_1 = \tan(\rho_1/2)\,\hat{\mathbf{n}} = R(\hat{\mathbf{n}}_1, \rho_1)$. Since the eye is in Listing's plane, the torsional component of $\mathbf{r}_1$ is zero, and the line of sight (which is eye fixed) points along $\hat{\mathbf{e}}_{1x}$.

A target at position $\mathbf{r}_2$ is foveated as soon as the new direction of the line of sight points at the target. This direction is given by the the gaze line in the new rotated eye-fixed coordinate system, $\hat{\mathbf{e}}_{2x}$.

The coordinates of the target, $\mathbf{s}_o = (s_{o,y}, s_{o,z})$ (i.e., the retinal error), with the eye in the initial position, $\mathbf{r}_1$, are given by the projection of the future line of sight onto the current gaze direction, i.e.:

$$
\begin{aligned}
s_{o,y} &= (\hat{\mathbf{e}}_{2x} \bullet \hat{\mathbf{n}}_{1y}) \\
s_{o,z} &= (\hat{\mathbf{e}}_{2x} \bullet \hat{\mathbf{n}}_{1z})
\end{aligned}
\tag{A14}
$$

When the eye fixates the new target, the gaze direction is described by a rotation vector: $\mathbf{r}_2 = R(\hat{\mathbf{n}}_2, \rho_2)$. By Listing's law Eqn. (2.33), the position axes of both eye positions are in LP so that they are perpendicular to the primary direction of the head-fixed primary Listing frame, $\hat{\mathbf{e}}_{ox}$:

$$
(\hat{\mathbf{n}}_1 \bullet \hat{\mathbf{e}}_{ox}) = 0 \quad \text{and} \quad (\hat{\mathbf{n}}_2 \bullet \hat{\mathbf{e}}_{ox}) = 0
\tag{A15}
$$

We now express the retinal error in terms of the initial and final eye positions, by using the third-order Taylor approximation of the rotation formula Eqn. (2.10), and the properties of vector triple-products: $\mathbf{a} \times (\mathbf{b} \times \mathbf{c}) = (\mathbf{a} \bullet \mathbf{c})\mathbf{b} - (\mathbf{a} \bullet \mathbf{b})\mathbf{c}$ and $\mathbf{a} \bullet (\mathbf{b} \times \mathbf{c}) = (\mathbf{a} \times \mathbf{b}) \bullet \mathbf{c}$. Substitution into (A14) (only including terms up to $\mathcal{O}(\rho^3)$) yields (with k=y, z):

$$
\begin{aligned}
s_{o,k} &= (R_2(\hat{\mathbf{n}}_2, \rho_2)\hat{\mathbf{e}}_{ox} \bullet R_1(\hat{\mathbf{n}}_1, \rho_1)\hat{\mathbf{e}}_{ok}) \\
&= (\hat{\mathbf{e}}_{ox} + \rho_2\hat{\mathbf{n}}_2 \times \hat{\mathbf{e}}_{ox} + \tfrac{1}{2}\rho_2^2\hat{\mathbf{n}}_2 \times (\hat{\mathbf{n}}_2 \times \hat{\mathbf{e}}_{ox})) \bullet \\
&\quad\quad (\hat{\mathbf{e}}_{ok} + \rho_1\hat{\mathbf{n}}_1 \times \hat{\mathbf{e}}_{ok} + \tfrac{1}{2}\rho_1^2\hat{\mathbf{n}}_1 \times (\hat{\mathbf{n}}_1 \times \hat{\mathbf{e}}_{ok})) \\
&= \rho_2\hat{\mathbf{e}}_{ok} \bullet (\hat{\mathbf{n}}_2 \times \hat{\mathbf{e}}_{ox}) + \rho_1\hat{\mathbf{e}}_{ox} \bullet (\hat{\mathbf{n}}_1 \times \hat{\mathbf{e}}_{ok}) \\
s_{o,y} &= (\rho_2\hat{\mathbf{n}}_2 - \rho_1\hat{\mathbf{n}}_1) \bullet \hat{\mathbf{e}}_{oz} = \quad 2 \cdot \mathbf{d}_z \\
s_{o,z} &= (\rho_1\hat{\mathbf{n}}_1 - \rho_2\hat{\mathbf{n}}_2) \bullet \hat{\mathbf{e}}_{oy} = -2 \cdot \mathbf{d}_y
\end{aligned}
\tag{A16}
$$

Thus, in good approximation, retinal error is linearly related to the difference vector in Listing's plane between the final and initial eye positions. For eye positions within the periprimary range of 15 degrees (the range within which the far majority of natural saccades is made), the approximation is better than 2.5%; but even up to 30 degrees the error is of the same order than the natural scatter in saccade accuracy: 12% [30].

2.8 REFERENCES

[1] E. Bizzi, R.E. Kalil, and V. Tagliasco. Eye-head coordination in monkeys: evidence for centrally patterned organization. *Science*, 173:452-454, 1971.

[2] L. Brand. *Vector and Tensor Analysis*. Wiley, New York, USA, 1948.

[3] R.H.S. Carpenter. *Movements of the Eyes*. Pion, London UK, 1988.

[4] J.D. Crawford, W. Cadera, and T. Vilis. Generation of torsional and vertical eye position signals by the interstitial nucleus of cajal. *Science*, 252:1551-1553, 1991.

[5] H. Collewijn, F. van der Mark, and T.C. Jansen. Precise recording of human eye movements. *Vision Res.*, 15:447-450, 1987.

[6] J.D. Crawford and D. Guitton. Visual-motor transformations required for accurate and kinematically correct saccades. *J. Neurophysiol.*, 78:1447-1467, 1997.

[7] S.C. Cannon and D.A. Robinson. Loss of the neural integrator of the oculomotor system from brain stem lesions in monkey. *J. Neurophysiol.*, 57:1383-1409, 1987.

[8] J.D. Crawford. The oculomotor neural integrator uses a behavior-related coordinate system. *J. Neurosci.*, 14:6911-6923, 1994.

[9] J.D. Crawford. Geometric transformations in the visual-motor interface for saccades. In M. Fetter, D. Tweed, and H. Misslisch, editors, *Three-Dimensional Kinematics of Eye-, Head-, and Limb Movements*, pages 85-99. Harwood Academic Publ., Amsterdam, 1997.

[10] J.D. Crawford and T. Vilis. Symmetry of oculomotor burst neuron coordinates about listing's plane. *J. Neurophysiol.*, 68:432-448, 1992.

[11] J.D. Crawford, M.Z. Ceylan, E.M. Klier, and D. Guitton. Three-dimensional eye-head coordination during gaze saccades in the primate. *J. Neurophysiol.*, 81:1760-1782, 1999.

[12] J.L. Demer, J.M. Miller, V. Poukens, H.V. Vinters, and B.J. Glasgow. Evidence for fibromuscular pulleys of the recti extraocular muscles. *Invest. Ophthalmol. Vis. Sci.*, 36:1125-1136, 1995.

[13] L. Ferman, H. Collewijn, and A.V. van den Berg. A direct test of listing's law. i. Human ocular torsion measured under static conditions. *Vision Res.*, 27:929-938, 1987.

[14] L. Ferman, H. Collewijn, and A.V. van den Berg. A direct test of listing's law. ii. Human ocular torsion measured under dynamic conditions. *Vision Res.*, 27:939-951, 1987.

[15] L. Ferman, H. Collewijn, T.C. Jansen, and A.V. van den Berg. Human gaze stability in the horizontal, vertical, and torsional direction during voluntary head movements, evaluated with a three-dimensional scleral induction coil technique. *Vision Res.*, 27:811-828, 1987.

[16] M. Fetter, T. Haslwanter, H. Misslisch, and D. Tweed, editors. *Three-Dimensional Kinematics of Eye-, Head-, and Limb Movements.* Harwood Academic Publishers, Amsterdam, 1997.

[17] M.A. Frens and A.J. van Opstal. A quantitative study of auditory-evoked saccadic eye movements in two dimensions. *Exp. Brain Res.*, 107:103-117, 1995.

[18] M.A. Frens and A.J. van Opstal. Monkey superior colliculus activity during short-term saccadic adaptation. *Brain Res. Bull.*, 43:473-483, 1997.

[19] E.G. Freedman, T.R. Stanford, and D.L. Sparks. Combined eye-head gaze shifts produced by electrical stimulation of the superior colliculus in rhesus monkeys. *J. Neurophysiol.*, 76:927-952, 1996.

[20] H.L. Galiana and D. Guitton. Central organization and modeling of eye-head coordination during orienting gaze shifts. *Ann. Acad. Sci.*, 656:452-471, 1992.

[21] H.H.L.M. Goossens and A.J. van Opstal. Human eye-head coordination in two dimensions under different sensorimotor conditions. *Exp. Brain Res.*, 114:542-560, 1997.

[22] H. Goldstein. *Classical Mechanics*. Addison-Wesley, Reading, MA, 1980.

[23] J.A.M. van Gisbergen, A.J. van Opstal, and J.J.M. Schoenmakers. Experimental test of two models for the generation of oblique saccades. *Exp. Brain Res.*, 57:321-336, 1985.

[24] J.A.M. van Gisbergen, D.A. Robinson, and S. Gielen. A quantitative analysis of generation of saccadic eye movements by burst neurons. *J. Neurophysiol.*, 45:417-442, 1981.

[25] D. Guitton and M. Volle. Gaze control in humans: eye-head coordination during orienting movements to targets within and beyond the oculomotor range. *J. Neurophysiol.*, 58:427-459, 1987.

[26] B. Glenn and T. Vilis. Violations of listing's law following large eye and head gaze shifts. *J. Neurophysiol.*, 68:309-318, 1992.

[27] T. Haslwanter. Mathematics of three-dimensional eye rotations. *Vision Res.*, 35:1727-1739, 1995.

[28] W. Haustein. Considerations on listing's law and the primary position by means of a matrix description of eye position control. *Biol. Cybern.*, 60:411-420, 1989.

[29] H. von Helmholtz. *Handbuch der Physiologischen Optik*. Voss, Hamburg, 1867. English translation 1962, Dover, New York.

[30] K. Hepp. On listing's law. *Commun. Math. Phys.*, 132:285-292, 1990.

[31] K. Hepp. Saccades and listing's law. In M. Arbib, editor, *Handbook of Brain Theory and Neural Networks*, pages 826-830. MIT Press, Cambridge, MA, 1995.

[32] E. Hering. Die raumsinn und die bewegungen des auges. In L. Hermann and F.C.W. Vogel, editors, *Handbuch der Physiologie*, volume 3, part 1, pages 343-601. Leipzig, 1879.

[33] B.J.M. Hess. Dual-search coil for measuring three-dimensional eye movements in experimental animals. *Vision Res.*, 30:597-602, 1990.

[34] K. Hepp, V. Henn, T. Vilis, and B. Cohen. Brainstem regions related to saccade generation. In R.H. Wurtz and M.E. Goldberg, editors, *The Neurobiology of Saccadic Eye Movements*, volume 3, pages 105-212. Elsevier, Amsterdam, 1989.

[35] P.E. Hallet and A.D. Lightstone. Saccadic eye movements toward stimuli triggered by prior saccades. *Vision Res.*, 66:88-106, 1976.

[36] V. Henn, W. Lang, K. Hepp, and H. Reisine. Experimental gaze palsies in monkeys and their relation to human pathology. *Brain*, 107:619-636, 1984.

[37] K. Hepp, A.J. van Opstal, D. Straumann, B.J.M. Hess, and V. Henn. Monkey superior colliculus represents rapid eye movements in a two-dimensional motor map. *J. Neurophysiol.*, 69:965-979, 1993.

[38] B.J.M. Hess, A.J. van Opstal, D. Straumann, and K. Hepp. Calibration of three-dimensional eye position using search coil signals in the rhesus monkey. *Vision Res.*, 32:1647-1654, 1992.

[39] T. Haslwanter, D. Straumann, K. Hepp, B.J.M. Hess, and V. Henn. Smooth pursuit eye movements obey listing's law in the monkey. *Exp. Brain Res.*, 87:470-472, 1991.

[40] V. Henn, D. Straumann, B.J.M. Hess, A.J. van Opstal, and K. Hepp. The generation of torsional and vertical rapid eye movements in the riMLF. In H. Shimazu and Y. Y. Shinoda, editors, *Vestibular and Brainstem Control of Eye-, Head-, and Body Movements*, pages 177-182. Japanese Scientific Societies Press, Tokyo/S. Karger, Basel, 1992.

[41] R. Jürgens, W. Becker, and H.H. Kornhuber. Natural and drug-induced variations of velocity and duration of human saccadic eye movements: Evidence for a control of the neural pulse generator by local feedback. *Biol. Cybern.*, 39:87-96, 1981.

[42] E.M. Klier and J.D. Crawford. The human oculomotor system accounts for 3D eye orientation in the visual-motor transformation for saccades. *J. Neurophysiol.*, 80:2274-2294, 1998.

[43] E.L. Keller, N.J. Gandhi, and J.M. Shieh. Endpoint accuracy in saccades interrupted by stimulation in the omnipause region in monkey. *Vis. Neurosci.*, 13:1059-1067, 1996.

[44] M. Lagally. *Vorlesungen über Vektor-Rechnung*. Akademische Verlagsgesellschaft, Leipzig, Ger., 1928.

[45] A.H.W. Minken and J.A.M. van Gisbergen. A three-dimensional analysis of vergence eye movements at various levels of elevation. *Exp. Brain Res.*, 101:331-345, 1994.

[46] W.P. Medendorp, B.J.M. Melis, C.C.A.M. Gielen, and J.A.M. van Gisbergen. Off-centric rotation axes in natural head movements: Implications for vestibular reafference and kinematic redundancy. *J. Neurophysiol.*, 79:2025-2039, 1998.

[47] A.H.W. Minken, A.J. van Opstal, and J.A.M. van Gisbergen. Three-dimensional analysis of strongly curved saccades elicted by double-step stimuli. *Exp. Brain Res.*, 93:521-533, 1993.

[48] D. Mok, A. Ro, W. Cadera, J.D. Crawford, and T. Vilis. Rotation of listing's plane during vergence. *Vision Res.*, 32:2055-2064, 1992.

[49] L.E. Mays and D.L. Sparks. Saccades are spatially, not retinocentrically, coded.. *Science*, 208:1163-1165, 1980.

[50] D.P. Munoz, D.M. Waitzman, and R.H. Wurtz. Activity of neurons in monkey superior colliculus during interrupted saccades. *J. Neurophysiol.*, 73:2313-2333, 1995.

[51] K. Nakayama. Coordination of extraocular muscles. In G. Lennerstrand and P. Bach-y Rita, editors, *Basic Mechanisms of Ocular Motility and their Clinical Implications*, pages 193-209. Pergamon Press, Oxford, UK, 1975.

[52] M.J. Nichols and D.L. Sparks. Component stretching during oblique stimulation-evoked saccades: The role of the superior colliculus. *J. Neurophysiol.*, 76:582-600, 1996.

[53] A.J. van Opstal and K. Hepp. A novel interpretation for the collicular role in saccade generation. *Biol. Cybern.*, 73:431-445, 1995.

[54] A.J. van Opstal, K. Hepp, B.J.M. Hess, D. Straumann, and V. Henn. Two-, rather than three-dimensional representation of saccades in monkey superior colliculus. *Science*, 252:1313-1315, 1991.

[55] A.J. van Opstal, K. Hepp, B.J.M. Hess, D. Straumann, and V. Henn. Experimental test of two models for the role of monkey superior colliculus in 3D saccade generation. In A. Berthoz, editor, *Multisensory Control of Movement*, pages 240-254. Oxford University Press, Oxford, UK, 1993.

[56] A.J. van Opstal, K. Hepp, Y. Suzuki, and V. Henn. Influence of eye position on activity in monkey superior colliculus. *J. Neurophysiol.*, 74:1593-1610, 1995.

[57] A.J. van Opstal, K. Hepp, Y. Suzuki, and V. Henn. Role of monkey nucleus reticularis tegmenti pontis in the stabilization of listing's plane. *J. Neurosci.*, 16:7284-7296, 1996.

[58] A.J. van Opstal, K. Hepp, Y. Suzuki, and V. Henn. Three-rather than two-dimensional burst generation for spontaneous saccadic eye movements. In M. Fetter, D. Tweed, and H. Misslisch, editors, *Three-Dimensional Kinematics of Eye-, Head-, and Limb Movements*, pages 73-84. Harwood Academic Publ., Amsterdam, 1997.

[59] A.J. van Opstal. Representation of eye position in three dimensions. In A. Berthoz, editor, *Multisensory Control of Movement*, pages 27-41. Oxford University Press, Oxford, UK, 1993.

[60] C. Quaia and L.M. Optican. Commutative saccadic generator is sufficient to control a 3-d ocular plant with pulleys. *J. Neurophysiol.*, 79:3197-3215, 1998.

[61] T. Raphan. Modeling control of eye orientation in three dimensions. i. Role of muscle pulleys in determining saccadic trajectory. *J. Neurophysiol.*, 79:2653-2667, 1998.

[62] A. Roucoux, D. Guitton, and M. Crommelinck. Stimulation of the superior colliculus of the alert cat. ii. eye and head movements evoked when the head is unrestrained. *Exp. Brain Res.*, 39:75-85, 1980.

[63] D.A. Robinson. Eye movements evoked by collicular stimulation in the alert monkey. *Vision Res.*, 12:1795-1808, 1972.

[64] D.A. Robinson. Oculomotor control signals. In G. Lennerstrand and P. Bach-y Rita, editors, *Basic Mechanisms of Ocular Motility and their Clinical Implications*, pages 337-374. Pergamon Press, Oxford, UK, 1975.

[65] Y. Suzuki, J.A. Büttner-Ennever, D. Straumann, K. Hepp, B.J.M. Hess, and V. Henn. Deficits in torsional and vertical rapid eye movements and shift of listing's plane after uni- and bilateral lesions of the rostral interstitial nucleus of the medial longitudinal fasciculus. *Exp. Brain Res.*, 106:215-232, 1995.

[66] C. Schnabolk and T. Raphan. Modelling three-dimensional velocity-to-position transformation in oculomotor control. *J. Neurophysiol.*, 71:623-638, 1994.

[67] T.J. Stanford and D.L. Sparks. Systematic errors for saccades to remembered targets: Evidence for a dissociation between saccade metrics and activity in the superior colliculus. *Vision Res.*, 34:93-106, 1994.

[68] D. Straumann, D.S. Zee, D. Solomon, A.G. Lasker, and D.C. Roberts. Transient torsion during and after saccades. *Vision Res.*, 35:3321-3334, 1995.

[69] D. Tweed, M. Fetter, S. Andreadaki, E. König, and J. Dichgans. Three-dimensional properties of human pursuit eye movements. *Vision Res.*, 32:1225-1238, 1992.

[70] D. Tweed, B. Glenn, and T. Vilis. Eye-head coordination during large gaze shifts. *J. Neurophysiol.*, 73:766-779, 1995.

[71] D. Tweed, T. Haslwanter, and M. Fetter. Optimizing gaze control in three dimensions. *Science*, 281:1363-1366, 1998.

[72] D. Tweed, T. Haslwanter, V. Happe, and M. Fetter. Noncommutativity in the brain. *Nature*, 399:261-263, 1999.

[73] D. Tweed, H. Misslisch, and M. Fetter. Testing models of the oculomotor velocity-to-position transformation. *J. Neurophysiol.*, 72:1425-1429, 1994.

[74] W.-K. Tung. *Group Theory in Physics*. World Scientific Publishing Co., Singapore, 1985.

[75] D.B. Tweed and T. Vilis. Implications of rotational kinematics for the oculomotor system in three dimensions. *J. Neurophysiol.*, 58:832-849, 1987.

[76] D.B. Tweed and T. Vilis. Rotation axes of saccades. *Ann. N.Y. Acad. Sci.*, 545:128-139, 1988.

[77] D.B. Tweed and T. Vilis. The superior colliculus and spatiotemporal translation in the saccadic system. *Neural Networks*, 3:75-86, 1990.

[78] D. Tweed. Kinematic principles of three-dimensional gaze control. In M. Fetter, D. Tweed, and H. Misslisch, editors, *Three-Dimensional Kinematics of Eye-, Head-, and Limb Movements*, pages 17-31. Harwood Academic Publ., Amsterdam, 1997.

[79] D. Tweed. Three-dimensional model of the human eye-head saccadic system. *J. Neurophysiol.*, 77:654-666, 1997.

[80] T. Vilis. Physiology of three-dimensional eye movements: saccades and vergence. In M. Fetter, D. Tweed, and H. Misslisch, editors, *Three-Dimensional Kinematics of Eye-, Head-, and Limb Movements*, pages 59-72. Harwood Academic Publ., Amsterdam, 1997.

[81] G. Westheimer. Kinematics of the eye. *J. Opt. Soc. Am.*, 47:967-974, 1957.

3

Integrating Anatomy and Physiology of the Primary Visual Pathway: From LGN to Cortex

K. Funke, Z. F. Kisvárday, M. Volgushev, and F. Wörgötter

ABSTRACT This chapter deals with the structure and function of the visual thalamus (lateral geniculate nucleus, LGN) and the primary visual cortex and aims to put this system into a computational perspective. We start with an overview of the basic structures of the primary visual pathway and the terminology used. Next, the organization of the LGN and its main functions are described: receptive field structure of LGN cells, excitatory and inhibitory influences, contrast gain- control, spatial summation, temporal structure of activity and influence of extra-retinal inputs. The section closes with models on three functional aspects of the LGN: 1) Switching between burst firing and tonic transmission modes of LGN cells, 2) Control of LGN function during the sleep-wake cycle, and 3) Involvement of LGN in gating visual signals. The section on the visual cortex starts with details of its morphological organisation: cortical layers, cell types, columnar structure and horizontal connections. This is followed by a description of the basic response characteristics of neurons, the organisation of receptive fields and their dynamic behavior. Here, mechanisms of establishing cortical orientation selectivity are considered in detail. Next, we focus on functional maps, e.g. distribution of orientation preferences of cells. The chapter closes with a section on basic models of the primary visual cortex, concerning: 1) Temporal firing patterns of neuronal assemblies, i.e. oscillations and synchronization, 2) Cortical cell characteristics, e.g. orientation specificity, and 3) Formation of functional maps, e.g. orientation map.

3.1 Introduction

The visual system is the most intensively studied neuronal structure in vertebrates. Since the pioneering studies of Hubel and Wiesel (Hubel and Wiesel, 1959; 1962) a vast amount of literature has accumulated, covering a broad range of anatomical and physiological aspects of the primary visual pathway.

This chapter gives an anatomical and functional description of the thalamic and cortical components of the primary visual pathway. It is intended to provide an overview of the main components of this pathway and, therefore, we leave out many of the fine details. We will try to put the described anatomical and physiological observations into a computational perspective. In this section, we will start with an overview of the primary visual pathway, naming the chief components which will then be described in more detail in the following sections. Every major section will be followed by a section on computational approaches, which were undertaken to model the respective substructure or processing stage of the primary visual pathway.

3.1.1   The primary visual pathway: An overview

The primary visual pathway consists of three hierarchically arranged main structures: retina; lateral geniculate nucleus (LGN), which is a part of the thalamus; and primary areas of the visual cortex. All three structures can be divided into subsystems, each of which is linked to a number of brain regions (e.g., superior colliculus). The visual cortex itself is subdivided into a multitude of areas (more than 20). Here the strict hierarchical arrangement of connections, which hallmarks the relationship between the first three stages (from retina to V1 via the LGN), is given up and replaced by a widely branching parallel arrangement. The retina is characterized by a sophisticated anatomical structure which carries out a preprocessing of the visually induced signals generated first in the photoreceptors. It differs from the other two main parts of the primary visual pathway in that it does not receive (significant) feedback from structures higher up in the visual hierarchy. For this chapter, we will adopt a rather simplified view and regard the retina only as the input structure which translates light energy into spiking patterns of the retinal ganglion cells. It will not be considered any further.

The LGN receives afferent input from the retina and a strong efferent connection from the visual cortex. Intra-LGN connections, many of which are inhibitory, are used to improve the signal-to-noise ratio of the LGN responses, for example, by increasing the mutual antagonism between excitatory receptive field center and inhibitory surround. Connections from the brain stem and from the visual part of the nucleus reticularis thalami (perigeniculate nucleus, PGN), are involved in controlling the signal transmission properties of LGN cells. Depending on these inputs, LGN cells can change their different response modes. The rich connectivity of the LGN with other brain structures explains the many different behavioral states of cells observed already at this level.

The visual cortex is a complex accumulation of interconnected areas. Only those early in the hierarchy are exclusively concerned with visual signal processing. Areas higher up in the stream start to integrate nonvisual sensory modalities and neuronal signals (limbic system, etc.). This chapter is only concerned with the primary (and secondary) visual cortex (monkey: V1,V2; cat: area 17, area 18), a discussion of other cortical structures has been omitted.

In general, cell responses become more and more complex higher up in the visual hierarchy. Retinal ganglion cells have a disk-shaped center-surround receptive field organization, with a weak antagonism between center and surround. This basic spatial receptive field structure persists in the LGN with an increased antagonism. In the cortex, the spatial shape of the receptive field changes; cortical receptive fields are elongated and selective for the orientation of a visual stimulus. Relay cells can alter their response characteristic in a nonlinear way such that their temporal behavior can vary strongly even when the same stimulus is presented. At the level of the cortex even more complex temporal behavior can be observed. Hence, modern theories of the primary visual pathway that try to describe and explain visual information processing must take into account spatial *and* temporal aspects of the observed cell behavior.

3.1.2   Definitions: The receptive field

The receptive field of a neuron in the visual system is the region of retina over which the firing of that cell can be influenced (Hartline, 1940; Hubel and Wiesel, 1962). Since in visual neurophysiological experiments the eye movements are eliminated, projection of the visual space onto the retina remains constant, allowing to translate coordinates of the visual space into retinal coordinates. The receptive field can be divided into subregions on the basis of a number of different criteria. *Discharge field* of a visual neuron is a region of the retina, or a region of the visual space which is projected onto that retinal region where presentation of stimuli evokes spiking of that neuron. *Receptive field surround* is a region of the visual space where presentation of stimuli does not lead to spiking of a cell, but can influence responses of the cell to other stimuli. Responses evoked by the onset of a light stimulus are called *On-responses*, and a region from which they can be induced is an *On-region* of the receptive field. Responses evoked by the offset of the light stimulus or by the onset of a dark stimulus are *Off-responses*, and a region from which they can be evoked is an *Off-region*. We note here, that On/Off-subregions are not equal to excitatory/inhibitory subregions of the receptive field. For example, suppression of the On-responses by light stimuli presented within the Off-subregion can be caused simply by a decreased background excitatory drive from the Off subregion. That region cannot be considered as On-inhibitory. Only if On-suppression is mediated through an inhibitory interneuron which receives On-inputs, the respective region is inhibitory. Conventional extracellular recordings do not allow to distinguish between these two possibilities. When visual stimulation produces excitatory postsynaptic potentials (*PSPs*) in a cell, the response of this cell is called excitatory, and the region from which excitatory responses are evoked is an *excitatory region*. Similarly, a region is called *inhibitory* visual stimulation evokes inhibiting PSPs.

3.2   The LGN

3.2.1   General view and functional anatomy

The main function of the LGN

The LGN is the major retinocortical relay for visual signals. It is thought to mediate the kind of visual information that leads to a conscious perception of the visual environment. Lesioning this pathway causes a severe deficit in visual perception. Nevertheless, some visually guided reflex responses persist if other visual pathways (e.g. via the midbrain) remain intact, a phenomenon known as *blindsight*. From a topographic point of view the LGN can be described as an *internal retina* because retinotopic organization is preserved and the spatial characteristics of retinal and geniculate receptive fields (RFs) are similar. However, remarkable differences in the strength and temporal course of retinal and geniculate visual responses exist. Therefore, the major function of the LGN seems to be the control of the visual information flow before it enters the visual cortex. This includes the generation of distinct temporal patterns of spike activity. Neural and biochemical mechanisms of the retina adapt to changes in the external world, e.g. during dark adaptation. In addition to this the LGN is remarkably sensitive to changes of the *internal world*, e.g. the changing level of arousal of the individual, and can modify retino-cortical transmission in a number of ways. In the following, we first give a short overview about the fundamental neuronal connectivity of the system and then provides a detailed description of the physiological properties of the LGN.

The retino-thalamo-cortical pathway

Following the general layout of the visual pathway the LGN receives roughly the same amount of input from both eyes. However, the afferents terminate in an eye-specific manner in the LGN. The cat LGN consist of two dorsally located A-layers (A,A1) and the C-complex (Fig. 3.1). The axons of the temporal aspects of the retinae (the lateral halves) stay on the same side of the brain (ipsilateral) and terminate in layer A1 (second layer from top) and in a sublayer of the C-complex. Those axons leaving the nasal hemiretina cross the midline of the brain at the *optic chiasm*, join the axons of the temporal hemiretina of the other eye, and terminate within the contralateral layers A and C of LGN. Guided axonal outgrowth during development leads to retinotopically matched maps of the visual field in all layers of the LGN which are in register. In this way, vertical columns across the layers receive separate inputs from either eye but sensory information

about the same point of the visual field (Fig. 3.1). Further descriptions
will also focus on the cat visual system and the pathways relayed by the
A-layers of the LGN.

The parallel channels through the LGN

The retinocortical pathway mediated via the LGN A-layers (A,A1) of the
cat is functionally separated into On, Off, magno-(Y) and parvocellular (X)
subsystems. Already at an early state of retinal processing, increments and
decrements in light intensity are handled by separate On and Off channels,
respectively (Schiller, 1992). A change in illumination leads to opposite
changes of activity in these two channels: On cells show an increased firing rate in response to light increments within their receptive field, while
Off cell firing rate is reduced at the same time; the opposite behavior is
found for a light decrement. On ganglion cells project exclusively to LGN
relay cells of the On-type and retinal Off-cells project exclusively to LGN
Off-type cells. Most of the LGN relay cells receive one dominant input
from one ganglion cell and a few weaker inputs from other ganglion cells
of the same type (Mastronarde, 1992). Based on morphological (Boycott
and Wässle, 1974; Wässle and Boycott, 1991) and electrophysiological criteria (see Casagrande and Norton, 1991) the primary visual pathway via
the LGN A-layers is further divided into magnocellular (α,Y) and parvocellular (β,X) channels, each containing roughly equal numbers of On and
Off cells. α(Y)-type ganglion cells and LGN Y-cells possess larger dendritic
trees than β(X)-type ganglion cells and LGN X-cells. The receptive fields
of Y-cells are about three times larger than those of X-cells at all eccen-
tricities. LGN Y-cells receive multiple input from α-ganglion cells while
the activity of LGN X-cells seems to be dominated by one X-type ganglion cell. About 5% of the retinal ganglion cells are of the Y-type, 40-50%
of the X-type, and the remainder belongs to the diverse group of γ-cells
(W-type). The latter primarily constitute the second visual pathway which
terminates in midbrain structures after being relayed by the C-laminae of
the LGN. This pathway will not be considered further. In the LGN, the ratio changes to about 60% Y-cells and 40% X-cells. Together with the more
widespread terminal field of Y-cell axons in LGN and cortex the spatial
influence of the magnocellular pathway increases stepwise on the way up
to higher processing areas. The spatial distribution of the dendritic trees
of the different ganglion cells overlap in a way that each point of the visual
field is represented by at least one Y-On, Y-Off, X-On and X-Off cell in the
ascending pathway. With the exception of the perigeniculate nucleus (see
below), these channels remain almost separate at the subcortical level.

The LGN circuitry

The complex connectivity of the LGN is schematically shown by the wiring
diagram in Fig. 3.2. The main vertical stream originating in retinal gan-

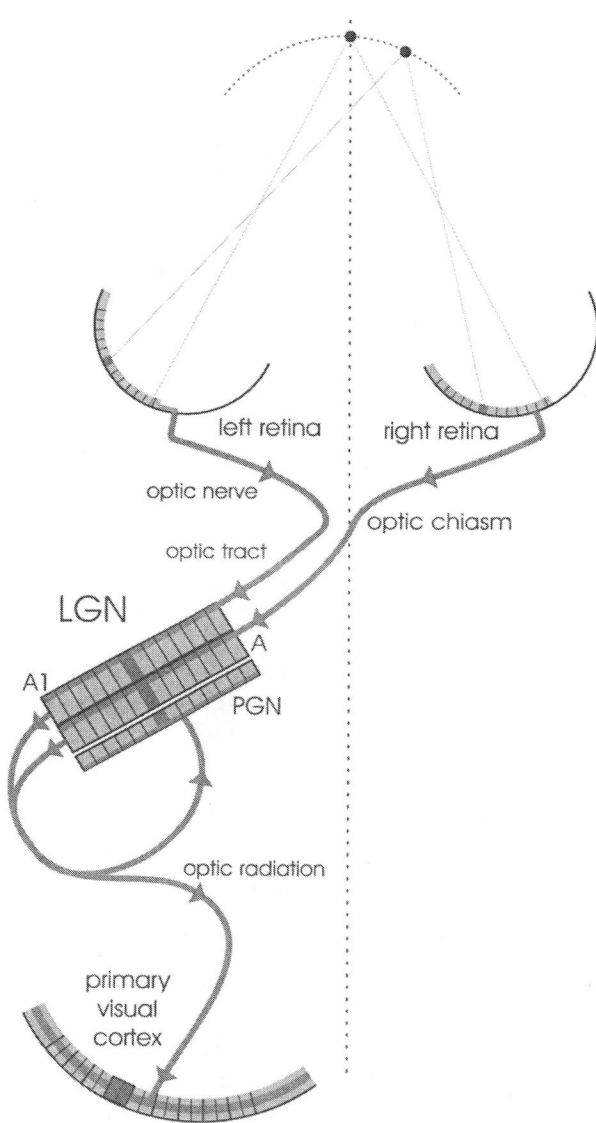

FIGURE 3.1. Basic organization of the primary visual pathway. Projections and topographic arrangements.

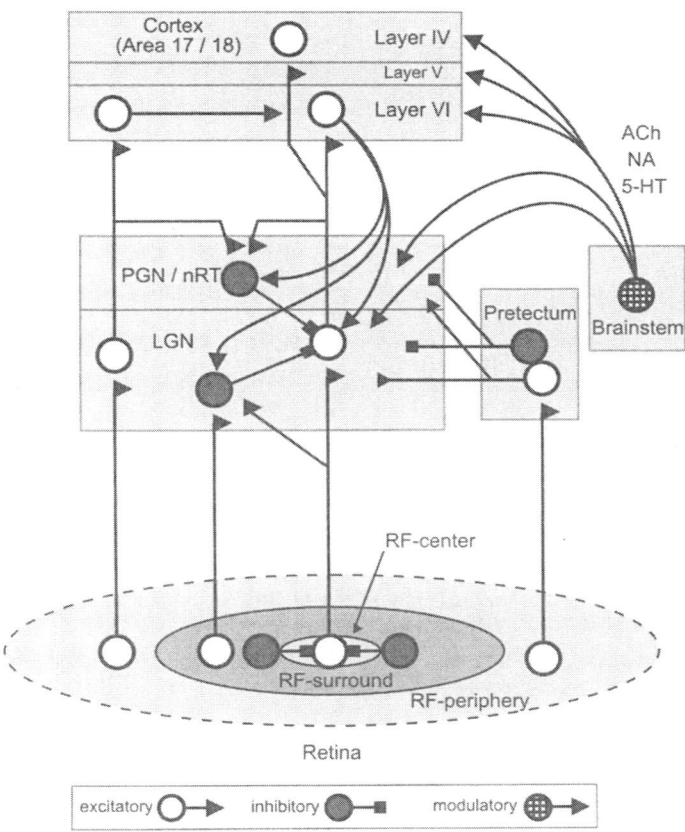

FIGURE 3.2. Wiring diagram of LGN connectivity. Upstream primary visual pathway, inhibitory connections, feedback from the primary visual cortex and modulatory inputs from pretectum and brain stem.

glion cells goes via LGN relay cells to the primary visual cortex. The visual cortex sends off excitatory back-projections to the LGN, which are gluta-matergic and form a positive feedback loop. Axon collaterals of the relay cells terminate on the neurons of the perigeniculate nucleus (PGN), which is part of the thalamic reticular nucleus (TRN) and joins the dorsolateral aspect of the LGN. The PGN neurons project back to the LGN, thereby establishing a negative feedback loop because they use the inhibitory trans-mitter, γ-amino-butyric acid (GABA). About 20-25% of the LGN neurons are local inhibitory (GABAergic) interneurons. They receive direct retinal input and project to LGN relay cells (feedforward inhibition). Their axons do not leave the nucleus and probably do not cross the layers (Wilson, 1993). Only 10-15% of the synapses on LGN relay cells are of retinal ori-gin. About 40% are cortical, about 20% can be related to local inhibitory circuits, including the PGN, and the remainder is made by diverse visual and nonvisual pathways originating in different brain stem structures. The latter use several different transmitters (acetylcholine, noradrenaline, serotonin, dopamine, GABA, and glutamate) and are summarized as so called *modulatory inputs.* Corticofugal and modulatory projections also terminate on local LGN interneurons and PGN neurons.

3.2.2   Physiological properties of the LGN

Organization of Receptive Fields

The spatial organization of LGN RFs is similar to their retinal counterparts. They are almost circular in shape and are composed of an excitatory center and an antagonistic (inhibitory) surround (Kuffler, 1953; Hubel and Wiesel, 1961). In mathematical terms the RF can be described as a combination of two centered Gaussian sensitivity profiles of opposite polarity (Rodieck and Stone, 1965; see Fig. 3.3A). The summation of both profiles results in a *Mexican-hat like* structure, the basis for a spatial filter. If the size of a spot of light flashed on the center of the RF is increased, first there is an increase in the response of the neuron and then when the spot extends beyond the center-surround border, there is a decrease in the response. Y and X cells differ with respect to their RF size and contrast sensitivity (the relative illumination to background to which the system is adapted). The RFs of Y-cells are about 3 times larger than those of X-cells and have higher sensitivity to luminance changes than X-cell RFs. In addition, the relative strength of the antagonistic RF surround seems to be weaker in Y-cells compared to X-cells.

Spatial frequency tuning

The spatial filter characteristic of the RF can be described by the contrast sensitivity function (CSF). Contrast sensitivity, the reciprocal of the contrast needed to elicit a criterion threshold response (a just significant change in ongoing activity), is plotted against the spatial frequency of a test grating moving across the RF at optimal velocity. The peak of the resulting curve represents the optimal spatial frequency for the cell and roughly corresponds to the RF center diameter. The curve shows a steep cutoff at higher frequencies which results from an incomplete coverage of the center by bright or dark bars. The sensitivity also declines to some degree at lower spatial frequencies as a result of center-surround interactions. This spatial tuning curve of the complete RF can be predicted by a difference-of-Gaussian (DOG) model based on the sensitivity profiles of RF center and surround (see Casagrande and Norton, 1991 and Fig. 3.3B). A comparison of the spatial frequency tuning curves of ganglion cells and LGN relay cells demonstrates a stronger rolloff at low frequencies for the LGN cells (Fig. 3.3C). This can be explained by additional inhibitory surround mechanisms originating in the LGN (see below). The optimal spatial frequency of cat X-cells close to area centralis is on average 1 cycle/degree, and that of Y-cells is around 0.3-0.5 cycles/degree (Fig. 3.3D). Due to some jitter in RF-diameter, both classes build a continuum of spatial filters that coincides with the behavioral range of spatial frequency detection.

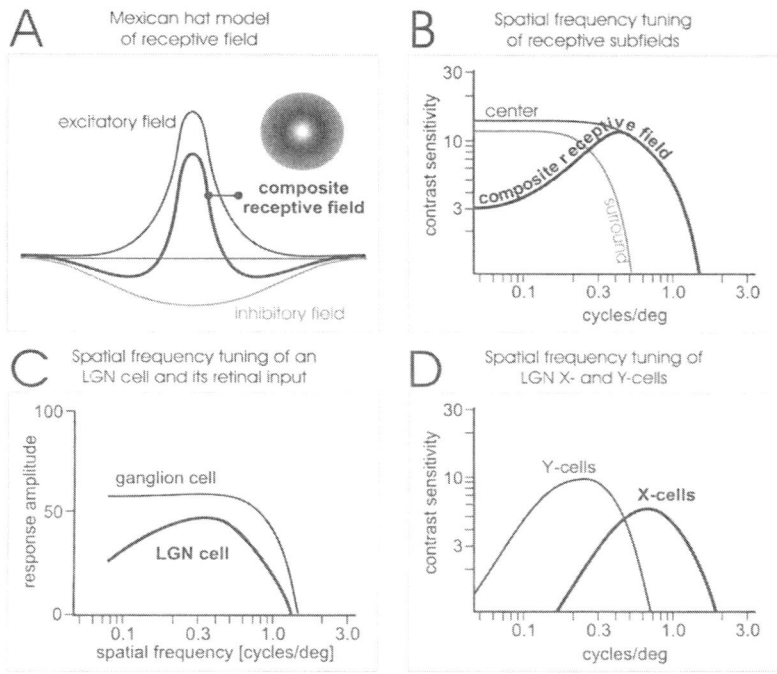

FIGURE 3.3. Spatial characteristics of retinal and geniculate receptive fields (RF). A) Composite RF of centered excitatory and inhibitory fields (Mexican-hat model, Rodieck and Stone, 1965). B) Contrast sensitivity functions of RF center surround and composite RF for different spatial frequencies of the stimulus in analogy to A). C) Comparison of spatial frequency tuning of retinal and geniculate RF. Drop of the LGN response at low frequencies due to the action of the additional antagonistic surround of LGN RFs. D) LGN X- and Y-cells differ in spatial frequency tuning.

Linear and nonlinear summation

The previously mentioned RF-composition is called the *linear summation field* of the RF. Linearity of spatial contrast integration is usually tested by flashing or counterphasing (contrast reversal) a grating of optimal spatial frequency at different positions (spatial phases) with respect to the RF center (Shapley and Hochstein, 1975). The luminance of the grating varies sinusoidally. A bright bar centered on an On-cell RF with flanking dark bars, covering parts of the surround causes a strong excitatory response. If the grating is reversed in contrast or shifted by 180° of spatial phase, the strongest inhibitory response is elicited. However, if the grating is shifted by 90°, both center and surround are covered by equal surface areas of bright and dark bars and excitatory and inhibitory responses are balanced. The result of a contrast reversal at this position is a *null-response* characterized by an almost absent modulation of activity (see Fig. 3.4A). This behavior is typical for X-cells. Y-cells, however, show a somewhat different kind of contrast integration. At low spatial frequencies (optimal frequency, matched to RF center size) they also show an almost linear integration of contrast similar to X-cells. At higher spatial frequencies, exceeding the spatial resolution of their RF center, the linear response shows the typical cutoff and the following type of response evolves: An excitatory response is elicited with each contrast reversal of the grating, resulting in two excitatory responses (one On and one Off response) for each complete stimulus cycle (Fig. 4B). The cell now responds with a frequency twice that of the stimulus frequency. This so called second harmonic response is a nonlinear kind of contrast integration suggested to be generated by the rectified (only excitation is passed) convergent activity of smaller On- and Off-subunits within the Y-cell RF. These subunits are of about the same size as the linear RF centers of X-cells. The input field of these subunits seems to spread out over the limits of the classical linear RF and also mediates the so called *periphery effect* or *shift-effect* (Fischer and Krüger, 1974).

Synergistic inhibition and spatial contrast

Different functional types of inhibitory interactions that shape the spatial and temporal properties of geniculate visual responses have been described. One of these is the *synergistic inhibition*. The term synergistic means that this kind of inhibition is initiated by the same type of neuron that also carries the excitatory response, e.g., light On elicits an excitatory response within the RF center of an On-cell but also an inhibition mediated by the same cell or other On-cells. Two spatially different components can be distinguished. One seems to be mediated by feedback inhibition within the center of the RF. On-center activity of an On-type ganglion cell drives an LGN On-cell. The LGN relay cell activates inhibitory neurons (local interneuron or PGN cell) and those inhibit the relay cell itself (pathway I in Fig. 3.5A). This circuit may cause the post-peak inhibitory response often visible between the early transient and the following tonic response of LGN relay cells. One important function of the feedback inhibition seems to be the control of the contrast gain (which is generally lower in LGN compared with the retina) and the temporal waveform of geniculocortical response volleys (see Funke and Eysel, 1998). The second type of synergistic inhibition works as lateral inhibition (pathway II, Fig. 3.5A) and amplifies the center-surround antagonism (Singer et al., 1972). LGN cells have been found to show a clearly stronger center-surround antagonism

Linear / non-linear spatial contrast integration

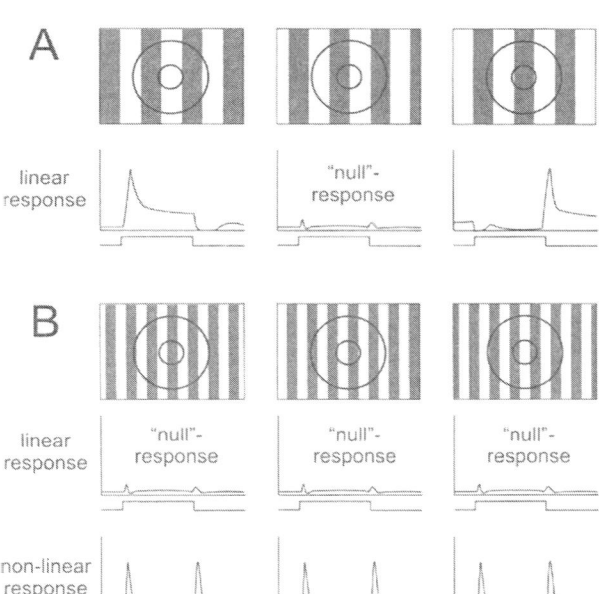

FIGURE 3.4. Linear and nonlinear spatial contrast integration. A) The strongest visual responses of the linear type are elicited by a contrast pattern of a spatial frequency that fits well to the diameter of the center of the receptive field (RF). The strength of the visual response depends on the spatial phase of the pattern (e.g. a grating). A maximal excitatory On-response is evoked if a bright section of the pattern covers the complete center of an On-center cell. By contrast, the strongest Off-inhibition is caused when the pattern is shifted by 180° in spatial phase so that the RF center is now covered by a dark bar. Graded responses can be elicited if the center is to some degree covered by bright and dark bars, and a balanced stimulation of the RF center (and surround) by bright and dark bars results in the *null-response* (middle) which is characterized by only small, if any change in activity. In this case of linear contrast integration, On-excitatory and Off-inhibitory activities are of equal strength and cancel each other. B) Linear visual responses become independent of the spatial phase of the grating if the spatial frequency is clearly higher than the spatial resolution of the RF center. A null-response is observed at many spatial phases because of the simultaneous stimulation of the RF center by bright and dark bars. However, Y-cells do not show a null-response but an excitatory On-Off response irrespective of the spatial phase of the grating. The idea is that there exist On- and Off-subfields which are smaller than the diameter of the classical RF center of a Y-cell. The excitatory activity of both, On- and Off-subunits is summed up probably without inhibitory interaction and results in a pure excitatory (rectified) On-Off response.

than retinal ganglion cells (Hubel and Wiesel, 1961), an indication for an additional inhibitory surround evolving at the thalamic level. One clear difference between retinal and geniculate surround inhibition is that retinal surround inhibition fades with dark-adaptation but geniculate lateral inhibition does not. Since lateral inhibition in the LGN has about the same latency as the excitatory center response, it has been suggested (Singer et al., 1972) that this inhibition of the feedforward type is driven by direct inputs of surrounding ganglion cells to local geniculate interneurons (pathway II in Fig. 3.5A). The main function of the feedforward type of lateral inhibition is to enhance spatial (or simultaneous) contrast discrimination. A further, more widespread, inhibitory surround seems to emerge via recurrent inhibition: axon collaterals of neighboring LGN relay cells converge onto PGN neurons and elicit a volley of feedback inhibition as mentioned above. However, because of the convergence of many LGN relay cells, the PGN cell has a large RF and can be responsible for a so called long-range lateral inhibition (Eysel, 1986).

Reciprocal inhibition and successive contrast

Another type of inhibition first visible at the geniculate level is the reciprocal inhibition (see Singer and Creutzfeldt, 1970). It reflects the interaction of different center response type ganglion cells at the level of the LGN. For example, an On-center ganglion cell directly excites an On-center LGN cell and an Off-center ganglion cell at the same retinal location inhibits the LGN cell via a local GABAergic interneuron (pathway I in Fig. 3.5B). Thus, an On-type relay cell is excited by a bright stimulus projected into the center of its receptive field, but also actively inhibited by a dark stimulus or the offset of a bright stimulus. The opposite wiring scheme is realized for an Off-center LGN cell at any location within the LGN (pathway II in Fig. 3.5B). This push-pull mechanism increases and sharpens the response to a change in contrast, assures the linearity of responses, and promotes successive contrast detection. Reciprocal inhibition seems to be mediated primarily by local interneurons in a feedforward manner. Nevertheless, PGN cells receive convergent input from On- and Off-relay cells, so that a recurrent component could also contribute to this phenomenon (pathway III in Fig. 3.5B). Inhibitory interactions of the push-pull type have been also postulated for the retinal network (Gaudiano, 1994).

Binocular (interocular) inhibition

A third functional type of geniculate inhibition is characterized by an interaction between the two ocular channels, which remains separate at the subcortical level. Interocular inhibition of an LGN relay cell is best visible when the so called *nondominant eye* is visually stimulated while the dominant eye is closed to prevent a direct modulation of activity by excitatory inputs from this eye. Excitation in one eye is mediated via the corresponding LGN A-layer and further passed to the PGN. The back-projection of the

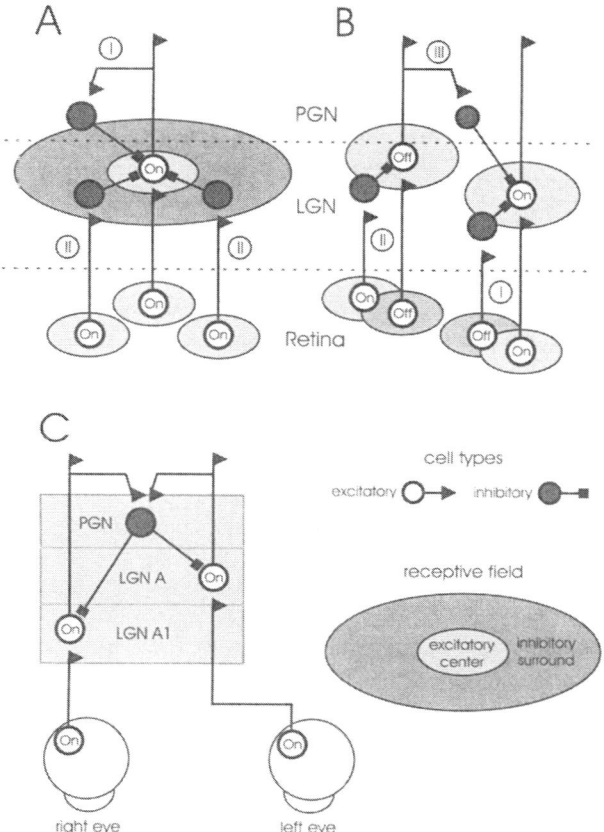

FIGURE 3.5. Synergistic, reciprocal and binocular inhibition. A) Synergistic inhibition is either a feedback inhibition within the same channel (On or Off, pathway I), or a lateral interaction between inputs of the same RF-center type (On-On, pathway II). The latter works as lateral inhibition, produces the additional antagonistic RF surround of LGN relay cells and enhances spatial contrast resolution. B) Reciprocal inhibition is an interaction between inputs of different RF-center type (On-Off, Off-On). These interactions are found at every location within the LGN. It is found within the center of an RF as well as within the RF-surround. Off-center inputs inhibit On-center cells (I) and On-center inputs inhibit Off-center cells (II) in a feedforward fashion. Reciprocal inhibition is also possible via the PGN-loop (III), because PGN cells receive both, On- and Off-center relay cell input. C) Binocular (interocular) inhibition is most likely mediated by the PGN cells because they receive convergent input from both LGN A-layers and also project back to both layers. However, the projection to one A-layer is usually stronger than that to the other layer, so that monocular and interocular inhibitory influences can vary.

PGN to both A-layers results not only in iso-ocular but also cross-ocular (interocular) feedback inhibition (Fig. 3.5C).

Contrast gain control

Retinal and geniculate cells respond to increasing stimulus intensity (or contrast) with an almost proportional increase in firing rate, so that the relative stimulus intensity is represented by the mean firing rate. A contrast response function can be established by plotting firing rate versus contrast on a logarithmic scale. A comparison between retinal and geniculate contrast response functions reveals that the geniculate function is characterized by a smaller slope (Kaplan et al., 1993). That means that the contrast gain is reduced in the LGN. The cellular mechanisms of geniculate contrast gain control are unknown, but possible candidates include intrageniculate feed-forward inhibition (Sherman and Koch, 1986), inhibitory feed-back via the PGN (Kaplan et al., 1993; Lo and Sherman, 1994), and - in a more sophisticated manner - the excitatory feed-back from the visual cortex (Cudeiro and Sillito, 1996). At a first glance, this reduction in contrast gain is surprising because the lowered gain diminishes the resolution of contrast differences. However, it has to be considered that many LGN relay cells converge onto a cortical cell. One idea is that the reduced gain prevents cortical cells from being overexcited and thereby loosing contrast resolution due to early saturation. The reduction of geniculate contrast gain should be compensated by the summation of multiple geniculate inputs at the cortical cell. A concomitant effect of such a convergence is that it improves the signal to noise ratio. A temporally structured activity will further promote this effect as described below.

Response latency

The latency of visual responses in retina and LGN is quite variable and depends on the magnitude of the change in contrast, the slope of the contrast change, the stimulus size and cell type (see Funke and Wörgötter, 1997). Usually, response latency declines with increasing contrast and with increasing stimulus size. The higher the amount of light energy, which hits the receptive field of a ganglion cell, the stronger and steeper the change of the membrane potential is, so that the firing threshold is reached faster. A large stimulus stimulating both center and surround of the RF elicits a smaller response amplitude than a stimulus confined to the center due to surround inhibition, nevertheless the response latency may be shorter because surround inhibition is lagging behind the excitatory center response by a few milliseconds (Enroth-Cugell and Lennie, 1975). Response latency exhibits some variability even when using identical stimuli. Spontaneous fluctuations of the membrane potential (noise) are the reason for variable spike timing. The standard deviation of the latency declines with decreasing latency (Bolz et al., 1982). The minimal latency of Y-cell responses is about 30 ms, that of X-cells is on average 10-15 ms longer. However, with suboptimal stimulation of the RF center by a small spot, the latency of a Y-cell can be longer than that of an X-cell. The response of ganglion cells precedes that of LGN relay cells by 2-4 ms. The intraretinal conduction velocity of ganglion cell axons is only 1.5-3.0 m/s because of the missing myelin shield. The myelinated part within the optic nerve has a considerably higher conduction velocity (30-50 m/s for Y-axons and 15-23 m/s for X-axons). Transmission of geniculate signals to the cortex takes 2-5 ms. Also in this case the Y-channel is about 1-3 ms faster than the X-channel.

Response dynamics

Depending on the temporal characteristics of the photopic input, the visual responses of retinal and geniculate projection cells show a distinct temporal waveform. For instance, a sinusoidal modulation of light intensity inside the receptive field is followed by an also almost sinusoidal change in firing frequency but with a phase difference which is depending on cell type and temporal frequency of the stimulus (Heggelund et al., 1989). On the other hand, a steep increment (for On-cells) or decrement (for Off-cells) in light intensity in the RF center produces a bimodal response with an initial phasic response (overshoot) and a following tonic response. The latter slowly adapts during standing contrast mainly due to adaptation processes in the retina. A comparison of the response of an LGN cell with its afferent retinal input shows distinct differences (Fig. 3.6). First of all, the geniculate response is generally smaller than its retinal counterpart indicating a transfer ratio less than 100% (Cleland et al., 1971; Coenen and Vendrik, 1972). This transfer ratio can strongly change in a state dependent way (Coenen and Vendrik, 1972), a matter described later on in more detail. Below saturation level both response components (phasic and tonic) are almost equally reduced as revealed by a comparison of the phasic-tonic-index (PTI, the ratio of phasic to tonic firing frequency) of retinal and geniculate responses. The initial overshoot (1) of LGN responses is often followed by a transient drop in firing rate (2) below that of the following tonic response, usually called *the post-peak-inhibition* which seems to evolve from intra- or perigeniculate inhibitory interactions. This inhibition is often followed by a *rebound* response (3) at the beginning of the tonic response (4). In addition, LGN cells also show a stronger decline in firing (5) when contrast changes in the direction opposite to RF center sensitivity (e.g., light Off for an On-center cell) which is the result of reciprocal inhibition (see above). This inhibitory Off-response also exhibits a multimodal time course. The strong and phasic inhibition after offset of stimulus is often followed by another rebound response (6) composed of a short burst of action potentials. The rebound is usually followed by a second inhibitory response (7) which is weaker and more sustained than the initial one (2). Additional bursts of action potentials can occur during the declining phase of the inhibition with variable latency (8). These rebound burst are not of retinal origin, are intrinsically generated by the LGN relay cell and are the only period during which retinogeniculate transfer ratio is higher than 1.0. X- and Y-cells slightly differ in their response dynamics. Y-cells were often called *the phasic or transient cells* because they exhibit a stronger initial overshoot and a less prominent sustained tail of their response when compared to X-cells (tonic cells; see Bullier and Norton, 1979; Cleland et al., 1971). So far, little is known about the significance of the different components of the visual response for higher level visual processing. The phasic and the tonic responses can be interpreted as two different messages about the visual stimulus: the slope of an intensity change is primarily transmitted by the initial phasic response, whereas the tonic part of the response carries information about the steady contrast difference between the new and the former intensity (Heggelund et al., 1989). Therefore, the phasic response might be used by the visual system to detect changes in the visual environment which are produced by fast eye or object motion. The tonic activity may be needed to analyze finer details like patterns and gradations in brightness and color.

Temporal structure of activity

The discovery of synchronized oscillatory activity patterns occurring during distinct stimulus situations (see Eckhorn, 1994; Heller et al., 1995; Singer and Gray, 1995; Theunissen and Miller, 1995) has revived the idea that sensory information is not only encoded by the rate but also by the temporal pattern of activity. A closer inspection of LGN activity reveals a temporal structure on different time scales. One is derived from the gross temporal waveform of the visual response and another is based on the pattern of spike intervals. McClurkin and colleagues (McClurkin et al., 1991a,b; Gawne et al., 1991) have studied the temporal waveform of monkey parvocellular responses to different spatial contrast patterns (Walsh patterns) using principal component analysis (PCA). The complex visual response waveform can be interpreted as a composite of a set of basic (principal) waveforms. It was found that the gross response waveform included more information about the spatial distribution of contrast than the mean firing rate, while the rate code largely reflected the mean luminance. A similar observation was made when the temporal structure of LGN responses was analyzed on a finer time scale by calculating the distribution of spike intervals (Funke and Wörgötter, 1997). When the spike interval distribution is determined for the entire response including the phasic and tonic response component, often an almost Poisson-like distribution can be seen. However, it is known that the spike interval distribution changes during the course of the response as shown in Fig. 3.7A,B. The time axis of these so called *intervalograms* runs from top to bottom and the summed light response is plotted as a peri-stimulus-time histogram along the left side of this axis. Spike interval distributions calculated from small time windows (100 ms)

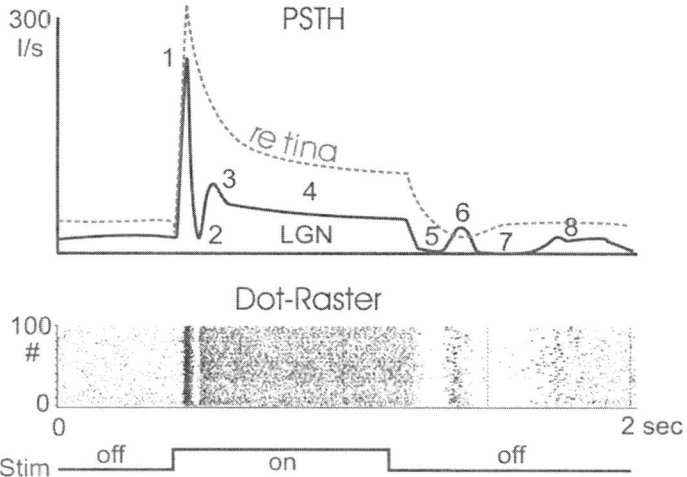

FIGURE 3.6. Typical LGN cell responses. The peri-stimulus-time histogram (PSTH) on top shows the gross temporal waveform of a geniculate (thick line) and retinal (broken line) visual response to a light spot flashed on and off within the center of the receptive field. The dot-raster diagram below shows 100 individual responses of an LGN relay cell to identical stimuli. Averaging the activity of these 100 traces results in the LGN response curve shown in the PSTH on top. The response to a sudden increment and decrement of RF illumination can show up to 8 components:1) initial transient response (overshoot, peak), 2) post-peak inhibition, 3) early rebound response, 4) tonic response, 5) stimulus off inhibition (off-response), 6) first post-inhibitory rebound, 7) late inhibitory response, 8) second post-inhibitory rebound. Strength and presence of these components depends on the type of stimulus and the polarization of the cell membrane. For comparison, the response profile of the retinal input is less complex. It is composed of an initial peak response, a following tonic response and a weak inhibitory off-response. At least these response components are also visible in the LGN.

are plotted on the right as gray scaled pixel rows. Thus, the intervalogram is similar to a sonogram but shows the spike interval distribution instead of frequency distribution over time. The interval distribution of the tonic part of the visual response of LGN relay cells is often multimodal (2 bands in the right diagrams of Fig. 3.7A,B). The individual modes are equidistant and thus represent integer multiples of a fundamental interval which is given by the leftmost mode of the diagrams (Fig. 3.7A,B,E). This multimodal distribution is first visible in the LGN. Simultaneously recorded retinal inputs do not show a multimodal distribution during the tonic response but only one dominating spike interval (left diagrams). The appearance of these multiple intervals, the relative number of integer multiples and their absolute length depends on the characteristics of the stimulus (brightness, size) and the amount of inhibition induced in the LGN. The underlying mechanisms are illustrated in Fig. 3.7C-E. The retinal ganglion cell generates an almost constant firing rate with a certain frequency (fundamental interval, as shown in 7A,B left side) and with the interval length depending on stimulus brightness (bottom traces in 7C). Thus, a bright spot stimulating the RF center elicits a higher firing rate than a less bright spot (compare bottom traces of left and middle column). This pattern of activity is transmitted by LGN relay cells in an almost one-to-one fashion (uppermost traces) unless individual spikes are canceled (dashed lines) by the activity of local inhibitory interneurons (middle traces). The surround inhibition induced by these interneurons is low for small stimuli but increases if larger stimuli which also excite neighboring ganglion cells (see right column). The increased inhibitory activity punches more holes into the spike train leading to a considerable number of double-sized spike intervals and a growing second and a new third peak in the interval histogram (see histograms in 7E). A large and bright stimulus can produce about the same mean rate of activity as a smaller but darker stimulus, however, the temporal pattern described by the distribution of spike intervals is different.

Electrical properties and activity modes

LGN relay cells can generate two principally different patterns of activity: sequences of single action potentials spaced by more than 5 ms and so called *burst* discharges which are groups of 2 to 7 action potentials with intraburst intervals generally shorter than 5 ms (200-500 Hz) and interburst intervals of several 100 ms. Due to the presence of voltage-dependent ion conductances, the predominance of burst-firing or tonic activity of an LGN relay cell is strictly related to the membrane potential. If the membrane potential is held close to the threshold for sodium/potassium action potentials (between -60 and -55 mV), a single suprathreshold EPSP elicits a single action potential. Similarly, a constant depolarization of the cell membrane, induced by current injection, elicits a train of single action potentials with the spike intervals shortening with increasing amplitude of the depolarizing step. The spike intervals are then usually longer than 5 ms. At a more hyperpolarized level (-60 to -70 mV), a small subthreshold depolarization, which could be an EPSP or a release of inhibition, elicits a so called *low threshold calcium spike* (LT calcium spike, Jahnsen and Llinás, 1984). The LT calcium spike is a large triangle shaped depolarization of 50 to 100 ms duration which triggers a high-frequency (200-500 Hz) burst of sodium action potentials when steeply crossing threshold (see Fig. 3.8A). The LT calcium spike is carried by a voltage-dependent calcium current which is inactivated if the membrane is held at potentials positive to -65 mV. A frequent occurrence of spontaneous bursts is found during periods of synchronized sleep (deep sleep), when geniculate relay cells are generally more hyperpolarized (Steriade and Deschênes, 1984; Steriade and Llinás, 1988; McCormick, 1992; McCormick and Bal, 1997; Steriade, 1991; 1997) than during periods of wakefulness or desynchronized sleep (REM-sleep). Repetitive spontaneous

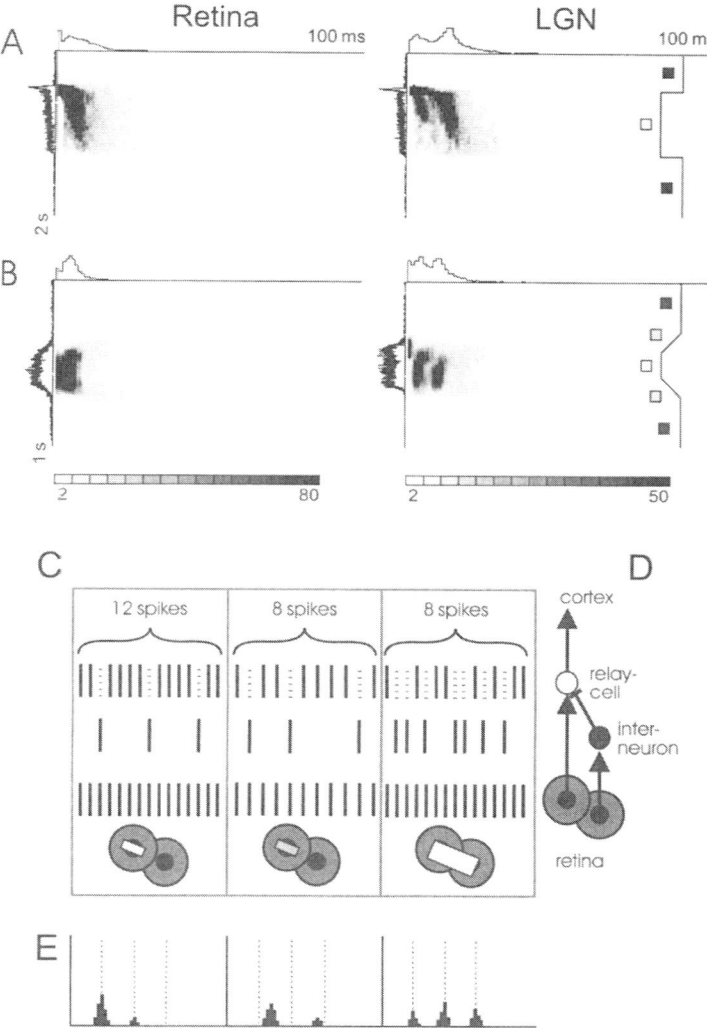

FIGURE 3.7. Temporal structure of spike patterns within the tonic visual response. (A,B) Examples of retinal (left) and LGN (right) spike interval distributions calculated for small (100 ms) consecutive time windows. The tonic visual response of the retinal ganglion cell is composed of only one dominating spike interval, the LGN response also includes intervals of double length. C) Schematic illustration of retinal and geniculate spike interval distributions resulting from a sustained stimulation of the receptive field with stimuli of different size and intensity. D) Connection scheme of retinal ganglion cells, LGN relay cells and LGN inhibitory interneurons. E) Spike-interval histograms of the relay cell spike trains shown in C. For further explanation see text. Parts A and B of the figure modified from Funke and Wörgötter, 1997.

bursts preferentially occur either with a frequency of 0.3 to 3 Hz (delta-rhythm) or in sequences with 7 to 14 Hz (alpha-rhythm/sleep-spindles). LGN relay cells are able to intrinsically generate rhythmic burst discharges at about 1 Hz due to the interaction of the LT calcium inward current with a calcium-dependent potassium outward current and a slowly inactivating sodium inward current which is triggered by a hyperpolarization. This is a process very similar to the pacemaker activity of heart muscle cells within the sinus node. The time course of the involved currents is also illustrated in Fig. 3.8A. The higher frequency oscillation of the sleep spindles (about 10 Hz) is depending on the reciprocal connections between LGN relay cells and PGN cells. Examples of spindle activity in simultaneously recorded PGN and LGN cells are shown in Fig. 3.8B (modified from Kim et al., 1997). The divergent projections between LGN and PGN cause a synchronized spread of activity as travelling waves of sleep spindles. Rinzel and coworkers (Golomb et al., 1994; 1996; Wang et al., 1991) could reliably reproduce these different patterns of thalamic oscillatory activity with a computer model (see below).

The different modes of activity affect the transmission of afferent input signals. An almost 1:1 transmission of retinal input spikes is found when the relay cell is in the depolarized single spike mode (transmission mode). Retinal EPSPs are then close to spike threshold and each retinal spike can elicit an LGN spike if no inhibition is active at the same time. The transfer ratio drops considerably with progressive hyperpolarization of the relay cell and slowing of the EEG. This results in a more phasic visual response because hyperpolarization prevents EPSPs elicited by the single spike activity during the tonic response to reach spike threshold while the high frequency input of the initial response can efficiently sum up. The first retinal input spike can also trigger a spike burst if the hyperpolarization is strong enough to de-inactivate the LT calcium current, leading to a retinogeniculate transmission ratio larger than 1. Because of this voltage-dependence thalamic relay cells can show two totally different responses to identical retinal inputs: an almost faithful reflection of the retinal activity in the depolarized state (transmission mode) and a stereotyped phasic response in the hyperpolarized state (burst mode). During transmission mode the cortex is informed about intensity, spatial distribution and temporal modulation of light contrast, during burst mode only a sudden change in contrast is reported.

3.2.3 Extra-retinal control of LGN function

Transmission of visual signals through the thalamus is controlled by a couple of nonretinal input systems which differ in origin, projection pattern and type of action. In this chapter, we will focus on the corticogeniculate feedback projection and the ascending reticular arousal system of the mesopontine brainstem (ARAS) which are examples of two principally different systems. The massive corticogeniculate pathway is topographically orga-

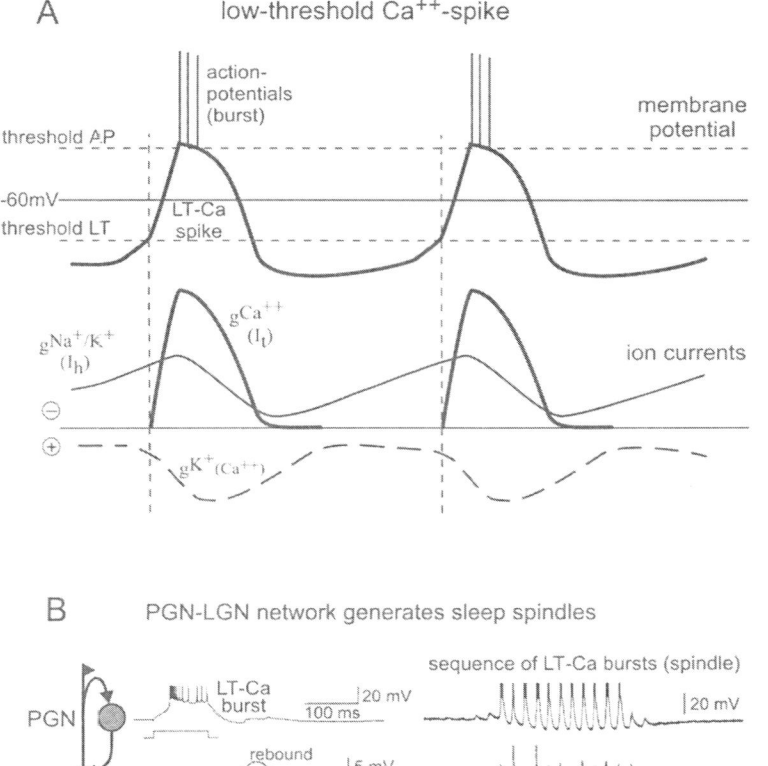

FIGURE 3.8. Oscillatory activity based on the low threshold Ca-spike mechanisms. A) Time course of repetitive LT-Ca-spikes (membrane potential trace, on top) and the involved ion currents (lower traces). For simplicity, action potentials riding on top of the LT-Ca-spikes are shown as short vertical lines. Positive inward currents $(g_{Na^+/K^+}, g_{Ca^{++}})$ are plotted upwards (negative currents), positive outward currents $(g_{K^+_{Ca^{++}}})$ are plotted downwards (positive currents). B) Simultaneous intracellular recordings of PGN and LGN activity. A LT-Ca burst of a PGN cell causes a strong IPSP in the LGN relay cell (lefthand traces). Repetitive, waxing and waning LT-Ca spikes and IPSPs consolidate a sleep-spindle. Part B of the figure modified from Kim et al., 1997.

nized with circumscribed terminal arbors of single fibers. Activity within this pathway is driven by visual stimulation or probably also during visual imagination. The projection of the ARAS is less strong but has a widely distributed diffuse field of termination missing any topographical arrangement. This pathway is made up by at least 3 different subsystems, a cholinergic, a serotonergic and a noradrenergic projection. The rate of activity of these subsystems co-varies with the state of arousal, e.g., the diurnal alternations of sleeping and waking (see Hobson, 1989; McCarley et al., 1995; Steriade, 1995).

Corticogeniculate feedback

The corticogeniculate projection originates in layer 6 of the primary and, in part, higher visual cortices. It is an excitatory, glutaminergic pathway that terminates on LGN relay cells, on local (inhibitory) LGN interneurons and on PGN cells. The corticogeniculate input constitutes the largest fraction of synapses within the LGN (40-50%) but their excitatory effect is small compared with that of the retinal input and appears to be subthreshold in the absence of retinal input. The cortico-geniculate projection seems to be composed of different subsystems because cortical activity arrives at the LGN with delays of less than 5 ms, 10-50 ms and more than 100 ms. The variable delay could be related to axons of different diameter and conduction velocity or to a contribution of different postsynaptic mechanisms. Corticofugal inputs are mediated not only by the glutamate receptors of the ionotropic AMPA or NMDA type but also by the so called metabotropic glutamate receptors. The latter control membrane conductances via intracellular second messenger systems and may account for long delays of corticothalamic modulation of activity (Salt and Eaton, 1996).

The function of the corticothalamic feedback in visual processing is still a bit puzzling but experimental findings indicate that there are at least 4 different functions:

1) gating of retinogeniculate signal transmission
2) improvement of temporal accuracy of spiking in the LGN
3) enhancement of the center-surround mechanism
4) synchronization of slow oscillations within thalamus and cortex

1) Gating of retinogeniculate signal transmission
The net effect of corticofugal activity is a facilitation of retino-geniculate signal transmission. The transfer ratio for retinal input signals is considerably reduced when the corticothalamic feedback is experimentally eliminated. However, the rather moderate corticofugal excitation of LGN relay cells does not affect all the visual responses in the same way. A strong phasic response (burst) elicited by a sudden change in contrast is almost independent of the corticofugal drive, but the tonic visual response is strongly

attenuated when corticofugal activity ceases. A drop in mean tonic firing rate is not caused by slowing of the fundamental firing frequency, it is the result of an increased failure of spike generation in LGN relay cells by retinal EPSPs. Retinogeniculate transmission is thus regulated by the combined action of intrageniculate inhibition and corticofugal excitation.

2) Improvement of temporal accuracy of spiking
The analysis of spike intervals during the tonic LGN response has shown that information about a stimulus is in part encoded by the temporal pattern of activity. For any analysis of the temporal pattern by higher visual centers it is therefore important that spike timing is as accurate as possible. However, neuronal activity is affected by noise and especially the alternation between electrical and biochemical transmission of signals at each chemical synapse will destroy the precision of spike timing. Feedback systems like the corticofugal projection in combination with a convergent-divergent connection pattern are in principle able to synchronize population activity and align spike timing to the mean firing time of the population. It has been demonstrated that the corticogeniculate projection reduces the jitter of spike intervals in the LGN. Individual peaks of the multimodal spike interval distributions mentioned above were on average 25% sharper when conditions with an active corticofugal feedback were compared with those during blockade of the corticofugal feedback (see Funke and Wörgötter, 1997).

3) Enhancement of center-surround mechanisms
Corticothalamic axons terminate both, on relay cells and inhibitory interneurons and are thus able to control also distinct inhibitory interactions, like lateral inhibition, within the LGN as well. Indeed, it has been found by cross-correlating the activity of topographically matched cells in LGN and cortex that cortical facilitation of geniculate activity is restricted to a circumscribed central region of 1 to 2 degree, while inhibition dominates beyond this limits (Tsumoto et al., 1978). This lateral interaction is not limited simply to light contrast. It has been demonstrated that the corticofugal pathway is also involved in the enhancement of local orientation contrast. An LGN cell is less excited when center and surround of its receptive field are stimulated by lines of the same orientation (a continuous grating) as compared to a situation when the orientation of a grating stimulating the surround is 90 degree different from the orientation within the center (Sillito et al., 1993; Cudeiro and Sillito, 1996).

4) Synchronization of slow oscillations within thalamus and cortex
A supportive role of the corticothalamic system in synchronization of activity is also evident for the spontaneous, slow oscillatory activity during sleep. Delta and spindle oscillations generated within the thalamus travel forth and back between LGN relay cells and cortical cells and become syn-

chronized over large areas because of the di- and convergence of upstream and downstream projections. The vertical projections are obviously more important in this process than lateral connections within the cortex (Steriade, 1997).

Mesopontine projections

The ascending reticular arousal system of the mesopontine brainstem (AR-AS) controls the activity of thalamic and other nuclei in a state-dependent way. Their rate of activity scales with the level of arousal, it is highest during wakefulness and lowest during deep sleep. A situation deviating from this scheme is found during the so called paradoxical or desynchronized sleep phases which are accompanied by rapid eye movements (REM). With a transition from deep sleep to REM-sleep noradrenergic (NA) and serotonergic (5-HT) neurons cease firing almost completely but at the same time cholinergic (ACh) neurons increase their firing rate even above the level during wakefulness. The thalamocortical system is then in a hyper-excitable state but activity is intrinsically generated, is barely modulated by peripheral sensory inputs. This seems to be the substrate of dreams (for review see Hobson, 1989; Steriade, 1991; McCormick, 1992). A sub-set of the cholinergic neurons shows periodic variations of activity which are phase-locked with rapid eye movements and cause the so called ponto-geniculo-occipital (PGO) waves of activity. Changes in the activity of this system are accompanied by correlated changes in thalamic and cortical activity patterns and are reflected by changes of the spectral composition of the EEG. The actions of ACh, NA and 5-HT at different types of thalamic target neurons differ considerably because of varying sets of receptors and different ion channel composition.

ACh excites LGN relay neurons via ionotropic nicotinergic receptors (Na^+/Ca^{++}-inward current) and muscarinergic receptors of type 1 (reduced K^+-outward current). The former mechanism causes transient excitations also in neurons of the perigeniculate nucleus (PGN), the latter results in a slow and long lasting depolarization of the membrane because of the involvement of the second messenger system. Local inhibitory LGN interneurons and PGN neurons are inhibited via muscarinergic receptors of the type 2 (increased K^+-outward current).

Stimulation of NA receptors of the alpha type activates the same second messenger system that reduces the K^+-outward current and NA beta-receptors increase the h-current (Na^+-inward current) and thus counteract hyperpolarizing influences. Both mechanisms switch a LGN relay cell from burst to tonic firing because the membrane potential is prevented from being shifted to a hyperpolarized state that de-inactivates the LT Ca^{++}-current.

Serotonin seems to increase the tonic firing of local interneurons and PGN neurons thereby causing an inhibition of the relay cells (see McCormick,

1992; Funke and Wörgötter, 1997). The concerted action of these 3 systems has two main effects: i) it controls the mode of activity in LGN relay cells and PGN cells (burst mode or tonic transmission mode) and ii) it regulates the ratio of retinogeniculate transmission in a global way. The former may be preferentially managed by the NA system, the latter seems to be controlled by the antagonistic actions of ACh and 5-HT on relay cells and inhibitory neurons, resulting in a reinforced synergistic action (see Fig. 3.9 and section 2.3.3).

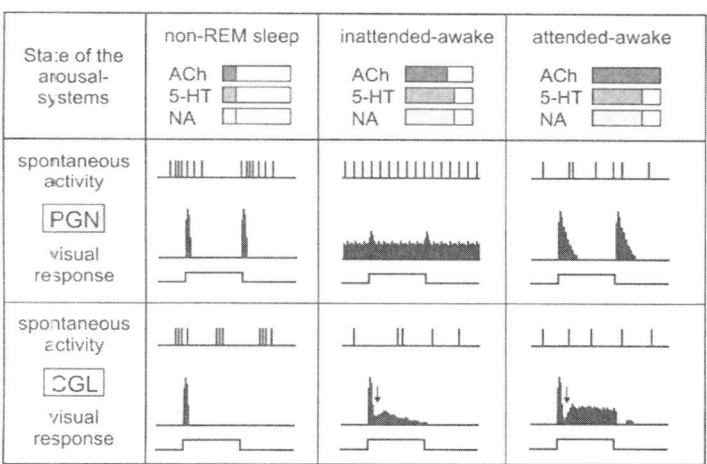

FIGURE 3.9. Model of state-dependent changes in PGN and LGN spontaneous and visual activity and the resulting LGN-PGN interactions during visually driven activity. Both cell types are hyperpolarized during deep sleep (non-REM sleep, synchronized EEG) when each of the 3 brain stem arousal systems is active at a strongly reduced rate. Burst discharges prevail during spontaneous and visually elicited activity. No tonic light response is observed. The activity of the arousal system is generally increased during wakefulness. However, during a drowsy state the effect of the 5-HT system might dominate over the effect of the ACh system, leading to an excitation of PGN cells. The high tonic activity of PGN cells has two consequences: i) a strong tonic inhibition of LGN cells, thereby reducing tonic visual responses more than phasic responses and, ii) a reduced signal-to-noise ratio for visual responses in the PGN and correlated inhibitory responses in the LGN (arrow). With increasing arousal the ACh system overcomes the action of 5-HT and now leads to inhibition of ongoing activity in the PGN. Released from tonic inhibition LGN cells now show a stronger tonic visual response and visual modulation of activity is improved in both, PGN and LGN.

The perigeniculate nucleus (PGN)

PGN neurons possess large binocular receptive fields because they receive convergent input from both main LGN layers (A, A1). In addition, they are typically excited by both, a sudden increment and decrement of light intensity, indicating that they receive convergent input from On- and Off-center cells. The back-projection of the GABAergic PGN neurons onto LGN relay cells generates a negative feedback loop that could, in principle, serve at least two different functions: 1) a direct control of LGN responses driven by visual activity, and 2) the modulation of the influences of other control systems.

1) Visually induced interactions between PGN and LGN
Simultaneous recordings of topographically matched PGN and LGN cells demonstrated a covariation of visual activity. Excitatory responses of PGN cells are temporally aligned with episodes of inhibition of LGN activity. For example, the transient response of a PGN cell to a sudden change in contrast matches with the phasic inhibition following the initial transient LGN response (see Fig. 3.6 component 2). The latency of both, the excitatory PGN response and the inhibitory LGN response are reduced in the same way when stimulus size is increased. This interaction is similar for monocular and binocular stimulation. It demonstrates that PGN activity is involved in the temporal modulation of thalamocortical visual activity.

2) State-dependent modulation of retinogeniculate transmission by the PGN
The ARAS originating in the brain stem innervates both, LGN and PGN cells. Therefore, it can be expected that global modulation of retinogeniculate transmission is also a matter of changing inhibitory interactions. Double recordings of PGN and LGN cells also demonstrated a state-dependent co-variation of LGN and PGN activity (Funke and Eysel, 1998). LGN relay cells switch from phasic-tonic visual responses to almost phasic and much less tonic responses when the EEG shows an increase in delta-wave activity, while at the same time tonic firing strongly increases in PGN cells. The high tonic firing of PGN cells during a synchronized EEG (which may be during light sleep, drowsiness, or an inattentive awake state, see Fig. 3.9) obviously results in a tonic inhibition of LGN relay cells leading to a reduction of tonic visual responses while strong phasic (burst) responses can overcome this inhibition. Geniculate information transmission is now altered as discussed in the section on *Electrical properties and activity modes*. The high tonic activity of PGN cells also diminishes their own responsiveness to visual inputs which could be an effect of saturation. Visually driven activity does not add to the tonic activity but is virtually submerged in it. As a consequence, a visually driven inhibitory modulation of LGN activity is missing. The situation is completely different during a less synchronized EEG (attended awake state). The low tonic activity of PGN cells now allows for a stronger tonic response of LGN cells and a better temporal modulation of PGN and LGN visual activity. Thus, retino-geniculate transfer ratio and the amount of sensory information is increased.

3.3   Models of the LGN

The LGN has long been regarded as a largely passive relay station between retina and visual cortex. This view started to change on the basis of experimental evidence showing that the LGN actively shapes visual signals on their way to higher areas (for an older review see Singer, 1977). Since then the visual thalamus has attracted theoreticians mainly to study the following functional aspects: 1) Its intriguing cell behavior where cells can switch between burst firing and tonic transmission mode, 2) its relation to the sleep-waking cycle, and 3) its involvement in early visual gating processes. The second and third aspects rely on the first.[1]

3.3.1   Basic membrane model of an LGN cell

The two different firing characteristics of LGN cells, described in the previous section rely on the membrane potential of the cells. When the cell membrane is hyperpolarized the cell will fire in burst mode, when the membrane is depolarized the tonic transmission mode is observed. McCormick and Huguenard (1992) have designed a model which includes the different membrane currents involved in this process. The model rests on a Hodgin-Huxley description of the individual neurons and faithfully reproduces the spiking behavior of the cells. According to this model the membrane potential changes can be described by a multitude of different currents:

$$\frac{dV}{dt} = -\frac{1}{C}[I_{Na}+I_{Nap}+I_L+I_T+I_C+I_A+I_{K2}+I_H+I_{K_{leak}}+I_{Na_{leak}}] \quad (3.1)$$

where Na stands for sodium, and K for potassium and C is the membrane capacitance. Concerning the firing behavior of LGN cells there are two groups of depolarizing and repolarizing currents which are of particular importance: The T- and the L-type currents will drive the membrane potential up, while the A-, C-, and K2-currents will push it to more negative values. The calcium-dependent T-current is inactive at resting membrane level. It becomes de-inactivated as soon as the cell becomes hyperpolarized for several tens of milliseconds. When a small depolarization occurs

---

[1] A large class of theoretical studies tries to understand LGN cell behavior by means of so called filter models. This model class simulates the cell behavior as the convolution between a spatiotemporal stimulus function with a spatiotemporal linear filter. This is a very successful model class (e.g., the *difference of Gaussian, DOG model for retina or LGN cells*) but basically all biophysical properties, which either derive from the intrinsic cellular characteristics or from the cells' connectivity within the neuronal network are removed when pursuing a filter approach. This chapter, on the other hand, tries to embed models in the framework of a description of the functional anatomy of the primary visual pathway. For this reason we will not discuss filter models here, even though this model class has in some parts lead to significant theoretical advancements.

the T-current sets in and produces a so called low-threshold calcium spike. Often this slow depolarization will also cross the firing threshold and elicits regular sodium spikes which will then occur as a burst sitting on top of the calcium spike (Fig. 3.10 A). The L-current is active only above -45 mV and adds to the depolarization. Repolarization during a calcium spike is attributable to the A-,C-, and K2-currents each of which has different dynamics and sets in after different latencies (Fig. 3.10 B). The comparison between real data and model simulation demonstrates that the model captures the basic features of a calcium spike quite accurately.

The H-current (also called *sag*-current) plays an interesting role in this process. It is activated during hyperpolarization but acts towards depolarizing the membrane. As a consequence, it can lead to the small depolarization which is necessary to trigger the T-current and, thus, elicit a calcium spike. The combined action of H- and T-current, therefore, leads to an oscillation of recurring calcium spikes.

The interplay of these currents describes faithfully the LGN cell behavior if the cells are in a hyperpolarized state. If they are depolarized the T-current is inactive and low-threshold calcium spikes are not anymore observed. Now the firing behavior is dominated by the more classical currents and regular stimulus-driven (contrast dependent) repetitive firing occurs, nonetheless the A- and K2-currents can still influence the firing frequency (Fig. 3.10 C).

3.3.2   Models of the hyperpolarized thalamic state

The NRT-TC network

Mostly from the work of the group of Steriade it is known that the intrinsic firing behavior of LGN cells determines the gross physiological states of the brain (e.g., sleep and wakefulness, see Hobson, 1989 for a review). In these processes, the cortex is also strongly involved, but the thalamus seems to play a key role in early phases of sleep where the EEG is interspersed by so called sleep spindles. Spindles are large and brief, burst like wave-cycles with a frequency in the alpha-range (10-14 Hz). During this time groups of neurons in the thalamus synchronize which leads to the observed rather strong spindle-shaped excursions in the EEG. The underlying mechanisms are understood to the degree that computer models of the spindling behavior could be made (Destexhe et al., 1994; Golomb et al., 1994; 1996).

The main driving force of spindling is the interaction between thalamic relay cells and cells in the nucleus reticularis thalami (NRT). NRT cells receive a major excitatory drive from the thalamic relay cells which they, in turn, inhibit. This induces a cyclic behavior where the NRT first inhibits its targeted relay cells which leads to a prolonged hyperpolarization and de-inactivates the T-current. Inhibition of the relay cells also leads to reduced excitation of the NRT cells which eventually seize to fire leading to a release of inhibition at the relay cells. Because the T-current is

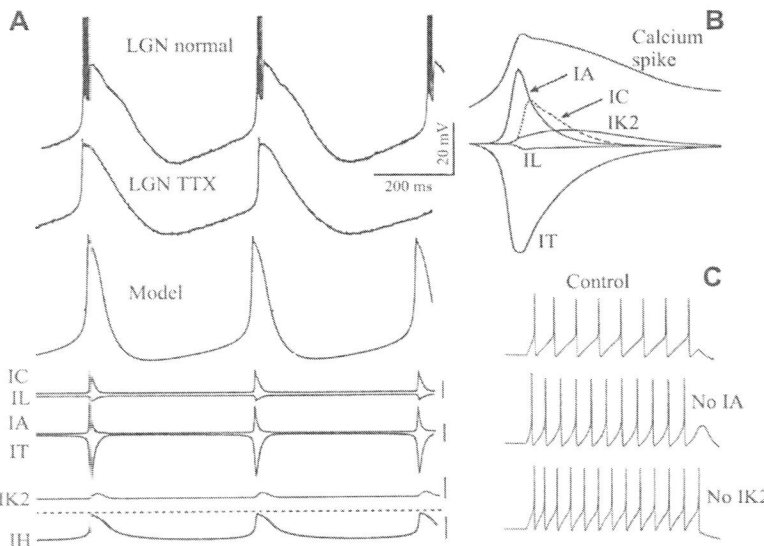

FIGURE 3.10. Experimental and modeling data showing the spiking behavior of LGN cells and the different currents involved. A) LGN cells are in a hyperpolarized state. One observes a repetitive occurrence of low-threshold calcium spikes in cat dLGN (top) with sodium spikes riding on the crest. Sodium spikes can be blocked by tetrodotoxin (TTX). The model simulates this situation. Bottom traces show the time-courses of the different currents involved in the generation of repetitive calcium spike activity. Part B shows a magnified plot of the different current traces. C) LGN cells are in a depolarized state. Low-threshold calcium spikes do not occur anymore and instead regular (sodium spike) firing is observed. The firing frequency can be attenuated by the I_A and I_{K2} currents. (Recompiled from McCormick and Huguenard, 1992).

de-inactivated and the H-current is still active, the relay cells will elicit a low-threshold calcium spike upon which a strong burst of regular sodium action potentials is generated. This burst of actions potentials then drives NRT cells very efficiently and the entire cycle starts again. In addition to the above qualitative description of the firing behavior of the thalamic cells Golomb et al., 1994 proposed a more detailed model (see also Wang and Rinzel, 1992, Destexhe et al., 1994).

Again thalamic cells are modeled as Hodgin-Huxley type neurons including many details. In particular they included $GABA_A$ and $GABA_B$ receptors mediated inhibitory connections between NRT cells and also between the NRT and the thalamic relay cell pool (Fig. 3.11 F). NRT cells are described by:

$$\frac{dV}{dt} = -\frac{1}{C}[I_T + I_{AHP} + I_{leak} + I_{GABA_A} + I_{GABA_B} + I_{AMPA}] \quad (3.2)$$

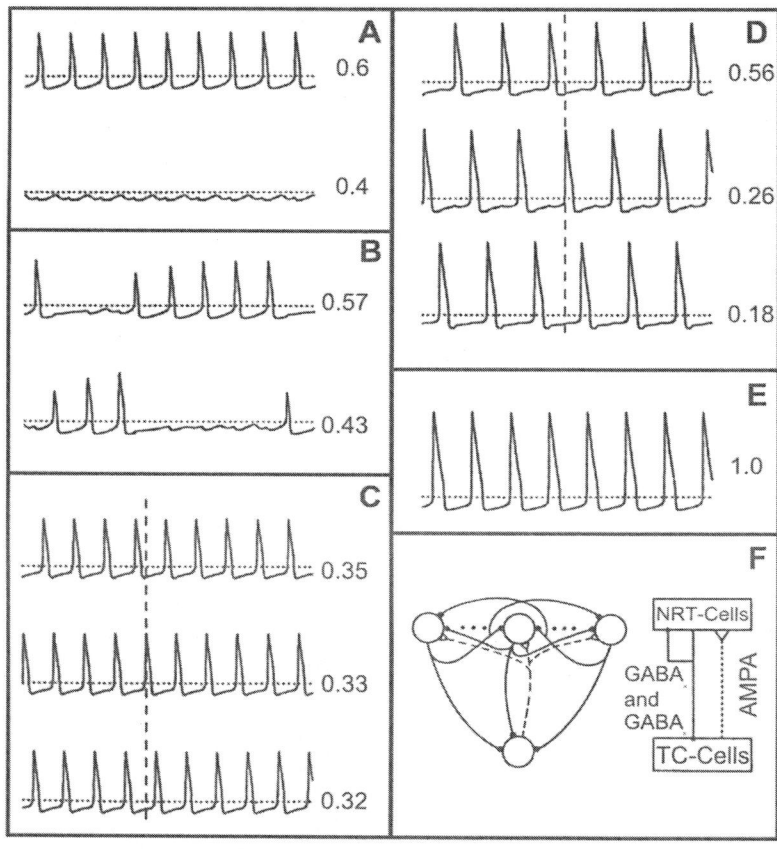

FIGURE 3.11. Model of the interactions between the nucleus reticularis thalami (NRT) and the thalamocortical projection neurons (TC). The wiring diagram is shown in F). Excitatory (AMPA) connection exist from TC to NRT. NRT cells project back onto themselves inhibitorily with $GABA_A$ and $GABA_B$ type connections. The same inhibitory connections also reach the TC-cell pool. Only low-threshold calcium spikes are modeled. Parts (A-E) show the behavior of the *isolated* NRT network without TC input. Depending on the input parameters (see text) different fractions of the NRT network will synchronize. As soon as the TC-input becomes active only the fully synchronized state E) is observed. (Recompiled from Golomb et al., 1994).

where I_{AMPA} describes the input current from the thalamic relay cells to the NRT cells (AHP = after hyperpolarization) and thalamic relay cells (TC-cells) are described by:

$$\frac{dV}{dt} = -\frac{1}{C}[I_T + I_H + I_{leak} + I_{GABA_A} + I_{GABA_B}] \tag{3.3}$$

Regular sodium spikes are not modeled because they are irrelevant in the context of these investigations.

The membrane equations together with the intrinsic characteristic of the different currents lead to a rich nonstationary (oscillatory) behavior even of the isolated NRT network (Fig. 3.11). The actually observed patterns depend on the initial conditions and the parameters of the cells. In the simplest case, the whole population synchronizes (E) or remains quiescent. More often, however, smaller groups of neurons will synchronize (A) and fire with different patterns (B). When the slow influence of $GABA_B$ is removed three synchronous clusters of cells with the same firing frequency but different phases appear (C). A similar situation occurs when the calcium conductance is increased (D). Frequencies are in general in the spindling range.

Destexhe and coworkers (Destexhe et al. 1994) used a similar approach and looked at the behavior of small NRT cell groups in detail. Fig. 3.12 shows different oscillatory patterns observed for either two mutually connected cells (A) or a group of five NRT cells with increasingly dense connectivity between them (B).

The situation changes as soon as thalamic relay cells are also taken into account. Almost in every case the NRT cell pool becomes fully synchronized when even a small excitatory drive from the thalamic relay cell pool enters the NRT (Golomb et al., 1994). Thus, the network simulation confirms the qualitative description of the antagonistic action of NRT and thalamic relay cells. Most interestingly the oscillation frequency drops from 10 to 4 Hz once the $GABA_A$ influences are removed. This phenomenon can be paralleled with experimental findings in which $GABA_A$ inhibition was blocked by bicuculline methiodide using a slice preparation (Bal et al., 1995).

Propagating waves in the NRT-TC network

In a second study Golomb et al. (1996) investigated the activation patterns in a spatially extended slab of their network. Only connections along one linear axis were included (Fig. 3.13 B), because this reflects the situation in a commonly used brain slice of the thalamic system.

The main result of this simulation is a lurching wave of spindle activity which propagates through the network at about 1mm/s engulfing more and more cells (Fig. 3.13 A). The reason for this slow propagation is that those thalamic relay cells which have just become inhibited must remain hyperpolarized for some time before the T-current is de-inactivated. Similar to the situation above, blockade of $GABA_A$ receptors leads to a strong

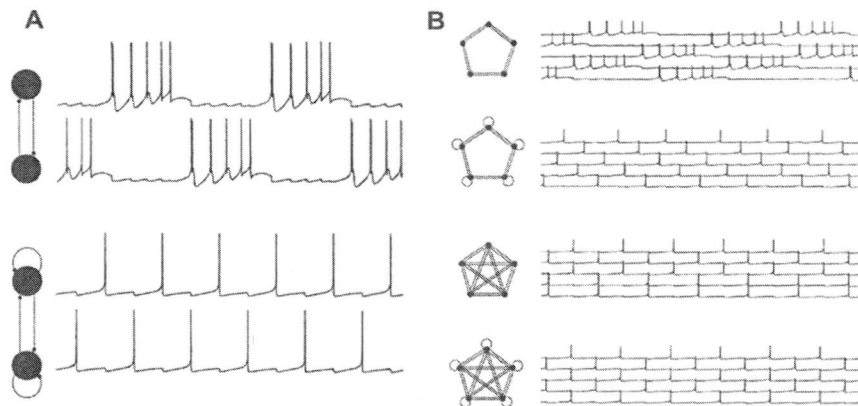

FIGURE 3.12. Simulation of different patterns of oscillatory behavior in a small, isolated NRT network. A) Only two cells are inhibitorily connected and the bottom trace shows the behavior as soon as self-inhibition is also included. B) Observed pattern for five connected NRT neurons. (Recompiled from Destexhe et al., 1994.)

frequency drop (Fig. 3.13, A, top right), because the inhibitory effect of $GABA_B$ receptors is smaller and has a much slower time-constant.

Simulations such as these, thus, can shed light on the involvement of the thalamus in the different sleeping patterns reflected in the EEG. It should be noted that so far we have only paid attention to the hyperpolarized state of the thalamic network. However, the LGN also plays a major role in the control of the information flow towards the cortex during the awake state when the thalamic network is in a more depolarized state.

3.3.3   Models of the depolarized thalamic state

The location of the LGN between retina and cortex and the integration of multiple input sources (including brain stem and corticofugal feedback), which takes place here, makes it ideally suited to control the throughput of information at a very early level.

Influences which generate multimodal LGN interval patterns

As described in the chapter dealing with the temporal structure of LGN activity (section 2.2.10) the multimodal interval pattern can be explained by the action of *deletion inhibition* (Fig. 3.7). Given all the different inputs to the LGN the situation, however, is much more complex. Fig. 3.14 summarizes the basic situation which has also been implemented in a biophysical realistic model (Suder and Wörgötter, 1999): LGN interval distributions can change from the control case (first peak of the interval distribution is highest) to a situation where the second (or higher order) peak is highest in at least three ways: 1) by increasing inhibitory inputs from LGN interneu-

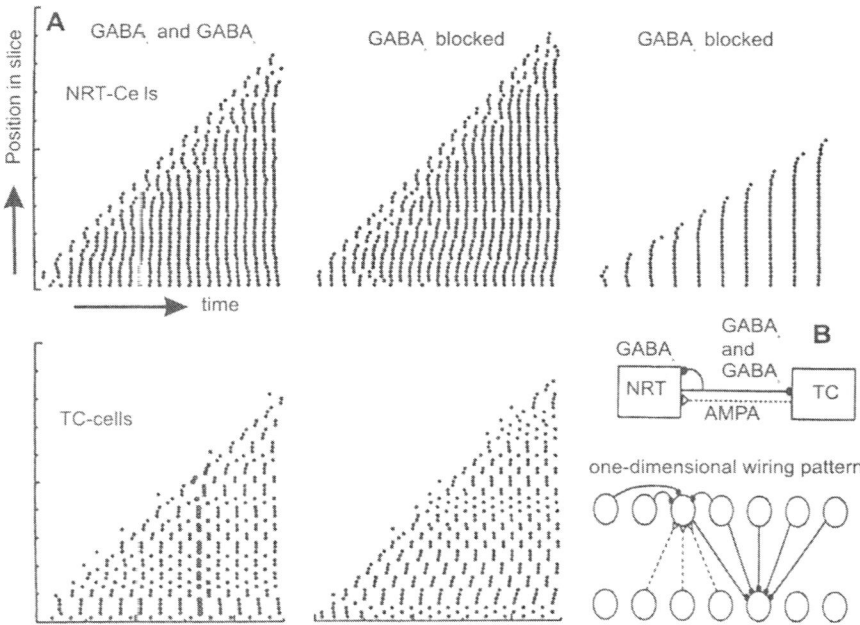

FIGURE 3.13. Spatiotemporal propagation of activity in a simulated thalamic slice. For abbreviations see also legend of Fig. 3.11. B) shows the wiring diagram. A one-dimensional structure is modeled and the connection types are slightly different than in the model shown in Fig. 3.11, because now the NRT cells receive only $GABA_A$ NRT connections (in Fig.11 also $GABA_B$ was present, which now is used only for connections onto the TC-cells). Part A shows how the activity spreads into the network. Every small dot represents a spike. NRT cells exhibit a rather regular firing pattern, TC cells are slightly more irregular. If $GABA_A$ influences are removed (top left) a strong drop in firing frequency is observed (also in the TC cells, not shown) (Recompiled from Golomb et al., 1996).

rons or from the PGN, 2) by removing corticofugal excitation, and 3) by a reduced activity of the ascending reticular arousal system (ARAS). The latter can ultimately lead to a hyperpolarized state of the LGN cells which can then enter the above described spindling patterns.

This shows that the corticofugal feedback itself is also involved in the gating of activity from the thalamus to the cortex (Crick, 1984, Mumford, 1991).

LGN firing influences cortical cell behavior

Recent findings indicate that in this process the initially existing activity of the thalamic cells has also an immediate influence. Even in the awake state LGN cells can change their firing characteristic from burst firing to tonic transmission mode (see aspect 1 in this section) and this can lead to a rather large difference in the point-spread of activity in the cortex already

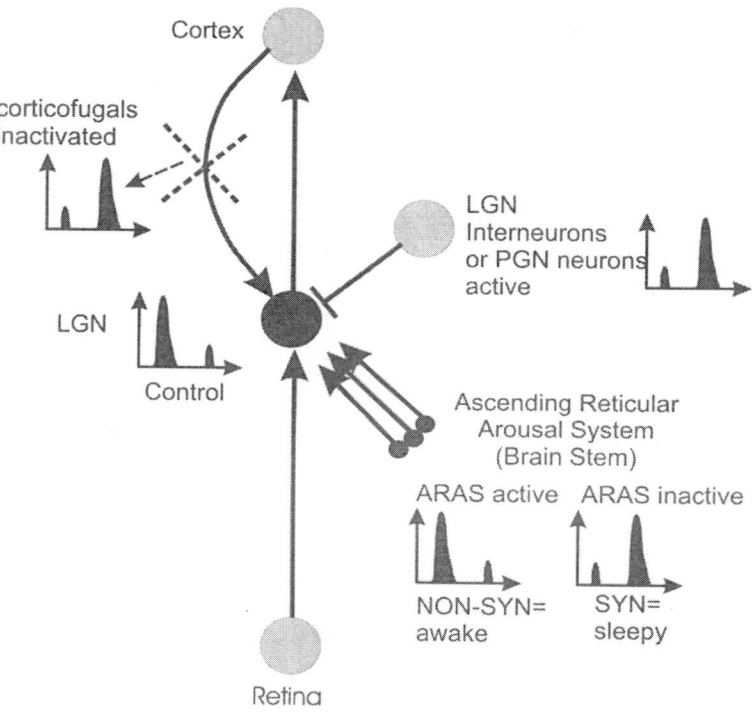

FIGURE 3.14. Schematic diagram of a biophysical model of the thalamocortical system including brain stem influences. The model focuses on the interspike interval patterns observed in the LGN. The little diagrams depict schematic interspike interval histograms (INTHs) of the LGN relay cell in the center. LGN interval distributions can change from the control case (first peak is highest) to a situation where the second (or higher order) peak is highest by: 1) by increasing inhibitory inputs from LGN interneurons or from the PGN, 2) by removing corticofugal excitation, and 3) by a reduced ARAS activity.

before the corticofugal feedback becomes active.

Burst firing in the LGN leads to a high degree of spatial and temporal summation in the cortical cells such that cells even further away from the central projection column will be driven. As a consequence a wide point spread of activity is observed (Fig. 3.15 A,C) and the fast and synchronous burst signals arrive with a rather short latency in the cortex where it will initiate a wave of activity in a chain of cortical areas (not shown in Fig. 3.15). Functionally this situation occurs during drowsiness in an inattentive situation such that a fast and wide-spread signal could signal a *wake-up call* for the system. On the other hand, if the LGN cells fire in tonic transmission mode their firing frequency is much lower and less temporal summation occurs at their cortical targets leading to a smaller point spread (Fig. 3.15 B,D) and a longer response latency. This situation is usually associated with alertness, attentive wakefulness and corresponds to a higher spatial resolution of receptive fields within the primary visual pathway. Such a functional distinction between low-resolution/short latency- and high resolution/long latency response could be demonstrated in a biophysical realistic model of the primary visual pathway (Fig. 3.15). There is experimental support for this concept showing that visual cortical receptive fields change their size in accordance with the EEG state. They are wider during sleep-like EEG (synchronized EEG) while they are smaller during nonsynchronized wake-like EEG (Wörgötter et al., 1998).

3.4   The Visual Cortex

The multitude of functional tasks that has to be completed in a fraction of a second in the visual cortex rests on the immense complexity of intracortical connections. At the single cell level this complexity is expressed by the different morphology of cells that can be appreciated by looking at a Golgi-stained section of the cortex, the method of choice for almost a century (Ramon y Cajal, 1899). Despite of the diversity of neurons in terms of anatomy, chemistry and physiology the cortex is astonishingly well organized. In this section we will describe some of the basic anatomical and physiological organization features upon which functional modules operate in the visual cortex.

3.4.1   Anatomical organization of the primary visual cortex

A cortical area, where the axons of LGN relay cells terminate is defined as the primary visual cortex. In primates, the primary visual cortex is restricted to area 17, which is also called striate cortex, or V1. In cat, the primary visual cortex includes, in addition to area 17, areas 18 and 19, because the LGN forms a systematic projection of the retina in these areas, too. Areas 18 and 19 are also called extrastriate cortex.

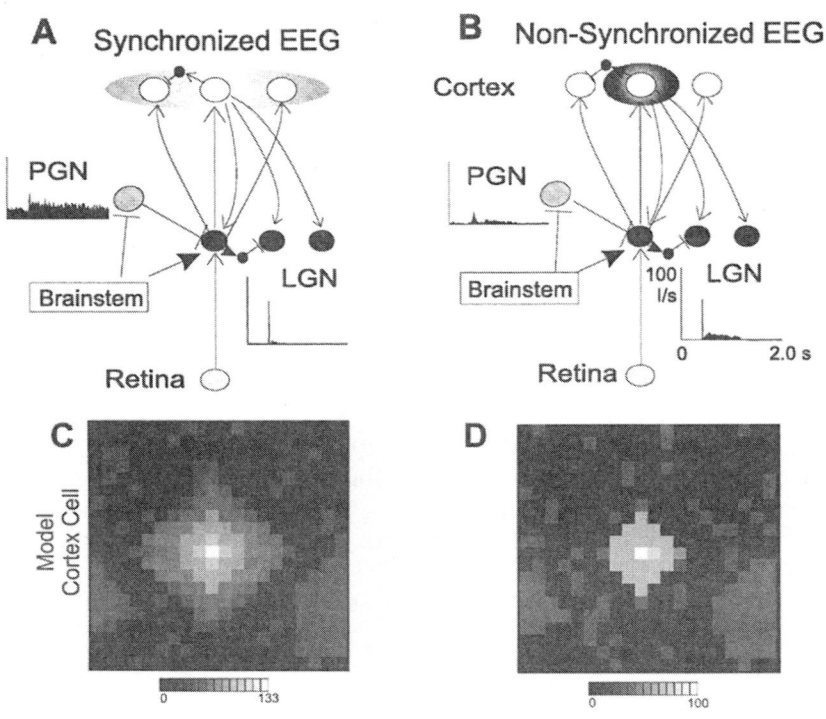

FIGURE 3.15. Schematic diagram of the cortical point spread of activity during a synchronized and a nonsynchronized EEG (A,B) and cortical model cell responses plotted as receptive field maps (C,D). Orientation selectivity has not been included in this model. Therefore the responses in (C,D) are spatially nonoriented. A) During a synchronized EEG the brain stem is less active but PGN cells fire strongly. The PGN in turn inhibits the LGN (see sections above). Inset PSTHs show real PGN and LGN responses. As a consequence LGN cells fire in burst mode. Bursts have a very high firing frequency and will therefore induce rather strong temporal summation at the cortex cells. This leads to a wide point spread of activity and consequentially to a wide receptive field (C). During nonsynchronized EEG (B), the brain stem is active, PGN is rather silent and the LGN cells fire in tonic transmission mode. Firing frequencies (after the initial burst) are lower and less temporal summation occurs at the cortical cells. The point spread of activity remains restricted and receptive fields are smaller (D).

Cortical layers

The gray matter of the cortex is a three-dimensional structure whose anatomical constrains can be best studied from sectioning planes perpendicular to the cortical surface or parallel to it. While the former view reveals the laminar composition and columnar organization of the tissue the latter is best suited to explore its relationship with laterally distributed functional representations. The visual cortex consists of 6 main laminae (Fig. 3.16 A), which are characterized by a particular size, type and density of neuronal cell bodies, fiber density, trajectory of the axon processes, sources of inputs and targets of projections. Each lamina contains cell bodies of both excitatory and inhibitory neurons of several types (see below). The major characteristics of different layers can be summarized as follows. In layer 1, neuronal cell bodies are sparse and the dominant structures are axon fibers and dendritic processes of cells located in deeper layers. Over 90% of the cells here are inhibitory containing GABA. In other layers, inhibitory cells represent about 20% of all neurons. Layer 2 consists of densely packed pyramidal neurons many of which emit a major axonal branch towards the white matter to reach other cortical structures (Fig. 3.16 A,B). Many of these cells project also to the opposite hemisphere via the corpus callosum. Layer 3 contains a variety of neurons of different sizes and types. The majority of cells are pyramids of which the largest are found close to the border with layer 4 (Fig. 3.16 A,C,E). Similar to layer 2 cells, many layer 3 pyramids have their main axon projecting to other cortical regions. Characteristic to layer 4 is a large number of densely packed medium size to small spiny stellate and star-pyramidal neurons. They have typical star-shaped dendritic tree and local axon projections. Layers 5 and 6 are composed largely of pyramidal cells which form the major output to subcortical structures. Some of layer 5 pyramids have large somata whose chief extracortical target is the superior colliculus (Fig. 3.16 A,D). In layer 6, many pyramidal cells are known to project to the LGN.

A good deal of anatomical and physiological data indicate that there is an orderly relationship between different laminae as one follows the information processing route from the input arriving from the thalamus to the output to other cortical and subcortical regions (Gilbert, 1977). In this sense layer 4, being in the middle, plays literally a central role because much of the thalamic input terminates in this layer. From here the information is relayed largely towards the upper layers 2 and 3 and, partly, towards deep layers via the apical dendrites of layer 6 cells. Layers 2/3 establish projections in the same layers and towards layer 5 which project back to the superficial layers as well as to layers 5/6. Finally, there is a strong projection from layer 6 cells into layer 4. Notice, that even this extremely simplified wiring diagram contains multiple loops via which the primary visual information could be influenced (see section: *First and second order cortical cells*).

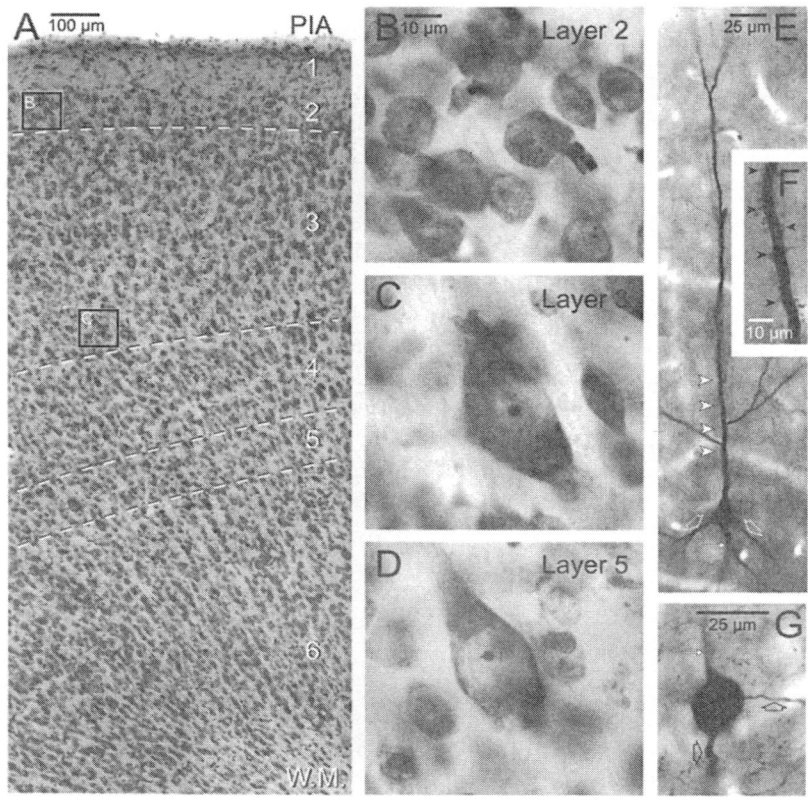

FIGURE 3.16. Neuronal composition of the primary visual cortex of the cat.
A-D:Nissl staining; E-G biocytin staining. In A, coronal section shows differences
in the packing density of neuronal cell bodies across the 6 cortical layers. B-D
are enlarged regions of A, demonstrating some morphological characteristics of
neurons of individual laminae. Layer 2 contains a high density of small, mainly
pyramidal neurons. In layers 3 and 5, there are small and large pyramidal cells. Of
the latter, examples are shown in C and D, respectively. In E, a biocytin stained
layer 3 pyramidal neuron demonstrates the major anatomical features of this cell
type; the pyramidal shaped soma, the apical (arrow-heads) and basal dendrites
(unfilled arrows), the axon (arrow) leaving from the lower pole of the soma,
and the dendritic spines (arrow-heads in F) which represent the chief target for
synaptic contacts. Inhibitory neurons typically have round or oval shaped somata
and spine-free (smooth) dendrites. In G, an example of a layer 5 interneuron is
shown whose dendrites (unfilled arrows) are smooth and the axon (arrow) is given
off from the upper pole of the soma. W.M. = white matter.

Neuronal cell types

Earlier anatomical studies spent heroic efforts in providing detailed description of neuronal cell types with some mixed successes. The major obstacle was the incomplete visualization of axons which we now know represent one of the key attributes in defining morphological categories. Modern anatomical techniques allow much of the axonal and dendritic arborization of individual cells or cell groups to be visualized. Many of them can be used in combination with functional approaches to explore the local as well as the global aspects of the functional organization of the cortex (e.g., in Somogyi and Freund, 1989).

At the most simplistic level there are two major morphological cell classes in the visual cortex (see Figs. 16-18 and Jones and Peters, 1984; White and Keller, 1989) . The first class which includes the majority of neurons represents pyramidal or principal cells (Figs. 16 E, 18 A). These cells are excitatory and have pyramidal shaped somata and spiny dendrites (Fig. 3.16 E, F). Characteristically, the soma gives off an apical dendrite directed towards the cortical surface, and basal dendrites directed laterally and towards deeper cortical layers (Fig. 3.16 E). The second class includes neurons with star-shaped somata and they are also called stellate cells. Some of these cells are excitatory and their dendrites are usually covered with spines. Other stellate neurons are inhibitory interneurons whose dendrites bare no spines (Figs. 16 G, 17 B). In the visual cortex, smooth cells are local neurons because their axons do not leave the cortex as opposed to many of the principal and spiny stellate cells which represent the output to other brain areas. Each class harbors a large variety of morphological types, and the basic anatomical differences between the cell types are often accompanied by functional differences (Kawaguchi, 1995). Notably, while virtually all spiny cells are excitatory in nature and use glutamate for signal transmission smooth dendrite interneurons express the inhibitory transmitter, GABA (see Mize et al., 1992). For clarity it needs to be added here that the name, interneuron, also applies to some of the spiny cells, in particular the spiny stellates, whose axons do not leave the cortex, nonetheless, their excitatory nature unambiguously delineates them from inhibitory interneurons. The cortex contains about four times as many excitatory cells and synapses as inhibitory ones (Gabbott and Somogyi, 1986; Beaulieu and Somogyi, 1990) . In order to compensate for the *numerical unbalance* between excitatory and inhibitory inputs the cell-to-cell connections have evolved an intriguing strategy. The location of the excitatory and inhibitory synapses along the surface of the cells shows a remarkable compartmentalization. Most of the inhibitory synapses occupy the soma and proximal dendritic regions as opposed to excitatory synapses which tend to influence neurons via dendritic shafts and dendritic spines. The functional implication of this kind of synaptic relationship between the inhibitory and excitatory input is best acknowledged when they are viewed in relation to the generation site of action potentials, the axon hillock. It is conceivable that proximally located inhibition is best suited to control the efficacy of the overall excitatory drive arriving from the dendrites whereas distally located inhibitory synapses ought to have a more specific role by influencing the excitatory input to particular dendritic branches. The proximity of the inhibitory input to the soma region suggests a powerful control of the excitatory drive. Thereby a smaller number of inhibitory cells and synapses are sufficient under normal conditions than of the excitatory types.

Most anatomical studies dealing with cell type characterization often face with the dilemma of answering the question as to how many neuron types exist in the visual cortex? Obviously, the answer depends on the number of criteria used in distinguishing cells from one another. Therefore, instead of providing a lengthy account on cell types and a take-home number, we show here some examples and demonstrate some major organization strategies of the neuronal network. Our first example is a layer 5 pyramidal cell shown in Fig. 17 A. The axon forms a characteristic clustered projection in layers 2/3 while avoiding all other layers. Pyramidal cells of layers 5 and 2/3 have been shown to establish 80-85% of the synapses with dendritic spines of other excitatory neurons and about 15-20% of the synapses with dendritic shafts of GABAergic, inhibitory neurons. The pyramidal cell in Fig. 3.17 A is interesting in two aspects. First, it demonstrates that interlaminar projections are layer selective. Second, it shows that remotely spaced cortical columns can be directly linked by the very same neuron (in many cases, pyramidal axons span 7-8 mm across, corresponding to 5-7 orientation hypercolumns or 3-5 ocular-dominance columns). Our second example is an inhibitory neuron of the large basket cell type (Fig. 3.17 B). Basket cells provide probably the majority of somatic and proximal dendritic inhibition and represent the most frequent inhibitory cell type, an estimated 50% of all GABAergic cells. Basket cells and many other GABAergic cells resemble excitatory neurons in possessing interlaminar axon projections. The lateral spread of axonal fields of some basket cells can reach 1.5-2 mm from the parent soma, corresponding to 2-3 orientation hypercolumns or 1-2 eye-dominance columns. However, most of the inhibitory interneurons project only locally often within a region of less than 300-500 μm in diameter.

Excitatory and inhibitory cell types with long-range lateral connections have been shown to form the basis of extensive corticocortical networks in an interlaced manner. The panels in Fig. 18 illustrate the succession of building up a horizontal network from individual pyramidal neurons. There are three important features to be noted. First, the connections are clustered, second, the majority of connections is reciprocal, third, local projections (intracolumnar) provide the richest connectivity.

Columnar structure

Another typical feature of the cortex is columnar organization. It derives from the spatial characteristics of the cellular constituents the dendrites and the axons. Typically, pyramidal cells emit a major apical dendrite

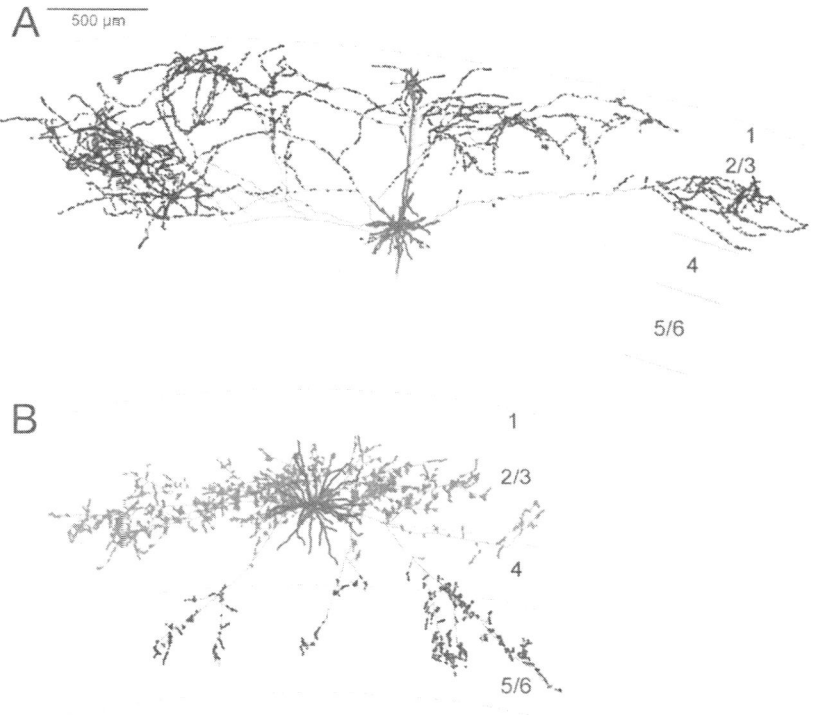

FIGURE 3.17. Demonstration of some of the main neuronal features in the cat visual cortex. A) Drawing of a layer 5 pyramidal cell. The soma (arrow) and the basal dendrites are located in layer 5. The ascending apical dendrite reaches the most superficial layers where it gives off a few branches. Characteristically, the axon arborises in a clustered manner providing patches of axon terminals (black dots) up to 1.5 mm form the parent soma. Interestingly, all terminals were found in layer 2/3 although in the 2-dimensional image some appear in lower layers. The main targets of pyramidal cells like this are the dendritic spines and shafts of other pyramidal neurons, nonetheless, to a lesser extent, somata and dendrites of inhibitory cells are also encountered. B) Inhibitory (GABAergic) large basket cell of layer 3. This cell type represents the largest inhibitory neuron in the visual cortex. The large soma, the dendritic field and the majority of the axon terminals (gray dots) are confined to the upper division of the cortex. Only a small proportion of the axon terminals (black dots) target deep layer neurons. Basket cells form the majority of axo-somatic contacts of all neuron types of the cortex. They are assumed to mediate powerful inhibition due to their proximity to the action potential generation site, the axon hillock. *(Drawing B is courtesy of Péter Buzás)*.

1 cell

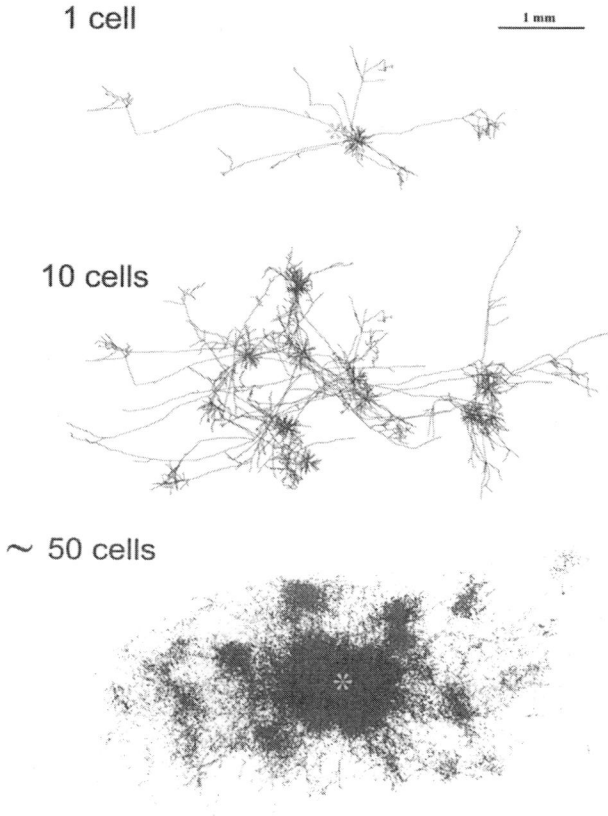

10 cells

~ 50 cells

FIGURE 3.18. The three panels demonstrate how a large horizontal neuron network is formed from individual pyramidal cells possessing clustered axonal fields. In all panels the reconstructions are viewed from the cortical surface. The upper panel shows a layer 3 pyramidal axon whose clustered axon spans 4.9 mm across the cortex. Asterisks mark the location of a small tracer injection that resulted in the labeling of this and other 9 cells. Middle panel shows a reconstruction of all ten cells labeled from the same injection site (asterisk). Notice the patchy appearance of the composite axonal field. The lower panel demonstrates the pattern of labeled axon terminals when an estimated 50 neurons are labeled from the same injection (asterisk). The resulting patchy pattern strongly resembles the pattern seen in the middle panel. Similar, interlaced networks are bound into *super-networks* that can mediate information across remote regions using only a few synapses.

which is directed at a right angle to the plane of the cortical surface. Furthermore, the majority of fibers that either leave or enter the gray matter form bundles whose trajectory runs parallel with the apical dendrites and perpendicular to the laminar borders. This polarization of the neuronal processes results in an overall columnar phenomenon (Fig. 3.19). Because the cortex comprises an unmatched complexity of overlapping axonal and dendritic networks, it is extremely difficult - if not impossible - to decipher what exactly a cortical column is, e.g., where is the border between neighboring columns, solely on the basis of morphology. A more meaningful approach is when structure and function are both taken into account. For example, in the primary visual cortex neurons lying one above the other share similar orientation preferences (as follows). Thus the definition of an orientation column could be approximated as the cortical cylinder within which the same orientation preferences are retained. However, considering the graded nature of change of orientation preferences from one cell group to another, and some variation of the preferred orientation even between neighboring cells, no absolute definition can be given.

Within the same column, communication between different layers is achieved via interlaminar connections. Based on available anatomical data virtually all neurons in the visual cortex, including excitatory and inhibitory types, send axon collaterals to other layers and, in turn, are exposed to input from other layers. This is particularly evident for cells in layers 3 and 5 which are known to have extensive intra- and interlaminar connections. The pattern of the axonal projections of cells of a single cortical column is demonstrated in Fig. 3.20. The graphs show the density distribution of processes (axon terminals) which were stained using small, focal injection of the neuronal tracer, biocytin into layer 3. Apparently, the terminal density at the injection site (highest peak) and in the immediate neighborhood overwhelms the labeling density found elsewhere, suggesting that a major processing takes place locally.

Horizontal connections

Communication between functional columns is a key feature of cortical organization. It is established via horizontal or long-range corticocortical connections mainly by pyramidal cells. Their axons emit laterally extended branches which run quasiparallel with the plane of laminae. In addition to pyramidal neurons there are inhibitory large basket cells which also emit long-range axons although their overall lateral extent is restricted to a few mm as mentioned above. From a functional point of view, long-range connections are well suited to integrate a broad range of effects over extensive regions of the visual space. Before providing exemplary data as to the functional specificity of long-range connections, we consider some basic response characteristics of cells in the visual cortex.

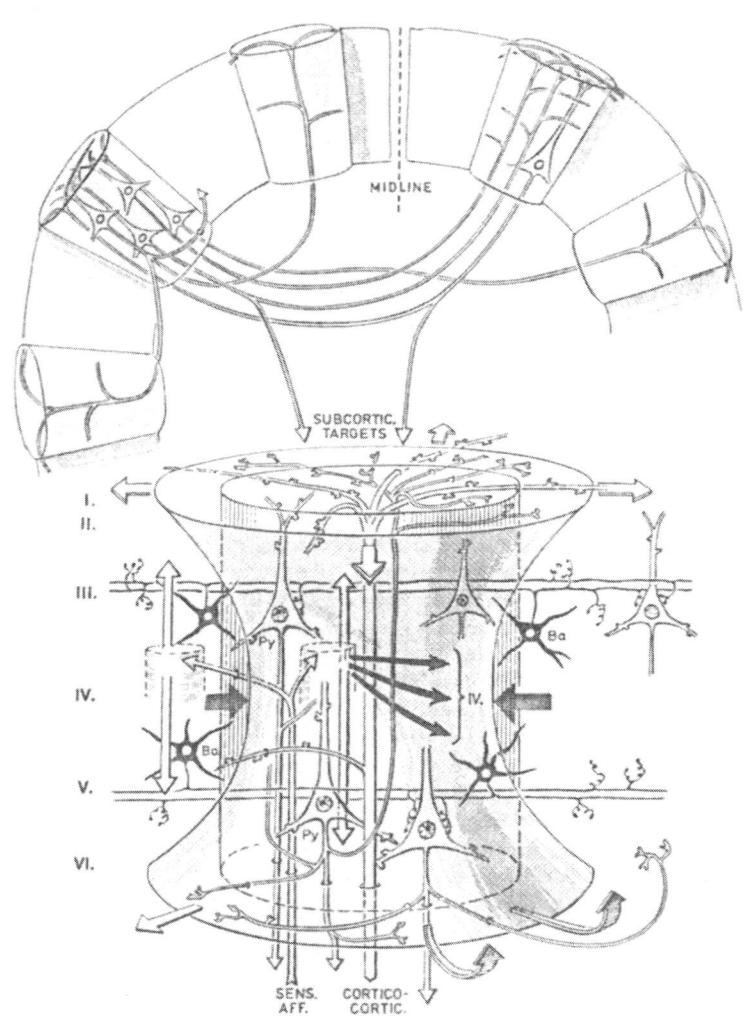

FIGURE 3.19.

FIGURE 3.19. Schematic view of the cerebral cortex organization pictured by the Hungarian anatomist, János Szentágothai. Here we provide the original caption of the figure:*Architectural principle of the cerebral cortex. Upper part of the diagram shows that the cortex can be envisaged as a mosaic of vertical columns, defined by the convergence of a group of cortico-cortical afferents. Only pyramidal cells are illustrated from which those situated in the outer three layers of the cortex project exclusively to other parts of the cortex, where fibers terminate as cortico-cortical afferents. Pyramidal cells of the lower two layers project mainly to subcortical targets, but their collaterals, or some of the main branches, reach the cortex of the opposite hemisphere. Lower part of the diagram gives a radically simplified diagram of the cortical column:a vertical cylinder containing about 5,000 nerve cells, ≈ 60% pyramidal (Py, projective = output cells), ≈ 40% are interneurons of various kinds, 20% of the total are now known to be inhibitory. The cylindrical column is defined anatomically by the cortico-cortical afferents (cortico-cortic.) placed into the axis of the cylinder. The specific sensory afferents (sens. aff.) are dominant only in the primary sensory areas, but secondary and further sensory areas have a similar input into lamina IV (lamination in Roman numerals at left margin) from the primary areas. The further local transmission of the specific input is indicated only by arrows (in outline if excitatory, and black if inhibitory). Two rows of large basket cells (Ba) at the border of layers III-IV and IV-V exercise inhibition over distances of up to 2-3 mm. This horizontal inhibition is assumed to narrow down the waist of the column in layers III-V (dark horizontal arrows), while excitation is conjectured to spread in radial direction both in lamina I and VI, either by the long terminal branches of cortico-cortical afferents in lamina I and pyramidal cell collaterals both in lamina I and VI. This distorts the original cylinder dynamically into the shape of an hourglass. The rich mutual local inhibitory and/or disinhibitory interconnections, oriented primarily in vertical direction, are not included in this diagram.* (Reproduced from: *The Brain-Mind Problem*, Philosophical and Neurophysiological Approaches, Gulyás B (ed.), p.71, (1987, with permission from Leuven University Press).

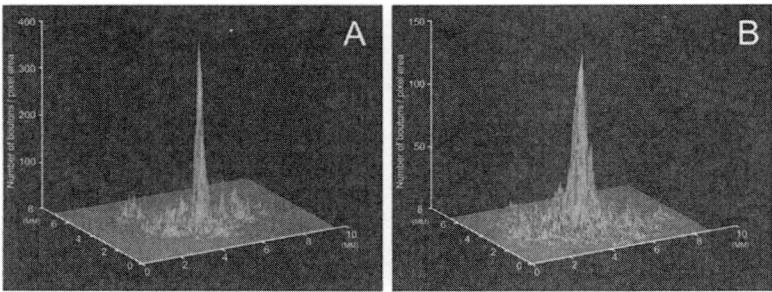

FIGURE 3.20. Density distributions of local (intracolumnar) and long-range lateral connections in the visual cortex revealed by tracer (biocytin) injections. The injections in A was smaller than in B. In both cases, locally labeled axon terminals (highest peak) far outnumber terminals in the surround representing patchy, long-range lateral connections.

3.4.2   Basic response properties of visual cortical neurons

Orientation and direction selectivity

An initial impulse for investigation of properties of neurons in the visual cortex came from the work of Hubel and Wiesel, who in the early sixties recorded spike activity of individual cells with an extracellular microelectrode placed in the primary visual cortex of cat (Hubel and Wiesel, 1962).

The responses of cortical neurons to the presentation of visual stimuli are quite different from the responses of cells in the retina and in the LGN. The most striking difference is that visual cortical neurons are orientation and direction selective. An elongated stimulus, such as a light or dark bar, presented in the receptive field of a cortical neuron, evokes a strong spike response only when it has a particular orientation, called the preferred orientation of that neuron. Stimuli of other orientations evoke much weaker responses or fail to produce spikes at all. In addition, the response of most neurons is stronger when an optimally oriented stimulus drifts across the receptive field in one (preferred) direction, but not in the opposite (null or nonpreferred) direction.

Further details of the mechanisms of orientation and direction selectivity were revealed with intracellular recordings. This method enabled to study membrane processes which underlie response selectivity, and to relate the visually evoked membrane potential changes (upper traces in Fig. 3.21) to the spike responses (lower traces in Fig. 3.21).

Receptive field structure, simple and complex receptive fields

The receptive fields of cortical neurons are very different from the receptive fields of cells at the earlier stages in the visual pathway. Cortical receptive

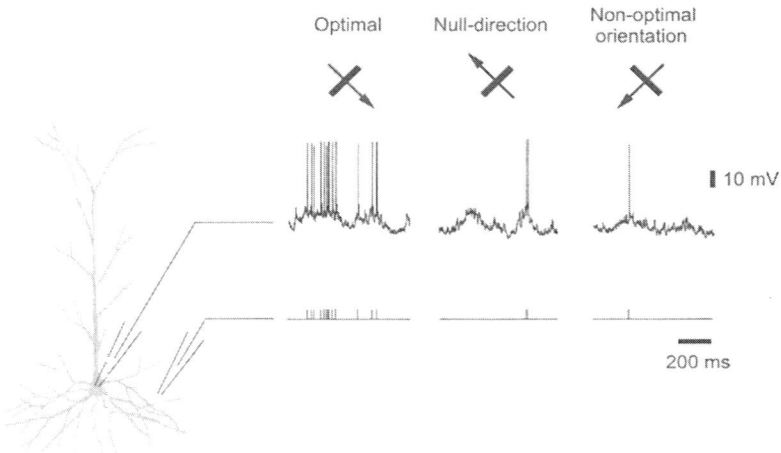

FIGURE 3.21. Orientation and direction selectivity of responses of a cortical cell. Responses of a cortical cell to an optimally oriented bar moving in optimal direction (Optimal), in the opposite direction (Null-direction) and to a bar of orthogonal to the optimal orientation (Nonoptimal orientation) are shown. The upper traces were recorded intracellularly and the lower traces show what would be recorded with an extracellular electrode *(Volgushev, Pernberg, Eysel, unpublished data)*. *Extracellular and intracellular recordings:* There are two basic techniques of recording electric activity of nerve cells. When a microelectrode is positioned close enough to a cell, action potentials produced by that cell can be detected as sharp deflections (spikes, discharges) in the recorded trace. Discharge rate can be influenced, either increased or decreased, by stimulation. However, since without visual stimulation most neurons in the visual cortex generate few irregular spikes (less than 5 per second), this technique has poor sensitivity for detecting inhibition or subthreshold inputs to a cell. Positioning a microelectrode inside a cell allows to measure the membrane potential, which is usually around -70 to -50 mV for visual cortical neurons and its deviation from the resting level in both positive and negative directions, as well as the action potentials. This enables detection of both subthreshold excitatory inputs and inhibitory inputs to a cell in a form of excitatory and inhibitory postsynaptic potentials. However, intracellular recordings are much more difficult technically, and therefore a majority of studies of the visual cortex exploit extracellular recordings.

fields have no central symmetry, but consist of one or several elongated subregions. From each subregion a response either to the onset of a light stimulus (On-subregion) or to the offset of a light stimulus (Off-subregion) could be evoked. When a receptive field consists of several subregions, they could be spatially separated or overlap. Receptive fields which consist of spatially separate On- and Off-subregions are called simple and receptive fields which consist of overlapping On- and Off-subregions are called complex.

The first hypothesis on the generation of cortical receptive fields

In the original scheme of Hubel and Wiesel (1962; 1965), both the spatial structure of the receptive field and orientation selectivity of a cortical neuron are due to the specific pattern of convergence of excitatory inputs. According to this scheme, a cortical cell which has simple-type receptive field receives input from LGN neurons whose circularly symmetric receptive fields are aligned in a row (uppermost scheme in Fig. 3.23). The dominant subregion of the receptive field corresponds to the area covered by the centers of LGN receptive fields, and subregions of an opposite sign located aside the dominant subregion and parallel to it, correspond to the periphery of LGN receptive fields. Orientation selectivity is then a direct and immediate outcome of the elongated shape of a simple receptive field. A cell with a complex receptive field receives its input from several cortical cells with simple receptive fields which are located close one to the other and have the same preferred orientation. As a result, complex receptive fields are usually larger than simple receptive fields and not elongated along the preferred orientation, but are orientation selective.

Cortical cells with simple and complex receptive fields were supposed to represent two hierarchical stages of information processing in the visual cortex. The simple receptive fields could be considered as local detectors of both position and orientation of fragments of visual objects, and complex cells could integrate information about the detected orientation over a certain area. These results and ideas produced a real breakthrough in the physiology of vision. First, they allowed a consideration of cortical neurons as detectors of particular properties of visual objects. Second, they related responses of individual cortical cells to something perceptually relevant, like orientation of fragments of contours. Third, they gave a hint on how visual information could be processed in the cortex, namely, by encoding more and more complex features of an image by progressively higher order neuronal detectors.

It should be noted that experimental data obtained by later studies demonstrated that excitatory convergence is not the only, and probably not even the main mechanism, which sculptures cortical receptive fields and orientation selectivity. There are alternative mechanisms, of which some are considered below, in this chapter (4.3).

First and second order cortical cells

Neurons in the visual cortex differ in their ordinal position relative to the thalamic input. Those cells, onto which the afferents from the LGN establish synapses are called first order cells. Those cells, which do not have direct thalamic input and receive specific visual information only from other cortical cells, are called second order cells. Most of the thalamocortical fibers terminate in layer 4, and some of them in layer 6, consequently, most of the first order cells are located in these layers. Nonetheless, neurons of other layers whose dendrites overlap with the termination zone of thalamic afferents could be first order, too. The majority of cells in the primary visual cortex are of first order type. Experiments in which recordings were made from visual cortical cells during electrical stimulation of the LGN demonstrated, that in about 75% of the cortical cells the electrical stimuli evoked orthodromic spikes (Singer et al., 1975; Bullier and Henry, 1979; 1980). For 60% of all cells, the short latency of orthodromic spikes clearly indicated the presence of a monosynaptic connection. It should be noted that this figure probably represents a lower estimate because it does not include those cells which do receive direct LGN input but their responses remain subthreshold for spike generation, or their input is mediated via slowly conducting fibers.

Electrophysiological recordings of single cells provided evidence that there is no strict correlation between first/second order and simple/complex dichotomy. Although most of the simple cells, and all those located in layer 4, receive direct thalamic input many of the complex cells are also first order. Furthermore, all cortical cells, even those receiving a direct thalamic input, are targeted by synapses (> 90%) from other sources than the LGN. Nevertheless, LGN afferents are likely to have, on the average, a more pronounced effect in activating simple cells than complex cells, as indicated by cross-correlation analyses of spike activity of simultaneously recorded LGN and cortical cells (Tanaka, 1983; 1985). About 10% of the spikes of simple cells were correlated with discharges of a single LGN afferent fiber whereas, for complex cells, this proportion was lower, about 3%. It is yet unclear as whether the above proportions reflect an effect of a single LGN fiber or of several fibers from a pool of thalamic relay cells firing synchronously.

The above considerations suggest a neural organization of the primary visual cortex that is not strictly hierarchical. Rather, each neuron is embedded into a number of intracortical loops of different degree of complexity. Given the massive corticogeniculate projection, relay cells of the LGN are incorporated in early cortical loops as well. From a functional point of view, multiple loops and massive feedback projections within the cortical circuitry have a number of consequences. First, they allow several iterations through computational mechanisms of the same circuitry and even of the same cell. Second, excitatory loops could prolong the responses of cortical cells and shape their temporal structure. Third, the reverberation of activity in a cortical network allows a cell to participate in several different assemblies. Fourth, feedback connections allow for top-down influences on visual information processing. Finally, the elaborated cortical circuitry provides a substrate for a multitude of plastic changes, for example, adaptation to extremely wide range of light intensities.

Receptive field dynamics

Responses of neurons to visual stimuli have a distinct temporal structure. The responses evoked from different parts of the receptive field have different latencies, time courses and strengths. As a result, the receptive field dimensions and topography must change with the development of responses (for a review see Wörgötter and Eysel 2000). To reveal the dynamics of that change, Shevelev and his colleagues used successive, short portions of responses evoked from different positions in the visual field for estimation of the topography of the receptive field (Shevelev et al., 1982; 1992). Their findings showed that the receptive field undergoes fast, systematic changes during the development of stimulus evoked responses, lasting from a few tens to several hundreds of milliseconds (Fig. 3.22). Initially, weak responses appear in the central receptive field area. Then, this area expands and the responses in the center become stronger. Finally, the receptive field breaks up and disappears. During certain time intervals, the receptive field relief differs significantly from the summary map. Therefore, in addition to position within the visual field, time has to be introduced as a second essential parameter for receptive field characterization. In fact, receptive fields of many neurons in the LGN and the visual cortex can be faithfully characterized only in a spatiotemporal domain (DeAngelis et al., 1995).

The dynamics of responses and of the receptive field structure can be due to several factors. The intrinsic properties of neurons and their synaptic connections are not suited to hold the firing activity of a cortical cell at a constant rate. When a visual stimulus evokes a burst of discharges in an LGN cell, then, at the level of the geniculo-cortical synapse the bulk of the transmitter pool is released during the initial 2-3 action potentials while later spikes cause only a little release. Consequently, the spike responses of a cortical cell decay even when the input is at a constant firing rate. In addition, reverberation of the input signal in the cortical network, accompanied by activation of some cells and suppression of others, might lead to redistribution of the activity at the cell's inputs. For example, early response phases could be dominated by geniculate while late phases by intracortical inputs.

Luminance and contrast adaptation

The visual system of higher vertebrates performs over a remarkably wide range of luminance conditions, from dim twilight to bright sunshine. While the differences between absolute luminance can be nearly ten orders of magnitude, the nerve cells are able to change their firing rate only by about

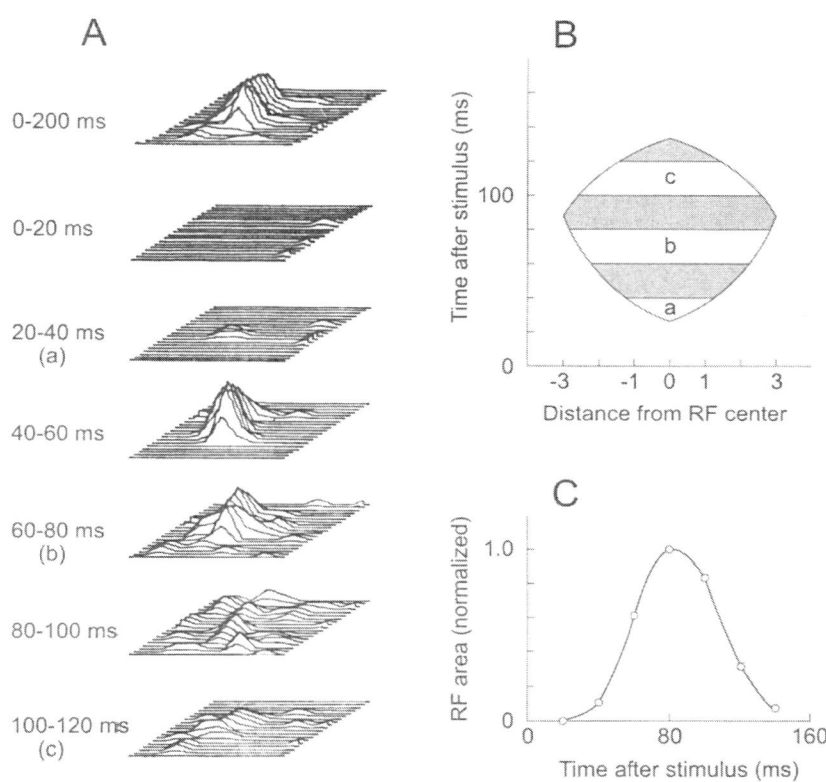

FIGURE 3.22. Dynamics of the receptive field of a cat visual cortical cell. A) Receptive field relief assessed from the number of spikes during the whole response (top, 0-200 ms) and from sequential 20 ms fragments of the responses (0-20 ms, 20-40 ms and so on; vertical scale is doubled relative to the summary map). The responses were evoked with 1° × 1° stimuli flashing in randomized order in 100 points (10x10 matrix) of the visual field. Plane of each map corresponds to coordinates in the visual field. The vertical axis shows response strength, measured as number of spikes. B) Scheme of dependence of temporal characteristics of responses to visual stimuli on location within the receptive field. The responses from receptive field center have the shortest latency and the longest duration. With increasing distance from the center, the latency of responses increases and duration decreases. a,b,c correspond to three temporal slices of the receptive field, as shown in A. C) Scheme of dynamics of the receptive field area. (Modified from Shevelev et al., 1992).

two orders of magnitude. During adaptation to luminance the actual working range of the visual system shifts in order to maintain resolution of grades of brightness at a sufficiently high level. Absolute luminance thresholds change with the changing level of light adaptation, and are lower in the dark adapted state than in the light adapted state. The process of luminance adaptation occurs on a time scale of minutes and the complete transformation from a light adapted state to a dark adapted state takes 20-40 min. Luminance adaptation is an associative effect of changes that takes place at all levels of the visual system: in the retina, in the LGN and in the visual cortex. Commonly, when adaptation level changes from light to dark, the responses of visual neurons become less phasic, the early phasic inhibitory component of the responses decreases or disappears, and the bulk of the responses shifts towards the long-lasting, sustained component (Lee et al., 1977; Virsu et al., 1977). In the dark, absolute luminance thresholds decrease and receptive field dimensions increase. These effects are probably due to decreased inhibition both in the center of the receptive field and in the periphery (Bear et al., 1971; Sasaki et al., 1971; Shevelev et al., 1974). The above processes occur at all levels along the visual pathway, and are likely to represent a trade off between sensitivity and spatial resolution. In the dark-adapted state, decrease of inhibition and increase of receptive field sizes enhance spatial summation resulting in an overall increase in the absolute sensitivity of cells. In the light adapted state, stronger inhibition and smaller receptive field sizes restrict spatial summation leading to an overall decrease in sensitivity but a better spatial resolution.

Another important feature of the visual system, contrast adaptation, occurs only in the visual cortex and is not present in the LGN and the retina. It is well known that the visual system can detect very fine relative luminance differences, but is poor for estimation of absolute luminance. Thus, stimulus contrast, that is usually defined as a difference between the luminance of the stimulus and the background divided by their sum, is a useful parameter for describing responses of visual neurons. Dependence of the response amplitude on stimulus contrast, the contrast gain, in the retinal ganglion cells and LGN relay neurons is affected little by contrast conditions that immediately preceded the test. The situation becomes different in the visual cortex. Contrast gain of most of the cortical neurons depends critically on test conditions. The contrast-response curve of a cell can shift with a certain time lag along the contrast axis so that the region of maximal gain (maximal slope of the curve) approximately corresponds to the contrast presented several seconds preceding the test stimulus. Thus, contrast adaptation keeps the response gain of the cortical neurons maximal around the mean contrast of the currently observed visual world, and it occurs without changes in the luminance adaptation level. Contrast adaptation is a relatively rapid, local and cooperative process (Fiorentini and Maffei, 1973; Ohzawa et al., 1985; Vidyasagar, 1990). It develops on a time scale ranging from a few seconds to tens of seconds, and is thus much faster than luminance adaptation. Contrast adaptation acts locally, so several retino-topical locations can be adapted to different contrasts, one independently from the other. Furthermore, contrast adaptation and luminance adaptation each has a different effect on the orientation selectivity of cortical cells. As long as the level of luminance adaptation remains the same, orientation selectivity is independent of the stimulus contrast (Sclar and Freeman, 1982). However, changing the luminance adaptation alters selectivity: in a dark adapted state orientation tuning curves of cortical neurons are broader and orientation selectivity is lower than in a light adapted state.

3.4.3 Mechanisms of selectivity of cortical responses: Orientation selectivity

Since the work of Hubel and Wiesel (1962), the transformation of receptive field properties that occurs at the transition from the LGN to the visual cortex has been in the focus of attention of experimenters and theorists. The most remarkable new property of cortical cells is orientation selectivity and we consider here the development of notions on the mechanisms of its generation.

The original hypothesis suggested by Hubel and Wiesel relies entirely on excitatory convergence of a number of LGN cells whose receptive fields are aligned in a row, along the preferred axis of orientation of a simple-type cortical cell (uppermost scheme in Fig. 23). Results of later experiments demonstrated that this mechanism alone cannot account for the sharp, final selectivity of cortical cells, and other, alternative mechanisms were suggested.

Intracortical inhibition

In a pioneering series of intracellular recordings from visual cortical neurons Creutzfeldt and colleagues (Creutzfeldt and Ito, 1968; Benevento et al., 1972; Creutzfeldt et al., 1974) found that excitatory input to a cortical cell originated from a slightly elongated region, but not from a highly elongated excitatory input field as suggested by Hubel and Wiesel (1962). The intracellular data of Creutzfeld and colleagues showed that in addition to excitatory inputs each cortical cell received inhibitory inputs as indicated by the presence of inhibitory postsynaptic potentials in responses to visual stimuli. This inhibition must be of intracortical origin since afferent fibers from the LGN form exclusively excitatory synapses. A logical interpretation of the intracellular findings was that inhibition plays a crucial role in creating receptive field properties by shaping largely unselective excitatory inputs. In the case of orientation selectivity, it was suggested that stimuli of nonoptimal orientations evoke stronger inhibition than those which are optimally oriented, and thus restrict spiking of a cell in response to presentation of nonoptimal stimuli. Since this inhibition was supposed to be maximal in responses to stimuli which were oriented orthogonal to the optimal, it was called cross-orientation inhibition (Fig. 3.23).

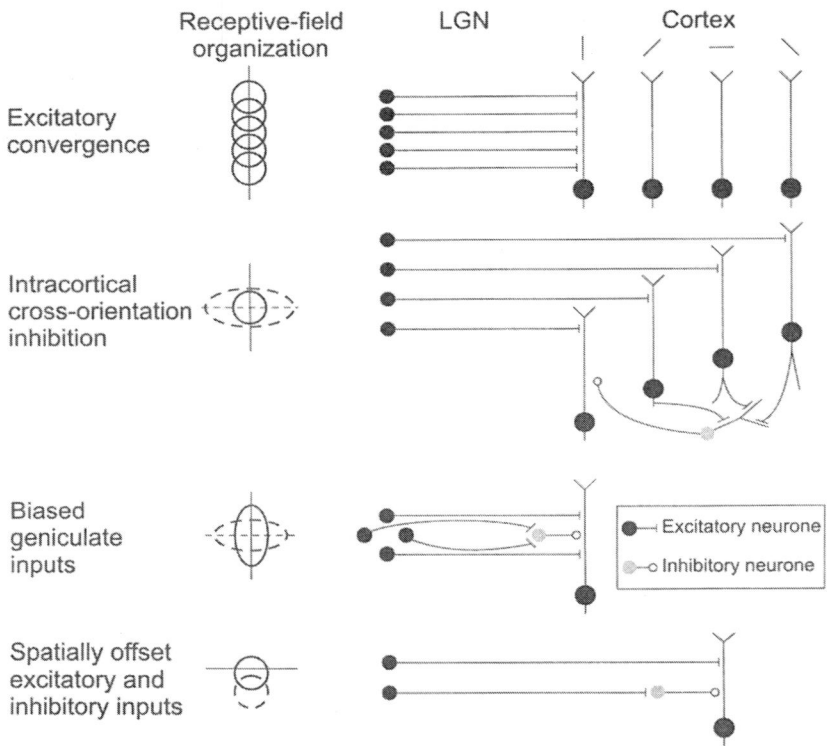

FIGURE 3.23. Different schemes, proposed to explain origin of orientation selectivity in the visual cortex. Spatial outlines of the LGN receptive fields converging onto a simple cortical cell (Receptive-field organization), and underlying wiring diagrams (LGN Cortex) are shown as proposed by different schemes. Receptive fields of excitatory inputs and the optimal orientation are represented by solid lines. Inhibitory inputs, via an interneuron, and their optimal orientation are shown as dashed lines. (Modified from Vidyasagar et al., 1996).

Once appeared, the idea of cross-orientation inhibition gave rise to several new experiments aimed to test its predictions. Generally, the obtained results supported the conjecture.

During application of $GABA_A$ receptor antagonists, which block intracortical inhibition, the majority of cortical cells loose or decrease dramatically their orientation and direction selectivity (Sillito, 1975; 1977; Sillito et al., 198C; Tsumoto et al., 1979; Eysel et al., 1998). Other neuropharmacologica̱ approaches showed that intracortical inhibition involved in the generation of orientation selectivity has a certain topographical organization (Wörgötter and Eysel, 1991; Crook and Eysel, 1992; Crook et al., 1997).

Increasing the firing rate of cortical cells by presenting a conditioning stimulus, or by the application of pharmacological agents allowed to uncover inhibitory influences by nonoptimally oriented stimuli (Blakemore and Tobin. 1972; Morrone et al., 1982; Ramoa et al., 1986; Heggelund and Moors, 1983). Cross-correlation analysis of spike activity of cell pairs recorded simultaneously revealed, in some cases, inhibitory interactions between cells preferring radically different orientations (Hata et al., 1988). Analysis of early components of postsynaptic potentials evoked in visual cortical ceḻs with flashing stimuli of different orientations showed that when excitatory and inhibitory response components could be separated, early inhibition was tuned to a different from the optimal orientation (Pei et al., 1994).

Taken together, these results obtained with various experimental approaches, provide converging evidence for the importance of intracortical inhibition in sharpening orientation selectivity of cortical neurons. It is important to note here that cross-orientation inhibition can sharpen orientation tuning of cortical cells but cannot create it anew from completely nonorientation selective input. Therefore, an already existing, initial orientation bias of the input is required right on arrival to the cortical cell that can be improved and sharpened by intracortical mechanisms.

Origin of initial weak orientation selectivity

To produce an initial weak orientation bias, the receptive field relief should contain some elongated components, which could cause some differences in the responses to different orientations. This nonhomogeneity of the receptive field profile could be brought about by a number of mechanisms. One possibility is an excitatory convergence of the input as suggested by Hubel and Wiesel (1962). However, the elongation of the receptive field does not need to be as marked as their scheme suggests since it should provide only an initial, weak selectivity which will be improved by additional mechanisms. Indeed, estimations of the geometry of the receptive field overlay of LGN fibers which arrive at a given orientation column (Chapman et al., 1991) or which converge onto a simple cortical cell (Reid and Alonso, 1995), demonstrated only a small degree of elongation of the excitatory input.

Another possibility is to exploit the orientation biases of the LGN receptive fields (Fig. 3.23). The receptive fields of some LGN relay cells are not strictly circular, but they show a 10-50% elongation in one of the directions and, in accordance with this, they display weak biases in responses to oriented stimuli. Transfer of these input biases onto either excitatory, or inhibitory, or both types of cortical cells would result in an initial, weak orientation selectivity of cortical neurons (Vidyasagar and Heide, 1984; Schall et al., 1986; Vidyasagar, 1987).

Finally, mild orientation selectivity of the input to a cortical cell could result from spatial displacement of excitatory and inhibitory receptive field regions (Fig. 3.23). In this scheme, suggested by Heggelund (1981), a simple cortical cell receives inputs from two pools of LGN relay cells of the same sign, with either On-centers or Off-centers. One pool of LGN cells makes direct excitatory synapses onto a simple cell, and forms the central, excitatory subregion of its receptive field. The other pool of LGN cells, with slightly displaced receptive fields, provides inputs to the cortical simple cell through an inhibitory intracortical interneuron, thus forming an inhibitory subregion of cortical receptive field. This scheme has two important advantages. First, the sum of the excitatory and inhibitory inputs produces an elongated, crescent-like relief of the receptive field and nonhomogeneity of the input in the orientation domain, even in the case of strictly circular subregions. Secondly, during development, it does not require elaborated rules to govern establishment of highly specific connections. Rather, spatial separation of excitatory and inhibitory subregions might be easily achieved on the basis of simple learning rules (Hebb, 1949).

Any of the above mechanisms, or their combination, could lead to a moderate elongation of the receptive field relief. In accordance with this assumption quantitative measurements of the shape of receptive fields of simple cortical cells did reveal a moderate degree of elongation of excitatory discharge regions (Heggelund, 1981; Reid and Alonso, 1995; Alonso and Martinez, 1998) and input fields of postsynaptic potentials (Pei et al., 1994; Volgushev et al., 1996; Hirsch et al., 1998). The ratio between the extension of the receptive field along and perpendicular to the optimal stimulus orientation was on average less than two. Thus, initial excitation to a cortical cell produced by an optimally oriented stimulus might be not more than two times stronger than initial excitation evoked by a nonoptimally oriented stimulus.

Intracortical amplification of the selectivity

An initial weak orientation selectivity of the input to a cortical cell needs amplification and improvement to achieve sharpness of the final tuning, typical for simple cells (Fig. 3.24). Improvement of the selectivity could start already during summation of the postsynaptic potentials, before any cortical spike is evoked. *In vitro* studies show that cortical neurons possess

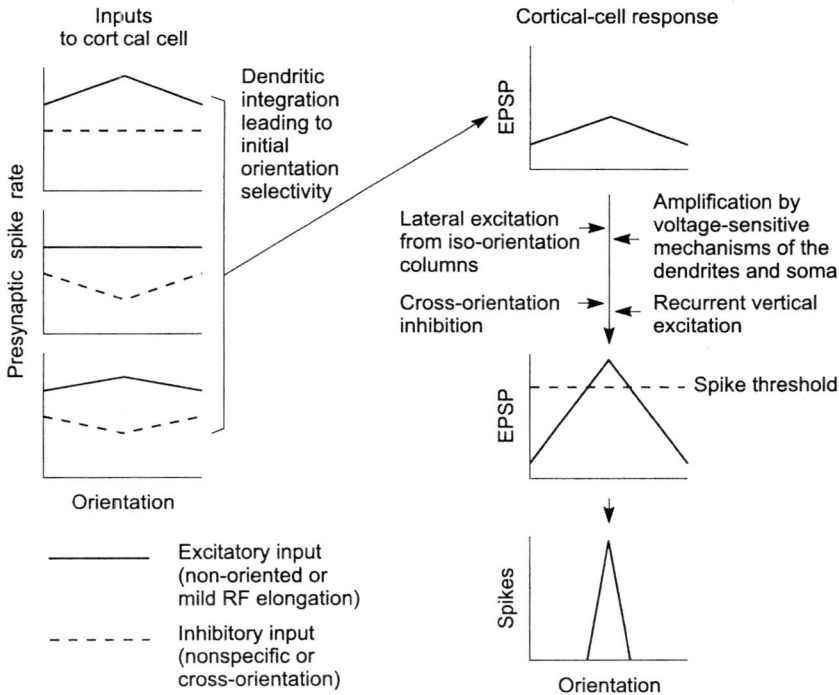

FIGURE 3.24. Generation of cortical orientation selectivity by multiple cooperative mechanisms. An initial mild selectivity of postsynaptic potential (PSP) responses in cortical cell can be produced by biases present in excitatory or inhibitory, or both inputs. A number of cortical mechanisms amplify this mild initial PSP selectivity to achieve final sharp tuning of spike responses. (Modified from Vidyasagar et al., 1996).

several voltage-dependent mechanisms, which are able to amplify excitatory postsynaptic potentials (Deisz et al., 1991; Huguenard et al., 1989; Magee and Johnston, 1995; Schwindt and Crill, 1995; Stuart and Sakmann, 1995; Nowak et al., 1984; Jones and Baughman, 1988; Thomson et al., 1988; Sutor and Hablitz, 1989). Activation of these voltage-dependent mechanisms *in vivo*, for example, during responses to visual stimuli, (Volgushev et al., 1992; Hirsch et al., 1995) can boost selectively the strong depolarizing responses but leave without amplification the small responses, thus leading to supralinear summation of large, visually evoked responses.

Further improvement of the selectivity can be achieved with the help of threshold mechanism for spike generation. A neuron receives its input in a form of excitatory and inhibitory postsynaptic potentials. The output of a neuron, an action potential, is generated only when the total input leads to depolarization of the cell membrane above a certain threshold. Only those orientations, which are capable of evoking suprathreshold excitation of a cell will be represented in the output of the neuron, while orientations, leading to excitatory responses that remain below the threshold will fail to produce any output signal. Picking up the uppermost part of the tuning curve and cutting off the rest by threshold mechanisms is called the *tip of the iceberg* effect. The actual contribution of this effect is determined by the relationship between the response amplitudes and the distance from the resting membrane potential to the spiking threshold.

Once spikes begin to occur in responses to visual stimuli, the orientation tuning can be further improved by the intracortical circuitry. Intracortical circuitry can both amplify the responses to the optimal orientation and suppress the responses to other orientations. The effectiveness of these mechanisms rests on the fact that the cells in the visual cortex receive all inhibitory inputs, and most of their excitatory inputs from other cortical cells (see below). It should be noted, that cortical cells respond to visual stimulation with different latencies. In many cells, generation of even the initial spike in the response could be influenced by intracortical inputs from other cortical cells which have shorter response latencies.

Intracortical inhibition could participate in sculpturing cortical orientation selectivity through a number of mechanisms. Of those, an obvious candidate is cross-orientation inhibition. It could play a decisive role in improving early orientation selectivity of the input and for defining a population of cells for which a particular stimulus is optimal and which generates spike responses. In this respect, cross-orientation inhibition restrains the spread of excitation in the orientation domain, and restricts spiking to those cells which prefer the same orientation. The other kind of inhibition, nonspecific to stimulus orientation, can also be instrumental for sharpening the tuning. This inhibitory system controls excitability of a cell and defines what amplitude an excitatory input should have to be sufficient to reach the threshold of spike generation (tip of the iceberg effect). Another kind of inhibition is of the recurrent type. Recurrent inhibition of a cell, or a population of cells, is mediated via inhibitory interneurons, which receive excitatory inputs from that cell or that cell population. This kind of inhibition is a necessary component of any system with a positive feedback and it protects the system from runaway excitation. Further, it provides the system with the possibility for response normalization over a broad range of stimulus strength and contrast (Carandini and Heeger, 1994; Somers et al., 1995; Ben-Yishai et al., 1995). The recurrent inhibitory system might be the strongest among the three since it should be able to restrict strong, full-blown excitation during the response to optimal stimulation (Douglas et al., 1991; Douglas and Martin, 1991).

Rich excitatory connections between cortical cells provide grounds for reverberation and amplification of the excitation in the cortical network. A bulk of intracortical excitation impinging upon a cortical cell originates from its close neighbors. Since neighboring cells prefer similar orientations (see figure 20 with related text and section 4.4.3 below), intracortical excitation must be highly iso-orientation selective. Long-range intracortical connections are also formed mostly between iso-orientation columns. Thus, once the cells preferring a particular orientation start to fire, excitation reverberates among only those neurons which prefer that orientation, and the overall excitation produced by a visual stimulus of that orientation becomes amplified. The importance of the intracortical excitatory feedback in creating strong responses to optimal orientations is evident from intracellular studies (Douglas et al., 1991; Douglas and Martin, 1991; Nelson et al., 1994). Capability of the above reverberating mechanisms in producing sharp tuning of the spike responses from an initial, weak selectivity was demonstrated in simulation studies (Douglas et al., 1995; Somers et al., 1995). In addition to intracortical mechanisms, excitatory corticogeniculate loops could also participate in the amplification of optimal responses (Sillito et al., 1994), although after a corticogeniculate loop the excitatory input that returns to the cortex may not be as selective as in the case of an intracortical loop. For both, intracortical and corticogeniculate loops, excitation returns to a cortical cell in a form of postsynaptic potentials, and the whole set of the above mentioned mechanisms of selectivity improvement is brought into play again.

The creation of the final sharp selectivity of cortical cells can be envisaged as a cooperation of multiple mechanisms (Vidyasagar et al., 1996; Fig. 3.24). The actual contribution of this or that mechanism can vary considerably from one cell to the other. It remains to be clarified whether this variability is related to some specific functional or morphological characteristics, or is simply guided by chance. Synergistic action of several mechanisms has a number of advantages. First, it reduces requirements for specificity of the mechanisms. Indeed, only low specificity of each particular mechanism is compatible with experimental data. In addition to this, computer simulations demonstrated that operation of several, not very specific but cooperative mechanisms could lead to a sharp, final selectivity of the cells' responses (Wörgötter and Koch, 1991). Second, cooperative mechanisms make the selectivity more robust against ongoing perturbations of cortical activity. Third, combination of the first two advantages might simplify establishment of the selectivity during development, and at the same time, it makes the process less sensitive to accidental disturbances.

3.4.4 Representations in the visual cortex

Retinotopy

In the visual cortex, many visual cues, such as position, orientation, direction of motion, disparity, eye-preference, and color, possess an orderly representation along the surface of the cortex. Neighboring cells usually have similar functional properties although representations are not strictly homogeneous. Take, for example, retinotopy. Electrophysiological mapping experiments showed that the visual cortex devotes a much larger area (and volume) of cortex to the central visual representation than to the periphery (Daniel and Whitteridge, 1961). As a result, the central 10 degrees of visual space, including the area centralis, occupy about half of the primary visual cortex in cat. In higher cortical areas, where the visual representation is often restricted to more central regions (few tens of degrees), the ratio between central and peripheral representations can be even higher. Receptive field position, receptive field size, point spread function, and magnification factor are all measures which depend on the physical extent of cortical tissue representing a given location in the visual space (see Orban, 1984). By determining these parameters over large regions of the visual cortex one clearly sees marked differences with regard to earlier levels in the visual pathway. For example, detailed electrophysiological mapping in the primary visual cortex of the cat shows that the smoothness of the retinotopic map present in the dLGN is regularly distorted at the so called orientation centers (Das and Gilbert, 1997). At these locations orientation selectivity jumps between neighboring neurons and so do the respective receptive field positions. Thus the inhomogeneity observed in orientation selectivity is in register with the inhomogeneity observed in the retinotopic representation. The significance of this joint anisotropy is not yet known.

Ocular dominance

The visual cortex is the first stage along the visual pathway where inputs from the two eyes converge. It has been shown in cat and primate that the visual thalamic input segregates into eye-specific domains in the cortex (Hubel and Wiesel, 1962; 1963; 1977). In the cat, convergence of the input from the eyes takes place on arrival of the thalamic afferents into layers 4 and 6 so that the majority of even the first-order neurons responds to stimulation of either eye. In the primary visual cortex of primates, binocular convergence is somewhat different because here the eye specific input remains largely segregated through many synaptic stages. Consequently, the so-called eye specific domains established by the afferents are maintained and thus, eye-specificity of neuronal responses is stronger than in the case of the cat. For demonstrating eye dominance in the cortex earlier studies used the 2-deoxyglucose technique (Hubel et al., 1977; Löwel and Singer, 1987) and more recent studies the optical imaging technique (Blasdel and Salama, 1986; Hübener et al., 1997). The results clearly showed that input from the two eyes segregate into large domains, each of which can house a minimum of one full cycle of orientation columns or as often called, an orientation hypercolumn.

Orientation selectivity

In the primary visual cortex, most cells are selective for the orientation of elongated visual stimuli such as a bright or dark bar moving across the corresponding retina location. Electrophysiological recordings from nearby cortical sites showed that neurons preferring similar stimulus orientations are clustered into columns and neighboring columns of cells have a strong tendency of responding to slightly different orientations (Hubel and Wiesel, 1962; 1963). The gradual change in the preferred orientation that occurs tangentially to the cortical surface is, however, occasionally interrupted by sudden orientation shifts. It was not until the introduction of the optical imaging technique that the global layout of orientation maps could be experimentally disclosed (reviewed by Bonhoeffer and Grinvald, 1995). As we will see the same technique has become one of the most powerful tools in exploring the organization features of a number of functional representations. Optical imaging experiments have provided elegant proof that orientation selectivity is mapped out into a two-dimensional spatial mosaic (Bonhoeffer and Grinvald, 1991). The map contains two main structural elements. Those regions, where the orientation gradient between neighboring locations is low, are called iso-orientation patches or domains. Those regions, where a full set of orientation domains converge and orientation gradients are high, are called orientation centers. Fig. 3.25 demonstrates optical images produced by moving gratings at each of four orientations. Notice that the central zones of iso-orientation domains (black regions) occupy distinct spatial locations. In order to visualize the entire range of orientations the iso-orientation maps are combined vectorially resulting in an orientation angle map such as the one shown in Fig. 3.31, where orientation ranges are coded with color. A prominent feature of such maps is that iso-orientation lines radiate out from orientation centers in a pinwheel-like manner. Orientation centers are particularly interesting because they denote locations where neurons can have either broad orientation tunings or, alternatively, closely spaced cells can be well tuned but their orientation preferences radically differ. Experimental findings support the second concept, demonstrating that a broad range of orientation specific neurons are intermingled in a small volume of tissue (Maldonado et al., 1997).

Direction selectivity

Direction selectivity is another emergent property of visual cortical cells which is not present at the level of the visual thalamus. Electrophysiological mapping experiments have suggested that visual cortical cells preferring different directions of movement are distributed in a systematic manner (Payne et al., 1981). In this respect, some predictions for the layout of direction representations can already be made on the basis of its relationship with orientation selectivity. Because the axis of movement of a stimulus must have a perpendicular trajectory to the preferred orientation of the bar stimulus, there are two possible directions along this axis. Assuming that direction selectivity is modularly represented in the cortex, a logical prediction would be a representation where the number of such

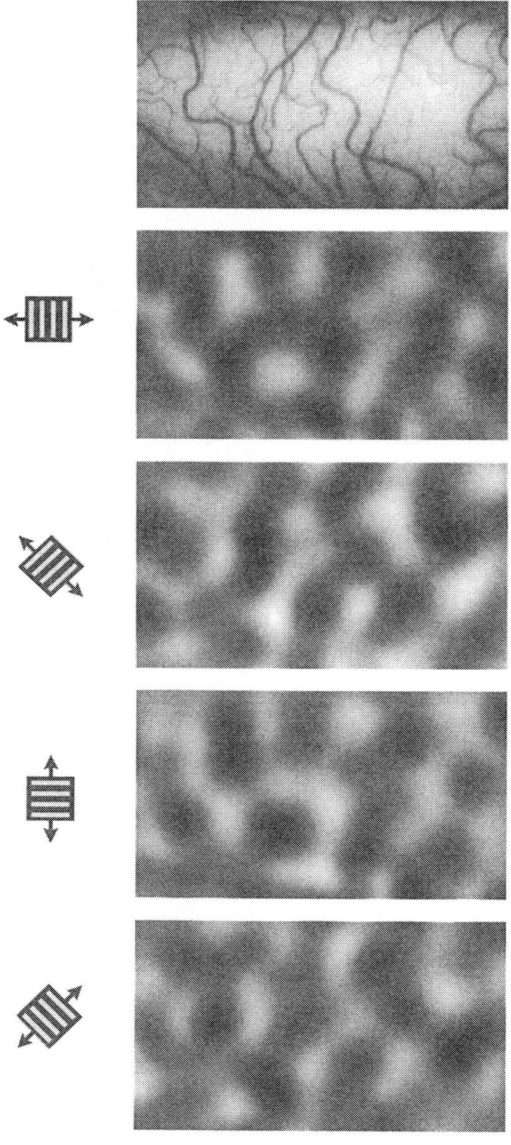

FIGURE 3.25. Optical images of intrinsic signals obtained in the cat visual cortex (area 18). The top panel shows the vascular pattern of the imaged region. Lower four panels show activity images, respectively, to visual stimuli of moving luminance gratings of four orientations (schemes). Black regions correspond to active zones and white regions to less active zones to the presented stimulus grating. Notice the complementary activity pattern between images where the stimulus orientation differed by 90 degrees (0-90 degrees; 45-135 degrees).

direction-modules is at least twice the number of corresponding orientation modules. Indeed, electrophysiological mapping data (Swindale et al.,
1987) and, more recently optical imaging experiments (Weliky et al., 1996;
Shmuel and Grinvald, 1996) showed that the actual number of isodirection
domains is 2-3 times that of orientation.

Spatial relationship between functional representations

Imaging techniques have been intensively used to explore the spatial relationship between functional representations. The actual search for a coherent view is rather complicated because in addition to the functional maps
described above, there are other modalities like spatial frequency, disparity and color each of which has its own form of representation. Recent
theoretical work has just begun to unravel the complex nature of multiple functional representations. Here we take an example from available
optical imaging data on the relationships between orientation preference
and eye-dominance in the monkey (Blasdel, 1992) and cat (Löwel et al.,
1988; Hübener et al., 1997; Crair et al., 1997). The results obtained in both
species point to a common characteristic, namely, iso-orientation lines tend
to intercept ocular dominance border regions close to right angles and orientation centers are most often found in the middle part of eye domains.
These data suggest a strong functional link between orientation and ocular
dominance maps.

Functional anatomy of horizontal connections

Horizontal connections are the chief candidates for mediating receptive field
interactions beyond the classical receptive field. In the visual cortex, the
extent of these connections allows integration of information over regions
which are 3-4 times larger than the receptive field discharge zones and, in
this sense, they link nonoverlapping receptive field locations. The orientation specificity of long-range connections, in particular, whether they link
cortical columns of similar or dissimilar orientations, has been investigated
by a number of research groups (reviewed in Salin and Bullier, 1995). Most
of the studies agree that the majority of connections is established between
similar orientation columns. Recent data on the orientation specificity of excitatory connections confirmed these findings but indicated also that there
is quite a significant proportion of the connections in layers 2/3, an estimated 40%, that links dissimilar orientation columns (Kisvárday et al.,
1997). To illustrate this phenomenon at the single cell level, Fig. 3.31 shows
an intracellularly stained pyramidal cell from a region of area 18 that had
been imaged for orientation preferences. The soma is located in an orientation column that prefers close to horizontal orientations indicated by ochre
in the map. To obtain a quantitative view of the orientation distribution

of the axon terminals, we counted each terminal in each orientation pixel of the map. The resulting bargraph, showing the orientation distribution of the terminals, could be interpreted as the orientation tuning of the connections established by that pyramidal cell. Notice that although the main output is devoted to similar orientations, many terminals of the same cell are linked to different orientations. What might be the significance of such a mixed orientation topography of horizontal connections? One possibility is that during cortical processing of natural visual scenes where all orientations could be present as part of the same object, coherent figure perception requires interactions between feature elements of a broad orientation range. Indeed, the results of experiments, in which the classical receptive field and its surround were simultaneously stimulated with particular stimulus combinations, support the above reasoning (Gilbert, 1992; Sillito et al., 1995). In these experiments, marked effects on the response characteristics of the recorded neurons were found when, for example, an optimally oriented grating in the receptive field center was presented together with a grating of orthogonal (nonoptimal) orientation in the surround region (Sillito et al., 1995). Often, a stimulus combination like this could elicit responses of even larger amplitude than evoked by an optimally oriented grating in the center alone. The above results are particularly interesting because they suggest some plausible link to psychophysical observations where the different effects could be mediated by the same basic circuitry. Following the same line of thought, it has been assumed that perceptual feature grouping and figure-ground segregation each is based upon different mechanisms. While grouping requires interactions between similar elements, segregation uses differences between the elements. A conceivable solution to such a diversity of demands using the same network is that the evolved long-range connections contain elements that can dynamically serve both tasks in accordance with the stimulus constellation. To translate this into functional anatomy, lateral connections should link a broad range of orientations, and recent findings provide evidence for that.

3.5   Models of the Visual Cortex

In general, models of the visual cortex can be subdivided into three groups: (1) Models of temporal firing patterns (e.g., oscillations and synchronization); (2) models of cortical cell characteristics (e.g., orientation specificity); and (3) models of map formation (e.g., orientation column maps). In these groups, subdivisions exist according to the level of abstraction with which the different models are implemented. For example a biophysical realistic model needs more details than a model which simulates a cortical cell by a filter function.

3.5.1   Models of the temporal structure of cortical responses

The binding problem

This chapter will discuss models of cortical specificities and of map formation in greater detail, because we are mainly concerned with the functional
anatomy of the visual system. Models of the temporal firing patterns of
cortical cells on the other hand are mostly concerned with higher cognitive functions, for example solving the *binding problem* (Millner, 1974;
v.d.Malsburg 1981). The binding problem can be ascribed as: Take a set
of neurons which detects different features of a stimulus (like its shape,
color, motion pattern, etc.) by rising the mean impulse rate. It is impossible to encode all arbitrary feature combinations using the impulse rate
alone. For example, a red triangle and a green disk will make the *red-* and
green-detectors fire as well as the *disk-* and *triangle*-detectors[2]. Without
additional information, however, it remains unclear that red belongs to
the triangle and green belongs to the disk. A solution to this problem can
be found when the temporal structure of the firing pattern is involved. If
those neurons, which encode the same object would fire always at the same
time then this synchrony would indicate that they are *bound* together and
represent a common object (Engel et al., 1992). There are other possible solutions to this problem (e.g., with super-detectors, sensitive for all objects
regardless of their intrinsic feature combinations, so called *grandmother
cells*) but all these other attempts introduce additional complications (e.g.,
combinatorial explosion). The temporal synchrony of neuronal firing as a
possible solution to the binding problem has generated an increase of experimental and theoretical interest, because it seems central to early cognitive
processes like segmenting a scene into its objects. There is a vast body of
literature covering these and related aspects, which, however, are not in
the core of this contribution.

Synchronous oscillations

In the context of this chapter we will only describe one model which deals
with the problem of synchronous oscillations. This model is currently the
largest biophysical model including 65000 cells in thalamus (NRT and
LGN) and several layers of cortex the (Lumer et al. 1997), modeled with a
large degree of realism. The goal of this study was to demonstrate under
which conditions synchronous oscillatory activity in the gamma-range is
generated. They also showed that their network exhibits oscillations very
generically in a wide variety of situations. Using partial lesioning of the
model circuitry they found that there are two long loops which play a
crucial role for this behavior: 1) the loop between the infragranular layers

---

[2]Obviously there are no such detectors in the visual cortex, and we just wanted to
give a descriptive example.

and the thalamus and 2) the loop between the supragranular layers and
the infragranular layers (Fig. 3.26). In addition to this, they found that
the short loop between LGN and NRT can also induce strong oscillations.
While the model of Lumer et al. is certainly very impressive with regard to
the implemented degree of realism, it has two drawbacks. The generic nature of oscillations seems to be overemphasized because they last too long
and are too strong. Furthermore, a major aspect of thalamic physiology is
not implemented in the model, namely, the different firing modes of the
thalamic relay cells (burst firing mode vs. tonic transmission mode). As
a consequence, the oscillation frequency which is generated by the LGN-
NRT loop is by far too high (approx. 50 Hz) and does not correspond to
the spindling activity which contains frequencies only about 10 Hz.

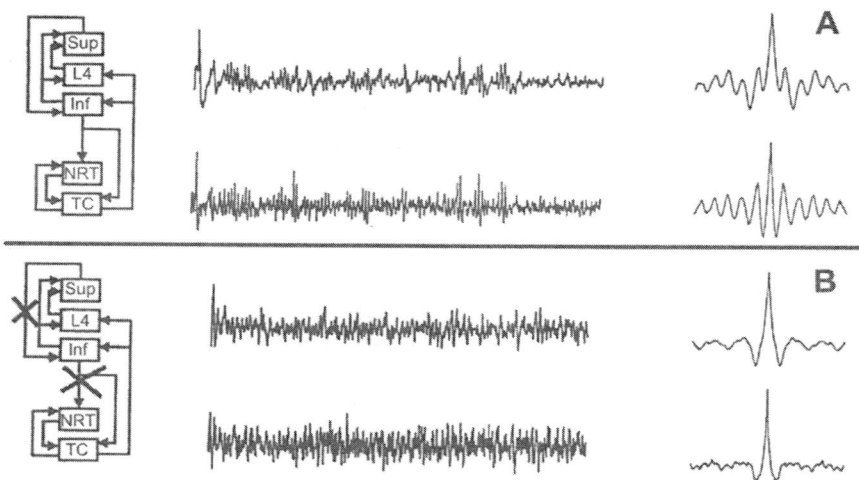

FIGURE 3.26. Large biophysical model (65000 cells) of the primary visual pathway. The wiring diagrams are shown on the left (Sup = supragranular layers,
L4=layer 4, Inf = infragranular layers, NRT = nucleus reticularis thalami, part
PGN, TC = thalamocortical relay cells.) Simulated evoked potentials are shown
in the center for the cortex and the thalamus. To the right, auto-correlation diagrams of the evoked potentials have been plotted. A) The complete network
exhibits pronounced oscillations in the gamma range (above 40 Hz). B) Intracortical feedback from Sup. to Inf. and the corticofugal feedback are both removed.
As a consequence the network stops oscillating. (Recompiled from Lumer et al.,
1997).

3.5.2   Models of cortical cell characteristics

Models of cortical orientation tuning including afferent bias and intracortical inhibition

From the viewpoint of cortical functional anatomy, cortical orientation selectivity was the first and probably most influential aspect of visual information processing in the brain which was modeled in biophysical simulations and successfully transferred to artificial computer vision systems. Almost all systems which utilize edge-detection for image analysis rely in one way or another on elongated detectors similar to cortical receptive fields. Computer vision approaches which regarded receptive fields as rather static spatial entities were introduced already in the early 80s (Marr, 1982). The first simple models of cortical specificities were designed around the same time (Nielsen, 1983), while detailed models of the primary visual pathway did not appear until 1991 (Wörgötter and Koch, 1991). It should be noted that all the above mentioned models excluded intracortical excitatory connections. The reason for this was, that at that time a major controversy existed which of the envisioned intracortical inhibitory circuits could best account for orientation specificity. The central result of the early biophysical models was that realistic orientation tuning can be obtained only by a *combination* of different inhibitory mechanisms. When only a single mechanism is employed the parameter settings have to be out of the physiological range in order to get orientation specificity (Fig. 3.27).

Analysis of recurrent excitation

The first theories which considered how to correctly embed intracortical excitation in a model appeared in the early 1990s. Figs. 29 A-C show the different stages of how to construct such a model. In the first stage afferent excitation enters and a regular EPSP is observed (A). Feedback excitation in the second stage drives the potential into saturation (B). Only by adding an inhibitory population (C) with a rather long time constant can a realistic intracellular response be modeled (Douglas and Martin, 1991). To mimic the cortical connectivity, the excitatory population more accurately is finally divided into pyramidal cells of the upper layers 2,3,(4) and those of the lower layer 5,6. Both populations have different cell characteristics. The model faithfully reproduces the responses of the superficial and deep layer neurons and it can also account for the changes in the response characteristic induced when eliminating $GABA_A$ components by bicuculline application (Fig. 3.28 E-H). This circuitry was called the *canonical microcircuit* (D), because it seems likely that such a wiring pattern can be found also in other cortical structures outside the visual cortex (Douglas et al., 1989; Douglas and Martin, 1991; Douglas et al., 1995).

Of great theoretical and physiological importance is the question whether such a circuit is stable or whether the neurons would be driven into saturation by the recurrent excitation. To this end Douglas et al. (1999) provided

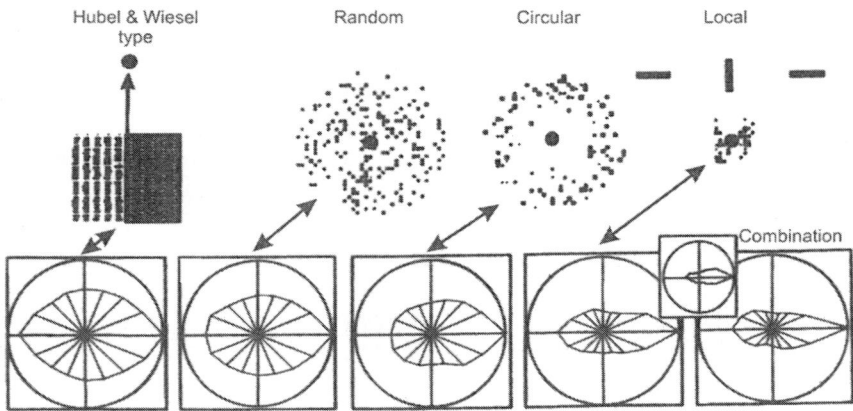

FIGURE 3.27. Results from a biophysical model of the primary visual pathway focusing on the orientation specificity of cortical cells. The top row shows the wiring pattern. The bottom row shows polarplots of the average cortical orientation selectivity for 55 cells. A polarplot depicts the normalized maximum response of a cell stimulated with a moving light bar with motion direction as given by the *wheel-spokes* in the polarplot. Hubel & Wiesel type connections represent purely afferent connections from LGN to cortex. On- (black dots on white ground) and Off-subfields of the cortical cells (white dots on black ground) are modeled. Every small dot represents the location of an LGN receptive field center with respective (On, Off) center dominance in the visual field. These LGN cells project onto the cortical target. As a consequence of the small elongation of the afferent projection field, a low orientation tuning is observed in the corresponding polarplot diagram. All other polarplots are obtained by adding certain inhibitory mechanisms (Random, Circular, Local and *Combination*) to the same Hubel & Wiesel afferent connectivity. The top row shows the topographical arrangement in the visual field of the inhibitory cells (small dots) terminating at the center cell (large dot). Bars above the Local connectivity diagram depict the initial (afferently induced) small orientation tuning across this patch of cortex which is the same for all situations. Random, Circular, and Local wiring patterns can improve the orientation tuning of the cells. However, only by a combination (*Combination*) of the Circular with the Local connection scheme (using different sets of weights) a realistic tuning is obtained. The small inset polarplot shows real data averaged from simple cell responses (Recompiled from Wörgötter, 1999).

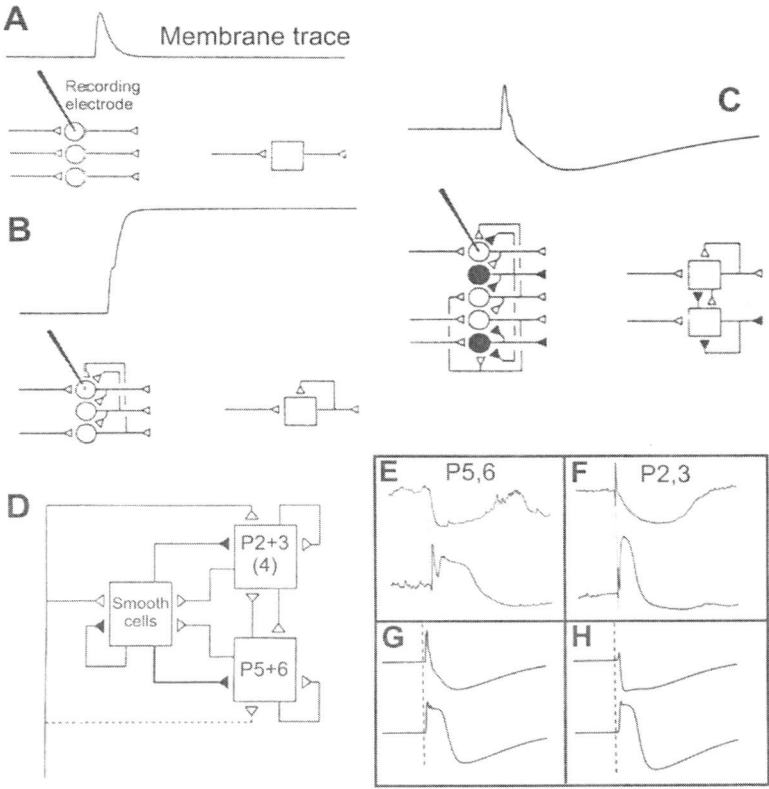

FIGURE 3.28. Construction and behavior of the *canonical cortical microcircuit.*
A-C) Constructing the circuit. A) Afferent excitation leads to an EPSP. B) recurrent excitation drives the membrane potential into saturation. C) long lasting
recurrent inhibition by $GABA_A$ and $GABA_B$ leads to a realistic response. D) The
final canonical microcircuit is obtained by subdividing the excitatory population
into two, simulating the behavior of the top layer pyramidal cells (P2+3,(4)) and
those of the bottom layer cells (P5+6), which have a different cell characteristic.
(E-H) Responses of real and model neurons to electrical stimulation of the afferents before (top traces) and during application of bicuculline (bottom traces). E)
Lower layer neuron, F) top layer neuron, G) model lower layer neuron, H) model
top layer neuron. In the model bicuculline application was simulated by reducing
the $GABA_A$ influence to 20% of its original value. A strong similarity between
real and model responses is visible. (Recompiled from Douglas and Martin, 1991).

an analysis of a neuronal network with linear neurons which showed that even a purely excitatory recurrent circuitry can be stable in a large parameter domain (Fig. 3.29 A). The output of such a neuron is given by the current flow I_g and represented as the voltage across G which ultimately leads to a firing frequency F. Two sources of input arise: I_{in} represents external (afferent) input and I_{rec} is the input via recurrent excitatory connections. If the network is homogeneous and has reached the stable state every neuron will fire with frequency F. Then, I_{rec} can be approximated by $I_{rec} = \alpha F$. The constant α must take the value of a conductance, because F is directly equivalent to a voltage. It is called the *network conductance* which numerically takes a negative value. The reason for this is that without recurrent excitation, the input-output relationship is linear with a slope of 1/G (Fig. 3.29 B). With recurrent excitation the same input I_{in} will lead to a higher output value F. Thus, the 1/G-curve appears to have a higher slope, now defined by $1/(G - \alpha)$. Thus, the network conductance α is negative. When the network is in the steady state, conservation of currents demands that the sum of all excitatory currents must equal the current flowing to ground across G. When the network is at a stable status it will be dependent only on the open-loop gain given by α/G. When, however, α/G is greater than one a small input change will lead to a larger change at I_{rec} (which turns into an input itself) and, consequently, the network will saturate. When α/G is less than one, I_{rec} will eventually fade and the network approaches stability.

This simplified analysis shows that even purely excitatory networks can be stable within large regimes. In a second step, Douglas et al. (1999) computed the closed loop in the presence of different types of inhibition. They found that inclusion of inhibition into the network increased the stability of the system.

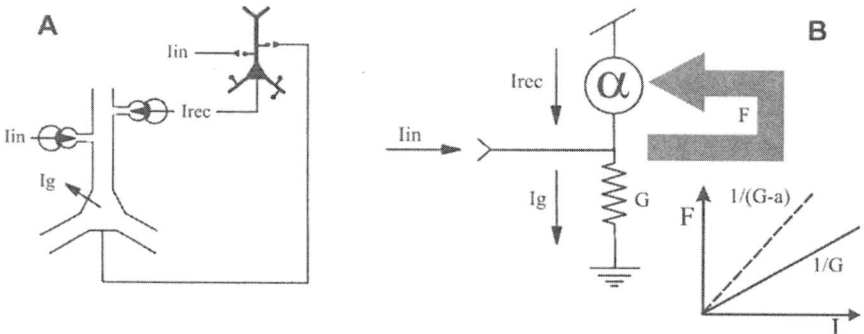

FIGURE 3.29. Linear analysis of recurrent excitation. A) Current flow in a recurrently connected cortical network. B) Schematic electric diagram depicting the same situation as in A. For abbreviations and explanations, see text. (Recompiled from Douglas et al., 1999).

Models of cortical orientation tuning including intracortical excitation

Together with a better understanding of the action of the excitatory circuitry more complete models of cortical orientation specificity were developed. Somers et al. (1995) demonstrated that a realistic orientation tuning could be obtained from a combination of weakly tuned afferent input, broad intracortical inhibition and narrowly tuned intracortical excitation (Fig. 3.30 A-E). The strong orientation tuning which appeared by using this model was an emergent network property. In particular, they could also demonstrate that a pure inhibitory model such as that of Wörgötter and Koch (1991) would produce realistic tuning only at the cost of a too low impulse rate of the cortical cells (Fig. 3.30 F). The approach of Somers et al. followed the tradition of typical biophysically realistic models containing many parameters, which made a rigorous analysis impossible. Carandini and Ringach (1997) were able to reduce the model of Somers et al. to a single differential equation (*neuronal field modeling approach*). They embedded afferent, intracortical inhibitory and intracortical excitatory influences at the same time. In this way, they could reproduce the findings of Somers et al. and provided a deeper analytical understanding of the underlying mechanisms. It leads to the conclusion that several aspects of this particular class of recurrent models for cortical orientation selectivity are probably still unrealistic, because very peculiar responses were predicted, for which currently no experimental support exists.

3.5.3   Models of cortical maps

In the previous sections we have seen that some cortical response features (like orientation selectivity or ocular dominance) are represented on the cortical surface in an orderly manner. The goal of a cortical map-model is to gain insight into the structure or the development of the spatial representation of these features. A wide variety of cortical map-models exists dating back as early as 1973 (von der Malsburg, 1973) many of which are abstract like those based on Kohonen maps (Obermayer et al., 1990) or those which use generalized correlation patterns between oriented *elements* (like dipoles, Swindale, 1982). Here we consider only the model of Erwin and Miller (1998) which tries to embed a larger degree of biological realism.

A model for the development of ocular dominance and orientation maps

In the paper of Erwin and Miller (1998), the joint development of the orientation and the ocular dominance map is studied and their model relates to an older study of Miller (1994) where he introduced the concept of On-Off correlations in order to develop an orientation map. Fig. 3.32 A shows the basic model setup. LGN On and Off cells from the left and the right eye are connected to the cortex with an arbor function that determines

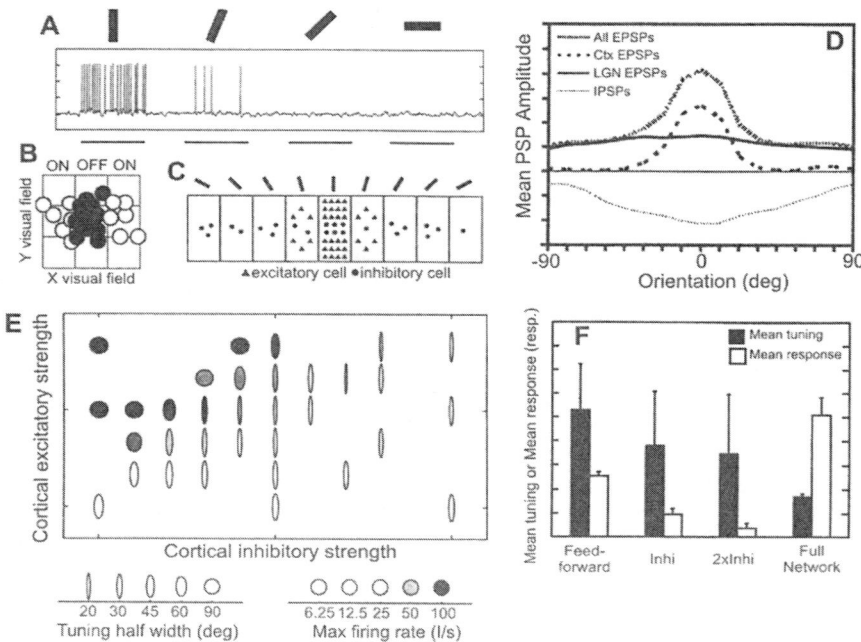

FIGURE 3.30. Biophysical model of cortical orientation selectivity which also includes recurrent excitation. A) Behavior of a single cell subjected to bar stimuli with different orientation (top). The cell responds maximally for a vertical bar. B) Afferent convergence pattern from the LGN to a single cortical cell. The cortex cell has an On-Off-On subfield structure. C) Intracortical connection pattern. The orientation selectivity of the source cells is depicted on top. Excitatory connection are narrowly tuned, inhibitory connection are broadly tuned. The target cell receives mainly iso-oriented (excitatory and inhibitory) connections. D) Orientation tuning of the different synaptic inputs to a cell. The afferent LGN induced tuning is very weak. Inhibitory tuning is a bit stronger, but only the intracortical EPSPs have a sufficient orientation tuning strength. This tuning must be generated by the total recurrent network activity, because the initial afferent tuning is very weak. This indicates that an amplification process takes place similar to that proposed by Douglas et al. (1999). E) Tuning strength and activity rates for differently strong cortical inhibition and excitation. F) Average tuning strength compared to the average activity of a population of cortical cells. Pure afferent connections (feedforward) lead to a too broad tuning and a reasonable activity strength. Inhibition reduces the activity very strongly (Inhi = single inhibitory strength and 2xInhi = double inhibitory strength) but the orientation tuning is still unsatisfactory. Only the full network produces a reasonable behavior. (Recompiled from Somers et al., 1995).

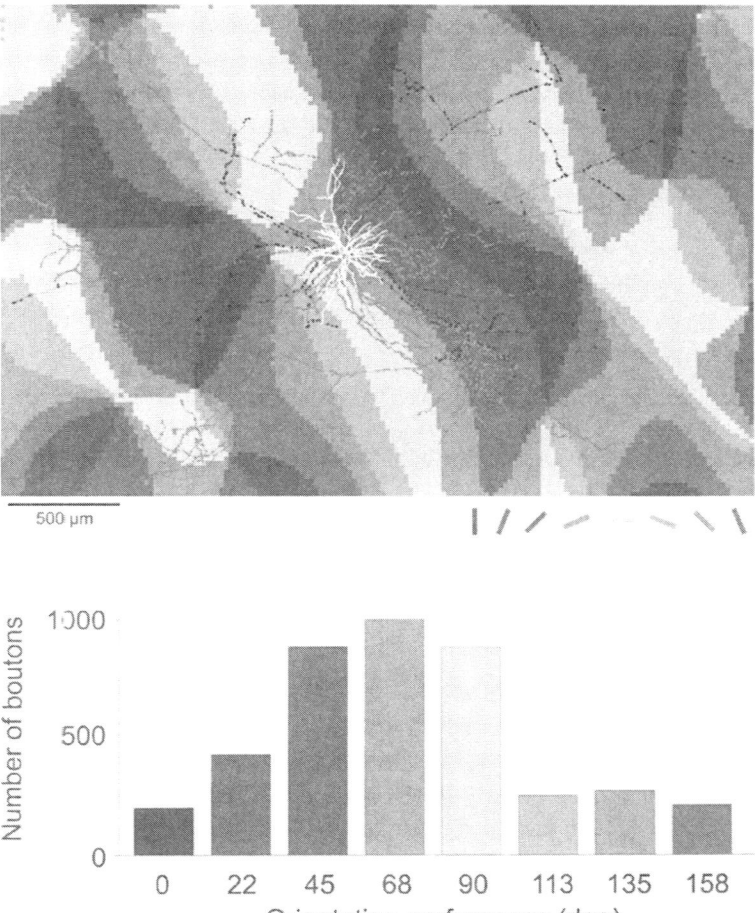

FIGURE 3.31. Functional anatomy of lateral connections in the cat visual cortex. Drawing of an intracellularly stained layer 3 pyramidal cell in superposition with an orientation map of the same cortical region. In the orientation map, orientation preferences are color-coded according to the lower right scheme. The soma and the dendrites of the pyramid are drawn in white and the axon in black. The axon terminals found in layers 2/3 are marked by grey dots and those in layer 5 by black dots. Some parts of the axon form terminal clusters while other parts show no clustering at all. The most obvious clusters occupy regions where the orientation preferences match that of the soma region (ochre-yellow) whereas other terminals are found at many different orientations. Bar graph shows the quantitative distribution of the axon terminals to the orientation map. Importantly, the orientation tuning of the cell is broad. An implication of the broad tuning is that although the majority of the connections prefer similar orientations to that of the soma region, other orientations are also encountered.

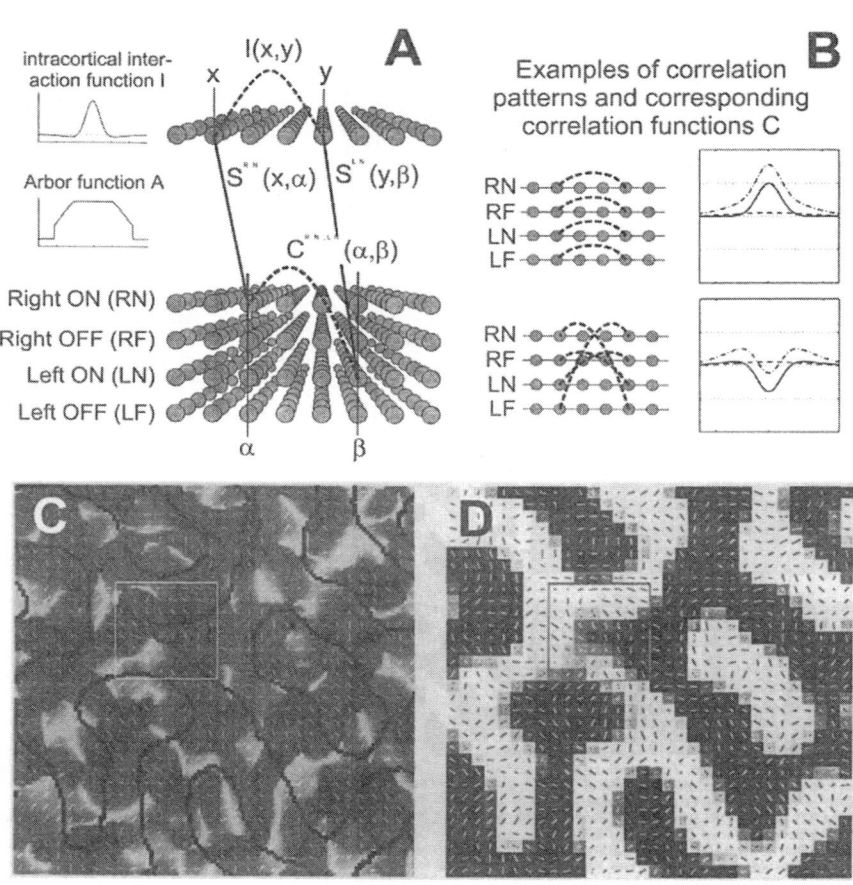

FIGURE 3.32. Model for the development of cortical ocular dominance and orientation maps. A) Structure of the model and shape of the intracortical interaction function I and the arbor function A. B) Examples of correlation patterns between the different LGN cell classes and the corresponding spatial correlation functions C. C) Color coded orientation column map. D) Ocular dominance column map (Bright = right eye, dark = left eye dominant). (Recompiled from Erwin and Miller, 1998).

their maximal connection ranges. The afferent connection strength is given by the synaptic weight function S, which changes for every connection following a hebbian learning rule. Intracortical connections are defined by an intracortical interaction function I, but intracortical connections will not change during development.

Hebbian learning of S is determined mainly by the firing correlations C between individual LGN cells. For example, adjacent cells from the same eye, which have the same response polarity (e.g., two On-cells from the left eye) will, with great likelihood, fire in a highly correlated manner when a stimulus appears. This situation is modeled by the correlation function shown in Fig. 3.32 B (top). A right On-cell, however, will very likely be negatively correlated to an Off cell at the same retinal position, regardless of whether this cell receives input from the right or the left eye (Fig. 3.32 B, bottom).

Accordingly, Erwin and Miller define the synaptic weight change as:

$$\frac{dS}{dt}(x,\alpha,t) = kA(x,\alpha)\sum_y I(x,y)\sum_{\beta,R,L,N,F}[C^{R,L,N,F}(\alpha,\beta)S(y,\beta,t)],$$

(3.4)

where k is a constant, A is the arbor function, I the intracortical interaction function and $C^{R,L,N,F}$ are the different correlation functions between the eyes and the response polarities, defined in a similar way as those shown in Fig. 3.32 B.

Using this approach the joint development of ocular dominance and orientation maps and their sequential development can be successfully modeled (Fig. 3.32 C,D). For the latter mechanism there is experimental evidence (Crair et al., 1997).

As opposed to many other map models the one of Erwin and Miller has the advantage of using very plausible correlation functions between the cells. It has been widely accepted that neuronal correlations determine the change of synaptic weights in learning and/or developmental processes. A still unresolved question, however, is whether the actual map structure is *pre-wired* very early on in development or whether it is determined in use-dependent manner. The model of Erwin and Miller favors the second view, however, cortical plasticity is still not included in their approach.

3.6   References

ALONSO, J.M., & MARTINEZ, L.M. (1998) Functional connectivity between simple cells and complex cells in cat striate cortex. *Nat. Neurosci.* 1:395-403.

BAL, T., VON KROSIGK, M., & McCORMICK, D.A. (1995) Synaptic and membrane mechanism underlying synchronized oscillations in the ferret dLGN in vitro. *J. Physiol. Lond.* 483:641-663.

BEAR, D.M., SASAKI H., & ERVIN, F.R. (1971) Sequential change in receptive fields of striate neurons in dark adapted cats. *Exp. Brain Res.* 13:256-272.

BEAULIEU, C., & SOMOGYI, P. (1990) Targets and quantitative distribution of GABAergic synapses in the visual cortex of the cat. *Europ. J. Neurosci.* 2:296-303.

BENEVENTO, L.A., CREUTZFELDT O.D., & KUHNT, U. (1972) Significance of intracortical inhibition in the visual cortex. *Nature* 238:124-126.

BEN-YISHAI, R., BAROR R.L., & SOMPOLINSKY, H. (1995) Theory of orientation tuning in visual cortex. *Proc. Natl. Acad. Sci. USA* 92:3844-3848.

BLAKEMORE, C., & TOBIN, E.A. (1972) Lateral inhibition between orientation detectors in the cats visual cortex. *Exp. Brain Res.* 15:439-440.

BLASDEL, G.G. (1992) Orientation selectivity, preference, and continuity in monkey striate cortex. *J. Neurosci.* 12:3139-3161.

BOLZ, J., ROSNER, G., & WÄSSLE, H. (1982) Response latency of brisk-sustained (X) and brisk-transient (Y) cells in the cat retina. *J. Physiol. Lond.* 328:171-190.

BONHOEFFER, T., & GRINVALD, A. (1991) Iso-orientation domains in cat visual cortex are arranged in pinwheel-like patterns. *Nature* 353:429-431.

BONHOEFFER, T., & GRINVALD, A. (1995) Optical imaging based on intrinsic signals: The methodology. *In: Brain Mapping: The Methods* (Toga, A., Mazziotta, J.C., eds.), 55-97. San Diego:Academic.

BOYCOTT, B.B., & WÄSSLE, H. (1974) The morphological types of ganglion cells of the domestic cat's retina. *J. Physiol. Lond.* 240:397-419.

BULLIER, J., & HENRY, G.H. (1979) Ordinal position of neurons in cat striate cortex. *J. Neurophysiol.* 42:1251-1263.

BULLIER, J., & HENRY, G.H. (1980) Ordinal positions and afferent input of neurons in monkey striate cortex. *J. Comp. Neurol.* 193:913-935.

BULLIER, J., & NORTON, T.T. (1979) Comparison of receptive-field properties of X and Y ganglion cells with X and Y lateral geniculate cells in the cat. *J. Neurophysiol.* 42:274-291.

CARANDINI, M., & HEEGER, D.J. (1994) Summation and division by neurons in primate visual cortex. *Science* 264:1333-1336.

CARANDINI, M., & RINGACH, D.L. (1997) Predictions of a recurrent model of orientation selectivity. *Vision Res.* 37:3061-3071.

CASAGRANDE, V.A., & NORTON, T.T. (1991) The lateral geniculate nucleus:a review of its physiology and function. *In: The Neural Basis of Visual Function* 4:41-84.

CHAPMAN, B., ZAHS K.R., & STRYKER, M.P. (1991) Relation of cortical cell orientation selectivity to alignment of receptive fields of the geniculocortical afferents that arborize within a single orientation column in ferret visual cortex. *J. Neurosci.* 11:1347-1358.

CLELAND, B.G., DUBIN, M.W., & LEVICK, W.R. (1971) Sustained and transient neurones in the cat's retina and lateral geniculate nucleus. *J.*

Physiol. Lond. 217:473-496.

COENEN, A.M.L., & VENDRIK, A.J.H. (1972) Determination of the transfer ratio of cat's geniculate neurons through quasi-intracellular recordings and the relation with the level of alertness. *Exp. Brain Res.* 14:227-242.

CRAIR, M.C., RUTHAZER E.S., GILLESPIE D.C., & STRYKER, M.P. (1997) Ocular dominance peaks at pinwheel centre singularities of the orientation map in cat visual cortex. *J. Neurophysiol.* 77:3381-3385.

CREUTZFELDT, O.D., & ITO, M. (1968) Functional synaptic organization of primary visual cortex neurons in the cat. *Exp. Brain Res.* 6:324-353.

CREUTZFELDT, O.D., KUHNT U., & BENEVENTO, L.A. (1974) An intracellular analysis of visual cortical neurons to moving stimuli:responses in a co-operative neuronal network. *Exp. Brain Res.* 21:251-275.

CRICK, F. (1984) Function of the thalamic reticular complex:The searchlight hypothesis. *Proc. Natl. Acad. Sci. USA* 81:4586-4590.

CROOK, J.M., & EYSEL, U.T. (1992) GABA-induced inactivation of functionally characterized sites in cat striate cortex (Area 18): Effects on orientation tuning. *J. Neurosci.* 12:1816-1825.

CROOK, J.M., KISVÁRDAY Z.F., & EYSEL, U.T. (1997) GABA-induced inactivation of functionally characterized sites in cat striate cortex:Effects on orientation tuning and direction selectivity. *Visual Neurosci.* 14:141-158.

CUDEIRO, J., & SILLITO, A.M. (1996) Spatial frequency tuning of orientation-discontinuity sensitive corticofugal feedback to the cat lateral geniculate nucleus. *J. Physiol. Lond.* 490:481-492.

DANIEL, P.M., & WHITTERIDGE, D. (1961) The representation of the visual field on the cerebral cortex in monkeys. *J. Physiol. Lond.* 159:203-221.

DAS, A., & GILBERT, C.D. (1997) Distortions of visuotopic map match orientation singularities in primary visual cortex. *Nature* 387:594-598.

DEANGELIS, G.C., OHZAWA I., & FREEMAN, R.D. (1995) Receptive-field dynamics in the central visual pathways. *TINS* 18:451-458.

DEISZ, R.A., FORTIN G., & ZIEGLGÄNSBERGER, W. (1991) Voltage dependence of excitatory postsynaptic potentials of rat neocortical neurons. *J. Neurophysiol.* 65:371-382.

DESTEXHE, A., CONTRERAS, D., SEJNOWSKI, T.J., & STERIADE, M. (1994) A model of spindle rhythmicity in the isolated thalamic reticular nucleus. *J. Neurophysiol.* 72:803-818.

DOUGLAS, R.J., & MARTIN, K.A.C. (1991) A functional microcircuit for cat visual cortex. *J. Physiol. Lond.* 440:735-769.

DOUGLAS, R.J., KOCH C., MAHOWALD M., & MARTIN, K.A.C. (1999) The role of recurrent excitation in neocortical circuits. *In:Cerebral Cortex*, Vol. 13 (Ulinski et al., eds.), 251-282, Kluwer Academic/Plenum Publ., New York.

DOUGLAS, R.J., KOCH C., MAHOWALD M., MARTIN K.A.C., & SUAREZ H.H., (1995) Recurrent excitation in neocortical circuits. *Science* 269:981-985.

DOUGLAS, R.J., MARTIN K.A.C., & WHITTERIDGE, D. (1989) A canonical microcircuit for neocortex. *Neural Comp.* 1:480-488.

DOUGLAS, R.J., MARTIN K.A.C., & WHITTERIDGE, D. (1991) An intracellular analysis of the visual responses of neurons in cat visual cortex. *J. Physiol. Lond.* 440:659-696.

ECKHORN, R. (1994) Oscillatory and non-oscillatory synchronizations in the visual cortex and their possible roles in associations of visual features. *Prog. Brain Res* 102:405-426.

ENGEL, A.K., KÖNIG, P., KREITER, A.K., SCHILLEN, T.B., & SINGER, W., (1992) Temporal coding in the visual cortex - new vistas on integration in the nervous system. *TINS* 15:218-226.

ENROTH-CUGELL, C., & LENNIE, P. (1975) The control of retinal ganglion cell discharge by receptive field surrounds. *J. Physiol. Lond.* 247:551-578.

ERWIN, E., & MILLER, K.D. (1998) Correlation based development of ocularly matched orientation and ocular dominance maps:Determination of required input activities. *J. Neurosci.* 18:9870-9895.

EYSEL, U.T. (1986) Spezifische Leistungen thalamischer Hemmungs-Mechanismen im Sehsystem. *Physiologie Aktuell* 2:159-175.

EYSEL, U.T., SHEVELEV I.A., LAZAREVA N.A., & SHARAEV, G.A. (1998) Orientation tuning and receptive field structure in cat striate neurons during local blockade of intracortical inhibition. *Neurosci.* 84:25-36.

FIORENTINI, A., & MAFFEI, L. (1973) Contrast perception and electrophysiological correlates. *J. Physiol. Lond.* 231:61-71.

FISCHER, B., & KRÜGER, J. (1974) The shift-effect in the cat's lateral geniculate neurons. *Exp. Brain Res.* 21:225-227.

FUNKE, K., & EYSEL, U.T. (1998) Inverse correlation of firing patterns of single topographically matched perigeniculate neurons and cat dorsal lateral geniculate relay cells. *Visual Neurosci.* 15:711-729.

FUNKE, K., & WÖRGÖTTER, F. (1997) On the significance of temporally structured activity in the dorsal lateral geniculate nucleus (LGN). *Prog. Neurobiol.* 53:67-119.

GABBOT, P.L.A., & SOMOGYI, P. (1986) Quantitative distribution of GABA-immunoreactive neurons in the visual cortex (area 17) of the cat. *Exp. Brain Res.* 61:323-331.

GAUDIANO, P. (1994) Simulations of X and Y retinal ganglion cell behavior with a nonlinear push-pull model of spatiotemporal retinal processing. *Vision Res.* 34:1767-1784.

GAWNE, T.J., McCLURKIN, J.W., RICHMOND, B.J., & OPTICAN, L.M. (1991) Lateral geniculate neurons in behaving primates. 3. response predictions of a channel model with multiple spatial-to-temporal filters. *J. Neurophysiol.* 66:809-823.

GILBERT, C.D. (1977) Laminar differences in receptive field properties of cells in cat primary cortex. *J. Physiol. Lond.* 268:391-421.

GILBERT, C.D. (1992) Horizontal integration and cortical dynamics. *Neuron* 9:1-13.

GOLOMB, D., WANG, X-J., & RINZEL, J. (1994) Synchronization properties of spindle oscillation in a thalamic reticular nucleus model. *J. Neurophysiol.* 72:1109-1126.

GOLOMB, D., WANG, X.,-J., & RINZEL, J. (1996) Propagation of spindle waves in a thalamic slice model. *J. Neurophysiol.* 75:750-769.

HATA, Y., TSUMOTO T., SATO H., HAGIHARA K., & TAMURA, H. (1988) Inhibition contributes to orientation selectivity in visual cortex of cat. *Nature* 335:815-817.

HEBB, D.O. (1949) The organization of behavior:A neuropsychological theory. J. Wiley and Sons, New York.

HEGGELUND, P. (1981) Receptive field organization of simple cells in cat striate cortex. *Exp. Brain Res.* 42:89-98.

HEGGELUND, P., & MOORS, J. (1983) Orientation selectivity and the spatial distribution of enhancement and suppression in receptive fields of cat striate cortex cells. *Exp. Brain Res.* 52:235-247.

HEGGELUND, P., KARLSEN, H.E., FLUGSRUD, G., & NORDTUG, T. (1989) Response to rates of luminance change of sustained and transient cells in the cat lateral geniculate nucleus and optic tract. *Exp. Brain Res.* 74:116-130.

HELLER, J., HERTZ, J.A., KJAER, T.W., & RICHMOND, B.J. (1995) Information flow and temporal coding in primate pattern vision. *J. Comp. Neurosci.* 2:175-193.

HIRSCH, J.A., ALONSO J.M., & REID, R.C. (1995) Visually evoked calcium action potentials in cat striate cortex. *Nature* 378:612-616.

HIRSCH, J.A., ALONSO J.M., REID R.C., & MARTINEZ, L.M. (1998) Synaptic integration in striate cortical simple cells. *J. Neurosci.* 18:9517-9528.

HOBSON, J.A. (1989) *Sleep.* Scientific American Libary, New York.

HUBEL, D.H., & WIESEL, T.N. (1959) Receptive fields of single neurones in the cat's striate cortex *J. Physiol. Lond.* 148:574-591.

HUBEL, D.H., & WIESEL, T.N. (1961) Integrative action in the cat's lateral geniculate body. *J. Physiol. Lond.* 155:385-398.

HUBEL, D , & WIESEL, T.N. (1962) Receptive fields, binocular interaction and functional architecture in the cat's visual cortex. *J. Physiol. Lond.* 160:106-154.

HUBEL, D.H., & WIESEL, T.N. (1963) Shape and arrangement of columns in cat's striate cortex. *J. Physiol. Lond.* 165:559-568.

HUBEL, D.H., & WIESEL, T.N. (1965) Receptive fields and functional architecture in two nonstriate visual areas (18 and 19) of the cat. *J. Neurophysiol.* 28:229-289.

HUBEL, D.H., WIESEL T.N., & STRYKER, M.P. (1977) Orientation columns in macaque monkey visual cortex demonstrated by the 2-deoxyglucose autoradiographic technique. *Nature* 269:328-330.

HÜBENER, M., SHOHAM D., GRINVALD A., & BONHOEFFER, T. (1997) Spatial relationships among three columnar systems in cat area 17. *J. Neu-*

rosci. 17:9270-9284.

HUGUENARD, J.R., HAMILL O.P., & PRINCE, D.A. (1989) Sodium channels in dendrites of rat cortical pyramidal neurons. *Proc. Natl. Acad. Sci. USA* 86:2473-2477.

JAHNSEN, H., & LLINÁS, R. (1984) Ionic basis for the electroresponsiveness and oscillatory properties of guinea-pig thalamic neurons in vitro. *J. Physiol. Lond.* 349:227-247.

JONES, E.G., & PETERS, A. (1984) *Cerebral Cortex* Vol.1: *Cellular Components of the Cerebral Cortex*, Plenum Press, New York.

JONES, K.A., & BAUGHMAN, R.W. (1988) NMDA- and non-NMDA-receptor components of excitatory synaptic potentials recorded from cells in layer V of rat visual cortex. *J. Neurosci.* 8:3522-3534.

KAPLAN, E., MUKHERJEE, P., & SHAPLEY, R. (1993) Information filtering in the lateral geniculate nucleus. In: *Contrast Sensitivity* (Shapley, R. and Man-Kit, D., eds.), Vol. 5:183-200, MIT Press, Cambridge MA.

KAWAGUCHI,Y., (1995) Physiological subgroups of nonpyramidal cells with specific morphological characteristics in layer II/III of rat frontal cortex. *J. Neurosci.* 15:2638-2655.

KIM, U., SANCHEZ-VIVES, M.V., & McCORMICK, D.A. (1997) Functional dynamics of GABAergic inhibition in the thalamus. *Science* 278:130-134.

KISVÁRDAY, Z.F., TÓTH É., RAUSCH M., & EYSEL, U.T. (1997) Orientation specific relationship between populations of excitatory and inhibitory lateral connections in the visual cortex of the cat. *Cerebral Cortex,* 7:605-618.

KUFFLER, S.T.W. (1953) Discharge patterns and functional organization of mammalian retina. *J Neurophysiol.* 16:37-68.

LEE, B.B., VIRSU V., & CREUTZFELDT, O.D. (1977) Responses of cells in the cat lateral geniculate nucleus to moving stimuli at various levels of light and dark adaptation. *Exp. Brain Res.* 27:51-59.

LÖWEL, S., & SINGER, W. (1987) The pattern of ocular dominance columns in flat-mounts of the cat visual cortex. *Exp. Brain Res.* 68:661-666.

LÖWEL, S., BISCHOF H.,-J., LEUTENECKER B., & SINGER, W. (1988) Topographic relations between ocular dominance and orientation columns in the cat striate cortex. *Exp. Brain Res.* 71:33-46.

LO, F.-S., & SHERMAN, S.M. (1994) Feedback inhibition in the cat's lateral geniculate nucleus. *Exp. Brain Res.* 100:365-368.

LUMER, E.D., EDELMAN, G.M., & TONONI, G. (1997) Neural dynamics in a model of the thalamocortical system. I. Layers, loops and the emergence of fast synchronous oscillations. *Cerebral Cortex* 7:207-227.

MAGEE, J.C., & JOHNSTON, D. (1995) Synaptic activation of voltage-gated channels in the dendrites of hippocampal pyramidal neurons. *Science* 268:301-304.

MALDONADO, P.E., GODECKE I., GRAY C.M., & BONHOEFFER, T. (1997) Orientation selectivity in pinwheel centres in cat striate cortex. *Sci-*

ence 276:1551-1555.

MARR, D. (1982) *Vision*, W.H. Freeman, New York.

MASTRONARDE, D.N. (1992) Nonlagged relay cells and interneurons in the cat lateral geniculate nucleus - receptive-field properties and retinal inputs. *Visual Neurosci.* 8:407-441.

MCCARLEY, R.W., GREENE, R.W., RAINNIE, D., & PORTAS, C.M. (1995) Brainstem neuromodulation and REM sleep. *Sem. Neurosci.* 7:341-354.

MCCLURKIN, J.W., GAWNE, T.J., RICHMOND, B.J., OPTICAN, L.M., & ROBINSON, D.L. (1991a) Lateral geniculate neurons in behaving primates. 1. Responses to two-dimensional stimuli. *J. Neurophysiol.* 66:777-793.

MCCLURKIN, J.W., GAWNE, T.J., OPTICAN, L.M., & RICHMOND, B.J. (1991b) Lateral geniculate neurons in behaving primates. 2. encoding of visual information in the temporal shape of the response. *J. Neurophysiol.* 66:794-808.

MCCORMICK, D.A. (1992) Neurotransmitter actions in the thalamus and cerebral cortex and their role in neuromodulation of thalamocortical activity. *Prog. Neurobiol.* 39:337-388.

MCCORMICK, D.A., & BAL, T. (1997) Sleep and arousal:Thalamocortical mechanisms. *Annual Rev. Neurosci.* 20:185-215.

MCCORMICK, D.A., & HUGUENARD, J.R. (1992) A model of the electrophysiological properties of thalamocortical relay neurons. *J. Neurophysiol.* 68:1384-1400.

MILLER, K., D. (1994) A model for the development of simple cell receptive fields and the ordered arrangement of orientation columns through activity dependent competition between On- and Off-center inputs. *J. Neurosci.* 14:409-441.

MILLNER, P. (1974) A model for visual shape recognition. *Psychol. Rev.* 816:521.

MIZE, R.R., MARC R.E., & SILLITO, A.M. (1992) *GABA in the Retina and Central Visual System*, Section III. Visual Cortex, Elsevier, Amsterdam, London; New York, Tokyo.

MORRONE, M.C., BURR D.C., & MAFFEI, L. (1982) Functional implications of cross-orientation inhibition of cortical visual cells.1 Neurophysiological evidence. *Proc. Roy. Soc. Lond. B* 216:335-354.

MUMFORD, D. (1991) On the computational architecture of the neocortex. I. The role of the thalamocortical loop. *Biol. Cybern.* 65:135-145.

NELSON, S., TOTH L., SHETH B., & SUR, M. (1994) Orientation selectivity of cortical neurons during intracellular blockade of inhibition. *Science* 265:774-777.

NIELSEN, D.E. (1983) A functional model of the wiring of the simple cells in the visual cortex. *Biol. Cybern.* 47:213-222.

NOWAK, L., BREGESTOVSKI P., ASCHER P., HERBET A., & PROCHIANTZ A., (1984) Magnesium gates glutamate-activated channels in mouse central neurons. *Nature* 307:462-465.

OBERMAYER, K., RITTER, H., & SCHULTEN, K. (1990) A principle for the formation of the spatial structure of cortical feature maps. *Proc. Natl. Acad. Sci. USA* 87:8345-8349.

OHZAWA, I., SCLAR G., & FREEMAN, R.D. (1985) Contrast gain control in the cat's visual system. *J. Neurophysiol.* 54:651-667.

ORBAN, G.A. (1984) *Neuronal Operations in the Visual Cortex.* Springer-Verlag, Berlin Heidelberg New York Tokyo.

PAYNE, B.R., BERMAN N.E.J., & MURPHY, E.H. (1981) Organization of direction preferences in cat visual cortex. *Brain Res.* 211:445-450.

PEI, X., VIDYASAGAR T.R., VOLGUSHEV M., & CREUTZFELDT, O.D. (1994) Receptive field analysis and orientation selectivity of postsynaptic potentials of simple cells in cat visual cortex. *J. Neurosci.* 14:7130-7140.

RAMOA, A.S., SHADLEN M., SKOTTUN B.C., & FREEMAN, R.D. (1986) A comparison of inhibition in orientation and spatial frequency selectivity of cat visual cortex. *Nature* 321:237-239.

RAMON, Y CAJAL S. (1899) Estudios sobre la corteza cerebral humana. *Revista Trimestral Micrographica* 4:1-63.

REID, R.C., & ALONSO, J.,-M. (1995) Specificity of monosynaptic connections from thalamus to visual cortex. *Nature* 378:281-248.

RODIECK, R.W., & STONE, J. (1965) Analysis of receptive fields of cat retinal ganglion cells. *J. Neurophysiol.* 28:833-849.

SALIN, P.,-A., & BULLIER, J. (1995) Corticocortical connections in the visual system:Structure and function. *Physiological Rev.* 75:107-154.

SALT, T.E., & EATON, S.A. (1996) Functions of ionotropic and metabotrop glutamate receptors in sensory transmission in the mammalian thalamus. *Prog. Neurobiol.* 48:55-72.

SASAKI, H., SAITO Y., BEAR D.M., & ERVIN, F.R. (1971) Quantitative variation in striate receptive fields of cats as a function of light and dark adaptation. *Exp. Brain Res.* 13:273-293.

SCHALL, J.D., VITEK D.J., & LEVENTHAL, A.G. (1986) Retinal constraints on orientation specificity in cat visual cortex. *J. Neurosci.* 6:823-836.

SCHILLER, P.H. (1992) The On and Off channels of the visual system. *TINS* 15:86-92.

SCHWINDT, P., & CRILL, W.E. (1995) Amplification of synaptic current by persistent sodium conductance in apical dendrite of neocortical neurons. *J. Neurophysiol.* 74:2220-2224.

SCLAR, G., & FREEMAN, R.D. (1982) Orientation selectivity in the cats striate cortex is invariant with stimulus contrast. *Exp. Brain Res.* 46:457-462.

SHAPLEY, R., & HOCHSTEIN, S. (1975) Visual spatial summation in two classes of geniculate cells. *Nature* 256:411-413.

SHERMAN, S.M., & KOCH, C. (1986) The control of retinogeniculate transmission in the mammalian lateral geniculate nucleus. *Exp. Brain Res.* 63:1-20.

SHEVELEV, I.A., SHARAEV G.A., VOLGUSHEV M.A., PYSHNII M.F., & VERDEREVSKAYA N.N. (1982) Dynamics of receptive fields of neurons in the visual cortex and LGB. *Neurophysiologia (Kiev)* 14:628-636.

SHEVELEV, I.A., VERDEREVSKAYA N.N., & MARCHENKO, V.G. (1974) Complete reorganization of detector properties of neurons in the cat visual cortex under different levels of visual adaptation. Dokl Acad Nauk SSSR 217:493-496.

SHEVELEV, I.A., VOLGUSHEV M., & SHARAEV, G.A. (1992) Dynamics of responses of V1 neurons evoked by stimulation of different zones of receptive field. *Neurosci.* 51:445-450.

SHMUEL, A., & GRINVALD, A. (1996) Functional organization of direction of motion and its relationship to orientation maps in cat area 18. *J. Neurosci.* 16:6945-6964.

SILLITO, A.M. (1975) The contribution of inhibitory mechanisms to the receptive field properties of neurons in the striate cortex of the cat. *J. Physiol. Lond.* 250:305-329.

SILLITO, A.M. (1977) Inhibitory processes underlying the directional specificity of simple, complex and hypercomplex cells in the cat's visual cortex. *J. Physiol. Lond.* 271:699-720.

SILLITO, A.M., CUDEIRO, J., & MURPHY, P.C. (1993) Orientation sensitive elements in the corticofugal influence on centre-surround interactions in the dorsal lateral geniculate nucleus. *Exp. Brain Res.* 93:6-16.

SILLITO, A.M., GRIEVE K.L., JONES H.E., CUDEIRO J., & DAVIS, J. (1995) Visual cortical mechanisms detecting focal orientation discontinuities. *Nature* 378:492-496.

SILLITO, A.M., JONES H.E., GERSTEIN G.L., & WEST, D.C. (1994) Feature-linked synchronization of thalamic relay cell firing induced by feedback from the visual cortex. *Nature* 369:479-482.

SILLITO, A.M., KEMP J.A., MILSON J.A., & BERARDI, N. (1980) A re-evaluation of the mechanisms underlying simple cell orientation selectivity. *Brain Res.* 194:517-520.

SINGER, W. (1977) Control of thalamic transmission by corticofugal and ascending reticular pathways in the visual system. *Physiol. Rev.* 57:386-420.

SINGER, W., & CREUTZFELDT, O.D. (1970) Reciprocal lateral inhibition of On- and Off-center neurones in the lateral geniculate body of the cat. *Exp. Brain Res.* 10:311-330.

SINGER, W., & GRAY, C.M. (1995) Visual feature integration and the temporal correlation hypothesis. *Ann. Rev. Neurosci.* 18:555-586.

SINGER, W., PÖPPEL, E., & CREUTZFELDT, O. (1972) Inhibitory interaction in the cat's lateral geniculate nucleus. *Exp. Brain Res.* 14:210-226.

SINGER, W., TRETTER F., & CYNADER, M. (1975) Organization of cat striate cortex: A correlation of receptive-field properties with afferent and efferent connections. *J. Neurophysiol.* 38:1080-1098.

SOMERS, D.C., NELSON S., & SUR, M. (1995) An emergent model of orientation selectivity in cat visual cortical simple cells. *J. Neurosci.* 15:5448-

5465.

SOMOGYI, P., & FREUND, T.F. (1989) Immunocytochemistry and synaptic relationships of physiologically characterized HRP-filled neurons. In: *Neuroanatomical-Tract tracing Methods 2* (Heimer, L. and Zaborszky, L., eds.), Plenum Publishing Corp.

STERIADE, M. (1991) Alertness, quiet sleep and dreaming. *Cerebral Cortex* 9:279-357.

STERIADE, M. (1995) Neuromodulatory systems of thalamus and neocortex. *Sem. Neurosci.* 7:361-370.

STERIADE, M. (1997) Synchronized activities of coupled oscillators in the cerebral cortex and thalamus at different levels of vigilance. *Cerebral Cortex* 7:583-604.

STERIADE, M., & LLINÁS, R.R. (1988) The functional states of the thalamus and the associated neuronal interplay. *Am. Physiol. Soc.* 68:649-742.

STERIADE, M., & DESCHÊNES, M. (1984) The thalamus as a neuronal oscillator. *Brain Res. Rev.* 8:1-63.

STUART, G., & SAKMAN, B. (1995) Amplification of EPSPs by axomatic sodium channels in neocortical pyramidal neurons. *Neuron* 15:1065-1076.

SUDER, K., & WÖRGÖTTER, F. (1999) The control of low-level information flow in the visual system. *Rev. Neurosci.* 11:127-146.

SUTOR, B., & HABLITZ, J.J. (1989) EPSPs in rat neocortical neurons in vitro. II. Involvement of N-Methyl-D-Aspartate receptors in the generation of EPSPs. *J. Neurophysiol.* 61:621-634.

SWINDALE, N. (1982) A model for the formation of orientation columns. *Proc. R. Soc. Lond. B* 215:211-230.

SWINDALE, N.V., MATSUBARA J.A., & CYNADER, M.S. (1987) Surface organization of orientation and direction selectivity in cat area 18. *J. Neurosci.* 7:1414-1427.

TANAKA, K. (1983) Cross-correlation analysis of geniculostriate neuronal relationships in cats. *J. Neurophysiol.* 49:1303-1318.

TANAKA, K. (1985) Organization of geniculate inputs to visual cortical cells in the cat. *Vision Res.* 25:357-364.

THEUNISSEN, F., & MILLER, J.P. (1995) Temporal encoding in nervous systems:A rigorous definition. *J. Comp. Neurosci.* 2:149-162.

THOMSON, A.M., GIRDLESTONE D., & WEST, D.C. (1988) Voltage-dependent currents prolong single-axon postsynaptic potentials in layer III pyramidal neurons in rat neocortical slices. *J. Neurophysiol.* 60:1896-1907.

TSUMOTO, T., CREUTZFELDT, O.D., & LEGENDY, C.R. (1978) Functional organization of the corticofugal system from visual cortex to lateral geniculate nucleus in the cat. With an appendix on geniculo-cortical monosynaptic connections. *Exp. Brain Res.* 32:345-364.

TSUMOTO, T., ECKART W., & CREUTZFELDT, O.D. (1979) Modification of the orientation selectivity of the cat visual cortex neurons by removal of GABA-mediated inhibition. *Exp. Brain Res.* 34:351-363.

VIDYASAGAR, T.R. (1987) A model of striate response properties based on geniculate anisotropies. *Biol Cybern* 57:11-23.

VIDYASAGAR, T.R. (1990) Pattern adaptation in cat visual cortex is a co-operative phenomenon. *Neurosci.* 36:175-179.

VIDYASAGAR, T.R., & HEIDE, W. (1984) Geniculate orientation biases seen with moving sine wave grating:implications for a model of simple cell afferent connectivity. *Exp. Brain Res.* 57:196-200.

VIDYASAGAR, T.R., PEI X., & VOLGUSHEV, M. (1996) Multiple mechanisms underlying the orientation selectivity of visual cortical neurons. *TINS* 19:272-277.

VIRSU, V., LEE B.B., & CREUTZFELDT, O.D. (1977) Dark adaptation and receptive field organization of cells in the cat lateral geniculate nucleus. *Exp. Brain Res.* 27:35-50.

VOLGUSHEV, M., PEI X., VIDYASAGAR T.R., & CREUTZFELDT, O.D. (1992) Postsynaptic potentials in the cat visual cortex:dependence on polarization. *NeuroReport* 3:679-682.

VOLGUSHEV, M., VIDYASAGAR T.R., & PEI, X. (1996) A linear model fails to predict orientation selectivity of cells in the cat visual cortex. *J. Physiol. Lond.* 496:597-606.

VON, DER MALSBURG, C. (1973) Self-organization of orientation sensitive cells in the striate cortex. *Kybernetik* 14:85-100.

VON, DER MALSBURG, C. (1981) The correlation theory of brain function. Internal report, MPI f. Biophys. Chem. Göttingen, Ger.

WÄSSLE, H., & BOYCOTT, B.B. (1991) Functional architecture of the mammalian retina. *Physiological Rev.* 71:447-480.

WANG, X.J., & RINZEL, J. (1992) Alternating and synchronous rhythms in reciprocally inhibitory model neurons. *Neural Comp.* 4:84-97.

WANG, X.J., RINZEL, J., & ROGAWSKI, M.A. (1991) A model of the t-type calcium current and the low- threshold spike in thalamic neurons. *J. Neurophysiol.* 66:839-850.

WELIKY, M., BOSKING W.H., & FITZPATRICK, D. (1996) A systematic map of direction preference in primary visual cortex. *Nature* 379:725-728.

WHITE, E.L., & KELLER, A. (1989) *Cortical Circuits. Synaptic Organization of the Cerebral Cortex / Structure, Function, and Theory.* Birkhäuser, Boston, Basel.

WILSON, J.R. (1993) Circuitry of the dorsal lateral geniculate nucleus in the cat and monkey. *Acta Anatomica* 147:1-13.

WÖRGÖTTER, F., & EYSEL, U.T. (1991) Topographical aspects of intracortical excitation and inhibition contributing to orientation specificity in area 17 of the cat visual cortex. *Europ. J. Neurosci.* 3:1232-1244.

WÖRGÖTTER, F., & KOCH, C. (1991) A detailed model of the primary visual pathway in the cat:comparison of afferent excitatory and intracortical inhibitory connection schemes for orientation selectivity. *J. Neurosci.* 11:1959-1979.

WÖRGÖTTER, F., SUDER, K., ZHAO, Y., KERSCHER, N., EYSEL, U.,

& FUNKE, K. (1998) State-dependent receptive field restructuring in the visual cortex. *Nature* 396:165-168.

WÖRGÖTTER, F. (1999) Comparing different modeling approaches of visual cortical cell characteristics. In: *Cerebral Cortex* (Ulinski et al., eds.), Vol. 13:201-249, Kluwer Academic/Plenum Publ., New York. 201-249.

WÖRGÖTTER, F., & EYSEL, U.T. (2000) Context, state and the receptive fields of striatal cortical cells. *TINS*, 23:497-503.

4

Neural Principles of Preattentive Scene Segmentation: Hints from Cortical Recordings, Related Models, and Perception

Reinhard Eckhorn

ABSTRACT Preattentive segmentation of visual scenes is a prerequisite of object recognition and effective visuomotor coordination. For this the visual system has to specify neural representations of contours and regions of potential relevance so that top-down acting mechanisms of attention, expectation and visual memory can interact with them. This chapter attempts to show how the largely unknown neural mechanisms of scene segmentation may be uncovered. Starting from principles of perceptual grouping, we present experimental results mainly of our own multiple microelectrode recordings from the visual cortex of awake monkeys. Perceptual and experimental hints at principles of scene segmentation are supported by related simulations of spike coding networks of minimal complexity. We follow the hypothesis that fast signal coupling and decoupling among visual cortical circuits define preattentively relations among scene segments. Two properties of scene segments and related signal coupling are differentiated: (1) Transient retinal changes evoke short simultaneous activations that are coarsely synchronized across the entire representation of the changing segments; (2) During ocular fixation stable retinal images induce fast cortical oscillations (FCOs; 30 − 90 Hz) that are phase-correlated along the cortical representation of contours and across segment regions. Phase coupling among FCOs in neighboring cortical populations is weak, distributions of phase differences are symmetrical to zero delay and their width increases with cortical distance. These properties explain the small size of cortical patches over which coherent FCOs have previously been reported. The representation of such a cortical patch in visual space (defined by the superimposed classical receptive fields (cRFs) of the synchronized neurons) is termed here the feature association field (AF). Arguments from experiments and simulations are developed showing that AF size at a lower level of processing can explain the larger cRFs at the next level and hence a stepwise increase in establishing relevant scene relations. In the light of our new data, the previous hypothesis of feature association by synchronization is modified to a hypothesis of coding region, contour and object continuity by phase-continuity, including single event stimulus-locked and rhythmic FCO processes. (This article contains some new unpublished experimental and modeling results.)

4.1   Introduction

4.1.1   Preattentive scene segmentation is a prerequisite for object recognition

Stimuli composed of coherent features are integrated by our visual system into perceptual entities. We can perceive a visual object as a perceptual whole even if various aspects of the object are occluded, obscured by the background, or are not present at all. The visual system can easily detect the correlations among an object's local stimulus features and is able to link, intensify, and isolate them. These capabilities of grouping, mutual facilitation, and figure/ground separation require neural mechanisms of self-organization that are able to construct reliable and unique percepts out of ambiguous sensory signals. Early processing of these relations is highly important because the infinite number of possibilities in which even simple "toy scenes" can be arranged has to be reduced substantially before visual associative memories can solve tasks of recognition.

The present work concentrates on neural mechanisms of preattentive scene segmentation. These are mechanisms mainly driven by the current visual scene (bottom-up) while influences of attention, expectation, visual knowledge, and recognition (top-down) are not included here, mainly because knowledge about them is still sparse. For visual scenes and its neural representations the following terms will be used:

Local features are characterized by the classical receptive field (cRF) properties of single neurons (retina) and by local clusters of neurons with similar cRFs (cortex). Nearly all cRF properties analyzed so far represent local feature gradients (contrasts). Examples are the concentric antagonistic On-Off cRFs of retinal ganglion cells and the orientation sensitive cRFs of simple cells in primary visual cortex. cRF properties are retinotopically represented in the lower visual system (e.g., retina, thalamus, and visual cortex including V1 to V5; for more details about the functional structure of the visual system see basic textbooks on visual neuroscience, e.g. [1, 2]).

Contours are a prominent property of a scene. They are present at elongated borders of feature contrasts defining segments, and hence delineate potential visual objects. The simple-cell orientation detectors of V1, for example, are particularly sensitive to short segments of oriented contours. Regions of a scene are separated by contours.

Regions are parts of, or as a whole, surfaces of potential visual objects (the latter are specified by rules of Gestalt psychology [3, 43]).

4.1.2   Principles of neural coding beyond the classical receptive field

A single object generally activates neurons in many visual cortical areas corresponding to a distributed representation of its features. While single neurons in the lower cortical areas, including V1-V5, represent a variety of local features, identified by their cRFs, it is still under debate how the distributed representation of an object is bound into a coherent whole and how unrelated features are separated. In other words, how does the visual system process its input to obtain a well segmented scene?

This topic has extensively been investigated by asking how the continuity of visual objects may be coded by visual cortical neurons (e.g. [4, 5]). Several solutions have been proposed for tasks of scene segmentation, the most prominent are continuity coding by spike-rate coherence and by spike-event coherence (e.g.[6, 7]). Coding a scene segment by spike-rate coherence means that neurons representing the same region are modulated in their spike rates in parallel. Coding by spike-event coherence states that these neurons discharge synchronized spike events and herewith specify binding of the features they are sensitive to. Two types of event coherence have been investigated intensively: Spike synchronization during fast cortical oscillations (FCOs, 30 − 90 Hz; reviews in [8, 9, 10]) and spike synchronization in isolated spike patterns occurring in a nonrhythmic way [10, 11, 12, 13, 14, 15, 16].

Coding by Rate and Event Coherence probably do not exclude each other. They may reflect activity changes at different time scales which can, in principle, be decoded by any neuron and may be simultaneously present. Fast postsynaptic time constants are necessary for extracting event coherence while spike rate modulations require slow mechanisms. Both codes become particularly effective with converging connectivities to a common target because then high activation probabilities are guaranteed by the coherence of converging signals (e.g.[17, 18, 114]). These common principles make it possible and probable that rate and event coding are simultaneously used for coding different stimulus aspects. With changing cortical activation states their relative contributions may vary depending on the network's state and its background activities [12], the states of its synapses [19], and other neural properties (e.g. [20]).

4.1.3   Coupling beyond the classical receptive fields defines association fields

Visual Association Fields of Local Assemblies. To bring mechanisms of synchronization into correspondence with the perceptual capabilities of feature associations and grouping, the concept of the "linking or association field" (AF) of a local neural assembly was introduced (see recent work [21, 22]). In the present context, the AF of a local assembly of visual neurones was defined as that area in visual space where appropriate local stimulus features induce synchronized activities within that assembly, and hence can support feature associations [21]. AFs are constituted both by the cRF properties of the assemblies' neurones and by the properties of their linking inter-connections (including type, strength, position of cRF). This implies that the AF of a given group of visual neurones is generally much broader than the cRF of single member neurones of the AF assembly. Recordings from primary and associative visual cortex areas of cat and monkey support this prediction [1, 2].

Partially complementary concepts of association (context) fields have been proposed by others on the basis of spike rate measures [23], psy-chophysical data [24, 25], intracellular potentials [26] and neural network models for visual tasks [27, 28]. Psychophysical and electrophysiological ex-periments to date support the hypothesis that visual networks for feature associations are closely related to the rules of Gestalt perception at least in the more peripheral areas of the visual cortex where they have been investigated in greater detail (for a recent review see [29]).

4.2   Properties of Synchronized Fast Cortical Oscillations (FCOs)

4.2.1   Sustained activation is required for the generation of FCOs

A general condition for the occurrence of FCOs is sustained cortical acti-vation and an absence of strong and fast excitatory or inhibitory response transients [30]. This condition is particularly given with stationary stimuli matching the cRF properties of the neurons. For example, stimulation with an oriented one-dimensional luminance grating sustainedly activates those neurons preferring the spatial frequency, orientation, and contrast of that grating [110]. Such states typically occur in situations with ocular fixation, i.e., when the retinal image of a fixated object is stable or slowly drifts. Model investigations support this experimental observation (e.g. [31, 109]).

4.2.2   Single neurons are differently involved in FCOs

Three types of coupling dynamics were observed between single cell spikes and oscillatory population activity (local field potentials LFP and multi-ple unit activity MUA; LFP and MUA are explained in Appendix) in cat [32] and monkey visual cortex [33] as well as in related neural network simulations [31]: 1. In *rhythmic states* single cell spike patterns are rhyth-mically modulated and many spikes are phase correlated with oscillatory population activity. 2. In *lock-in states* rhythmic modulation is not visible in single cell spike patterns (and not in their auto-coincidence histogram) while spikes are significantly phase-coupled to the oscillatory activity of the surrounding local population. 3. In *non-participation* states rhythmic modulation is absent in spike trains as well in addition, the spikes are not correlated with the current oscillatory population activity (4.1).

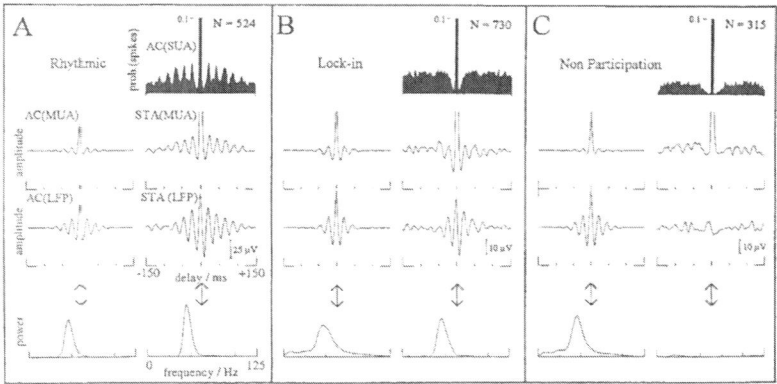

FIGURE 4.1. Three different states of single cell couplings with oscillatory population activities in the primary visual cortex. A: rhythmic, B: lock-in, and C: non-participation states of three different neurons. AC: auto-coincidence histograms of single cell spikes (SUA), auto-correlation of multiple unit activity (MUA), and of local field potential (LFP). STA denotes spike-triggered averages of multiple unit activity or local field potentials. According to the classification, STAs have oscillatory modulations in the rhythmic and lock-in states, and lack such modulation in the nonparticipation state. Note that in the rhythmic state (A) the single cell correlogram (top) is clearly modulated at 44 Hz, while in the lock-in (B) and the nonparticipation states (C) rhythmic modulations in the range 35 − 80 Hz are not visible (by definition). Lowest row of panels: power spectra for the above row of correlograms (modified from [32]).

State transitions of single cell coupling with oscillatory population activities depends on stimulation and cRF properties. For example, a weak lock-in state can change into a strong synchronized rhythmic state in a cortical (directionally sensitive) neuron when reversing the movement direction of an oriented contour stimulus [32]. Stimulus movement in the preferred direction causes higher levels of average sustained activation (also in other neurons of the same directional preference) and this results in a rhythmic synchronized state.

4.3   Coding Contour Continuity

Sharper Orientation Tuning with FCOs than with Other Response Measures. Essential for perceptual scene segmentation are elongated contrast edges defining contours between segments (i.e., between potential objects). For a precise representation of a contour's local position and orientation probably population coding is used instead of relying on a single neuron matching the orientation best. For the motor cortex Georgopulous [34] proposed coding by a population vector based on spike rates. He succeeded by showing close correlation of the population vector's direction with the movement direction of a monkey's arm. Corresponding measures have been applied to the coding of visual features by cortical neurons (e.g. [35]). Shadlen and Newsome [36] review evidence for population coding in sensory systems, including the visual modality, and came to the conclusion that population coding by spike rates is sufficient for explaining behavioral performance. This view is disputed here.

For this we will test the hypothesis of feature binding by synchronization by first looking at the conditions under which FCOs occur and second, under which conditions they synchronize (for "synchronization" see Appendix 4.10). As the primary visual cortex is specially equipped for the detection of oriented contour segments by its simple-cells we expect coherence of FCOs in local populations (with similar cRF properties) that are well tuned to orientation in their population amplitudes (MUA and LFP, see Appendix 4.10). In addition, if FCOs display even sharper orientation tuning than spike rates calculated from the same recording, the functional importance of FCOs against rate coding is favored for this task. Indeed, recent experiments revealed significantly sharper orientation tuning of population FCOs compared to population rates (Fig. 4.2).

Under what conditions might FCOs of local populations reveal sharper orientation tunings than those measured by population spike rates? No difference in orientation tuning would result if, for example, the number of spikes is kept constant while shifting them to certain phases (so that they become synchronized fast oscillations). In contrast, sharper orientation tuning with FCOs would result if some nonlinear facilitation is present among the local mechanisms generating FCOs and if mutual inhibition is acting among neigboring populations of different orientation preference (e.g. [38]). Sompolinsky and Shapley [39] reviewed the potential origin of orientation selectivity. They weighted experimental and theoretical results in favor and against its thalamic and cortical origin. Their final conclusion supports a substantial contribution of cortical feedback circuits to the sharp orientation tuning of neurons in striate cortex while the thalamic input was insufficient to explain it.

The investigation of Frien and coworkers [37] also favors a strong cortical involvement in the processing of sharp orientation selectivity. The main argument is that FCOs display a sharper orientation tuning than other response measures and that FCOs are probably generated in cortical circuits and are not injected by thalamic afferents [40, 112]. Against the latter possibility Sillito and coworkers [41] obtained direct experimental evidence by showing that fast oscillations in visual thalamus were suppressed by inactivation of striate cortex, indicating its cortical origin. In addition, synchronized FCOs occur in local populations of neurons with similar orientation preference (which is a specific cortical property) and can synchronize over several millimeters of cortical distance in a stimulus specific way (see Sec. 4.4). These and other arguments support a cortical origin of fast oscillations and thus agrees with the hypothesis that sharp orientation tuning is supported by coupling among local neurons with similar cRF properties.

In summary, the main and new finding of this investigation [37] is that FCOs of local V1 populations of awake monkeys show sharper orientation tuning than any other response measure tested. These included response components in low and medium frequency ranges and the average spike rates of local populations. As the comparisons were all based on the same set of raw data, differences could well be established. In addition, since FCOs occur synchronized in the awake monkey's striate cortex we assume that they have enhanced probability of activating successive stages of visual processing and hence contribute to the perception of orientation.

4.4   Coding Region Continuity

Event Coherence Supports Region Coding but Rate Coherence does not. In this paragraph it is argued that the degree of coherence among the different neural signals of a cortical visual representation may indicate how much these components can contribute to the coding of region, and thus object surface, continuity [42]. Functional coupling was quantified by calculating spectral coherence among population signals (MUA, LFP, see Appendix 4.10) recorded in striate cortex of an awake monkey, activated by a large field grating texture. During visual stimulation the spectral coherence was mostly dominated by fast $35 - 50$ Hz oscillatory components, and often by additional slow $2-15$ Hz components. In the intermediate range ($15-35$ Hz) the amplitudes, the coherence, and their changes with stimulation were generally small. Comparison of low and high frequency bands revealed a higher stimulus selectivity and higher significance levels for the coherence of FCOs compared to slow components, suggesting a stronger contribution to stimulus representation by spike-event than by spike-rate coherence, at least for the present experiments (Fig. 4.3).

Another new finding in the experiments by Frien and Eckhorn [42] is the form of the association fields (AF) during stimulation with a texture region. The AF is determined here by the superposition of cRFs in visual space (and hence is directly related to the neural distribution in V1) of neurons engaging in coherent FCOs [21]. When we used large grating textures, AFs were typically round patches in which the coherence declined homogenously with distance. This means, that the coherence at a given distance had equal values at parallel, oblique or coaxial relative cRF positions.

Coherence of FCOs across the AFs was strongly depending on the relative orientation preferences of the coupled neurons and other properties. Coherence was the higher the more similar the orientation preference was, the better the neurons were driven by the grating and the nearer they were in cortical and visual space, resembling Gestalt rules of proximity and

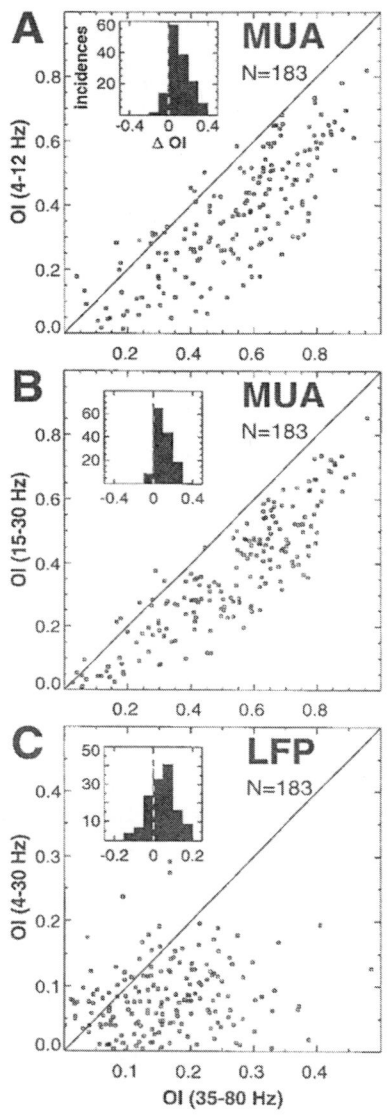

FIGURE 4.2. Sharper orientation tuning for FCOs than for slow components: Comparison of the orientation index OI from different frequency ranges for multi unit (MUA) and local field potential (LFP). A: MUA low over high frequency range (FCO). B: MUA medium over high frequency range; C: LFP alpha (around 10 Hz) over high frequency range. Insets show the distributions of differences in OI. Positive values correspond to larger orientation indices for the high frequency range (Fig. modified from [37]).

Polar direction tunings of MUA peak coherence

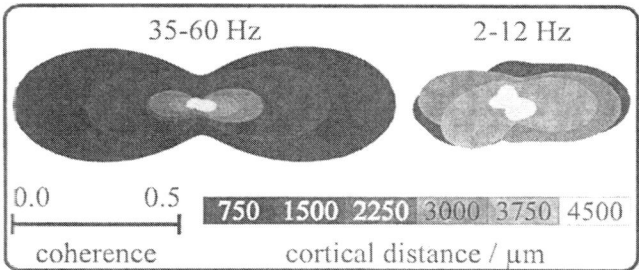

FIGURE 4.3. FCO-coherence shows stronger tuning for the orientation of a slowly moving texture (left) than spike rate coherence (right). Polar plots of the coherence of multiple unit activity recorded from V1 of awake monkey. Results support previous proposals that neurons of similar orientation preference engage in synchronized fast oscillations when stimulated close to this orientation (Figure modified from [42])

similarity [3, 43, 45].

In contrast, previous work on fast oscillations in striate cortex found generally highest values of coherence among neurons with coaxially aligned cRFs [5, 46, 47, 48, 49, 50, 51, 52, 53] (for a review see [54]), which means AFs were elongated along a contour. This difference may be explained by the different stimuli. While Frien et al. used a large texture grating resembling a scene's segment (e.g., the surface of an object) previous work found coupling among aligned neurons with stimuli resembling elongated object contours. This suggests that coupling of fast oscillations can flexibly match to spatial aspects of stimulus context. In other words, the cortical network may engage in elongated one-dimensional AFs resembling object contours and it is also able to couple neurons representing object surfaces establishing two dimensional AFs.

From the same experiments [42] coherence plots for pair recordings of neurones having about orthogonal or oblique orientation differences were calculated (not shown here). They revealed also broad elongated blobs of increased coherence, but these were spatially more restricted and their orientation fell half way between the preferred orientations at the two recording positions (e.g. at about 45° with 0° and 90° "detectors").

4.5 Coding the Separation of Adjacent Regions

If we continue to follow the temporal coding hypothesis, adjacent or overlapping scene segments of different visual objects should be coded by temporally separable signals. Discrimination would be possible if, for example, the neurons representing adjacent objects are activated at different relative phase shifts at the same frequency, or at different uncorrelated frequencies, or more generally, with uncorrelated signals that might even be stochastic (see Sec. 4.8 about stimulus-locked synchronization).

To our knowledge, there are only few experimental examples reporting uncorrelated FCOs and these used two stimulus objects moving independently in different directions [46, 54, 55]. Not a single experiment is known reporting FCOs at fixed phase differences at two adjacent representations of scene segments. In contrast, many models were developed in which scene segmentation is performed by phase shifts at a common oscillatory frequency (e.g. [56, 57, 58, 59, 60, 61]).

Our group therefore performed experiments that tested signal relations among adjacent, stationary regions as they occur during ocular fixation. It was found that adjacent scene segments (separated by a contrast edge) do not generate phaseshifted FCOs in their cortical representations but decouple their FCOs [62, 63]. In other words, the same local populations engage in synchronized FCOs if activated by a single coherent stimulus but decouple their FCOs when an object's boundary seperates the populations (Fig. 4.4). Accordingly, the frequencies became different, typically higher in the neurons representing the smaller of two stationary regions (for stimulus influences on average FCO frequency see 4.7.1).

4.6 Spatially Restricted Synchronization Among FCOs

4.6.1 Average zero-phase correlation within a cortical area

Reported phase differences between FCOs at separate locations of the same visual cortical area (e.g., V1 or V2) and among different areas are narrowly distributed around zero [4, 50, 33]. The average range in which FCO coherence was observed above noise level is 4-6 mm in visual cortex areas V1 and V2 of anesthetized cat and awake monkey [20, 64, 65, 66, 67, 68]. This was established by recordings by linearly arranged arrays of equally spaced microelectrodes and the calculation of coherence among FCOs at different separations. Spatial decline of coherence is less steep in positions where neurons represent similar visual features (for example similar local orientation of a contrast border) compared to those with dissimilar features (Fig. 4.5).

4.6.2 Average zero-phase correlation among two visual cortex areas

Average zero-phase correlation is also present among FCOs recorded in directly connected visual areas, including cat V1, V2, and V3 [4, 50, 69] as well as monkey V1 and V2 [33]. It is important to know, that phase coupling in these experiments was spatially restricted to neurons representing overlapping or directly neighboring cRFs in both cortical areas (signified

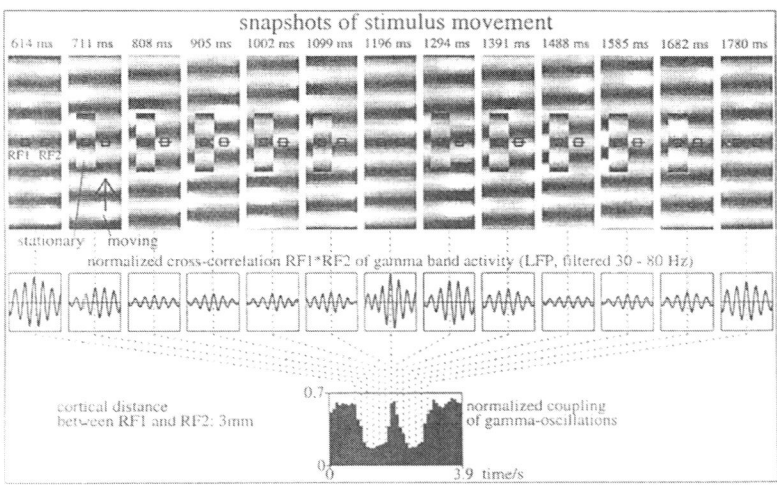

FIGURE 4.4. Figure-ground segregation defined by the reduction of FCO co-herence across region contours. LFP recordings from upper layers of V1 in 3 mm lateral cortical separation (while the monkey kept fixation within 0.5°). The measure of signal coupling utilized was similar to spectral coherence which was calculated with the dominant oscillations in the range 35 − 100 Hz. Stimulus: sinusoidal grating (the "background": spatial period 0.7° visual angle) was pre-sented stationary for 0.8 s and then moved perpendicular to its orientation at a velocity of 0 65°/s while a patch of it (the "object": 4° x 4° visual angle) remained stationary. RF1 and RF2 (left upper display) indicate the cRF positions of the recording pairs in- and outside the "object". Since the contour of the object was only visible when the gratings were out of spatial phase, we were able to examine a dynamical figure-ground separation (modified from [62]).

by the size of AFs).

Zero-phase correlation was particularly not expected among monkey V1 and V2 because these areas are known as serially arranged in the visual processing stream suggesting a delay of V2 against V1 signals. However, the average phase differences are narrowly distributed around zero with a standard deviation of less than 1 ms (Fig. 4.6). This result seems coun-terintuitive because the average conduction velocity among V1 and V2 is relatively slow, even though single V1-to-V2 projection fibers have trans-mission delays as short as 2-3 ms [70, 71, 72]. However, if only these fastest fibers are operational in V1-V2 coupling the 2-3 ms delay were clearly measurable in data as those of Fig. 4.6. The same argument holds for the correlation delays measured within a single visual cortex area.

Explanation of Zero-Delay Correlation is Difficult at Present. For finding at least a single plausible explanation of average zero-delay phase shift we concentrate at the conditions under which they occur. And we ask whether zero-delay correlation is restricted to FCO-epochs and if so, is it present only with certain FCO-frequencies?

We found average correlation delay among monkey V1 and V2 remaining near zero, independent of the frequencies of FCOs and even independent of

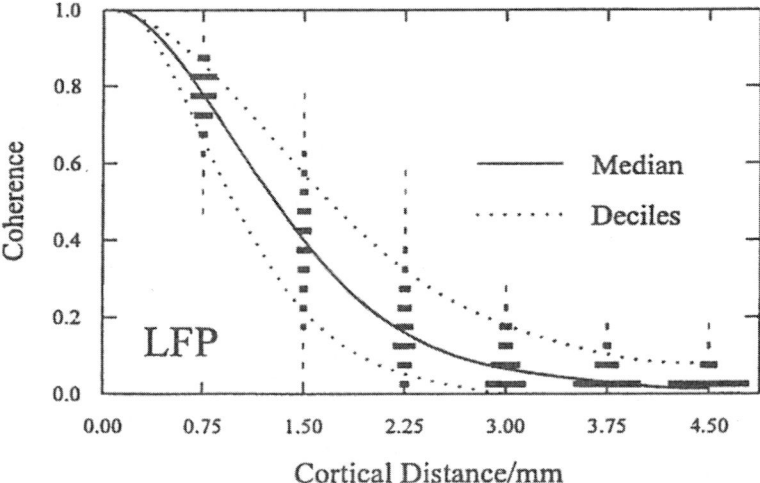

FIGURE 4.5. Coherence of FCOs declines with cortical distance. Local field potentials (LFP) recorded in primary visual cortex of awake monkey with linear multiple μ-electrode array from upper cortical layers. Horizontal bars are proportional to relative frequency of occurrence. Continuous line indicates mean, dotted deciles, of coherence. High values of coherence were obtained with similar, low with different, receptive field properties in the recording locations (modified from [68]).

their presence (non-rhythmic maintained activity also has a narrow peak at the center of V1-V2 cross-correlograms, if the cRFs of V1 and V2 recording positions overlap [33, 72, 73, 74, 113]).

A possible explanation of zero correlation delay is common input at equal delay from a single source because its occurrence does not depend on oscillation frequencies and type of activity. It has been argued that common input sources may be subcortical [75, 76]. At present, however, it seems more probable that common inputs are from cortical sources because the appearance of synchronized FCOs mostly shows stimulus specificities typical for cortical neurons (e.g., orientation sensitivity). Connections of equal activation delay may consist of inter- and intra-areal projections by the same neurons as suggested by anatomical findings [77]. They report single pyramidal neurons projecting, on average, with half of their synapses to distant targets of another single cortical area and with the second half to neurons of its own and nearby assemblies in the same visual area. Even though the distances of far and near targets are generally very different this may well be compensated to about equal delays because long range projections are rapidly conducting myelinated fibers while the local ones are unmyelinated and slowly conducting. If this is so, correlated activity at zero delay would show up in simultaneous recordings at the respective far and near target independent of the activation delays.

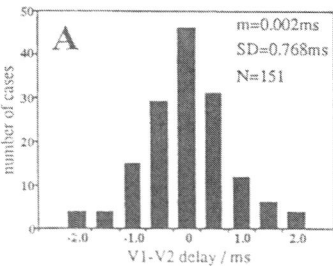 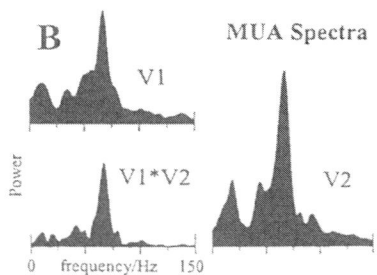

FIGURE 4.6. Average zero phase correlation among fast oscillations in two visual cortex areas (V1-V2). A: Distribution of average phase differences between oscillatory events in V1 and V2. B: Example of power spectra and cross-power spectra of V1 and V2 multiple unit activities (modified from [33]).

However, this explanation leaves some unsolved questions: (1) Is V1 the sole driving source of FCOs for all other cortical and subcortical targets? As V1 is connected to a large variety of cortical and subcortical targets this cannot be excluded to date. Inactivation of V1 neurons then should extinguish FCOs in all other parts of the visual system. (2) Is FCO synchronization at different frequencies due to selective changes in coupling with different conduction delays and/or due to modulation of neural time constants? These explanations can also not been ruled out at present, because (intra- and inter-) corticocortical connections span a wide spectrum of conduction velocities [70, 72] and neural time constants are variable over a broad range, in particular depending on synaptic activations [78].

4.6.3   Why declines FCO coherence with cortical distance and what are possible consequences for coding object continuity?

Correlation Analysis of Time- and Space-Resolved Single Responses. Conventional cross-correlation analyses revealed average phase differences among pair recordings of FCOs that did not increase with cortical distance, neither within the same nor among different visual cortex areas [4, 20, 68]. Instead, their average correlation delay remained narrowly distributed around zero while correlation strength declined to noise level within 3-5 mm cortical distance in V1 and V2 of anesthetized cat and awake monkey (Fig. 4.5). As the cortical representations of visual objects often are larger than the range over which synchronized FCOs occur, a common zero delay phase relation seems not suitable for coding the feature associations across the total representation of an object, at least not for larger ones. It has therefore been asked whether other measures of correlation among FCOs may support the coding of object continuity [79]. For this LFP and MUA correlations of single responses were calculated, resolved in space and time (bandpassed at 30-70 Hz) with a sliding time window of 20 ms duration, resulting in a two dimensional function of delay and cortical distance (Fig. 4.7). With linear regression (among the FCOs correlation maxima relative to the time axis) a criterion for linear phase relation and phase velocity was derived with respect to the cortical recording positions. So the dynamics of single FCO epochs could be measured along the recording positions as a function of time.

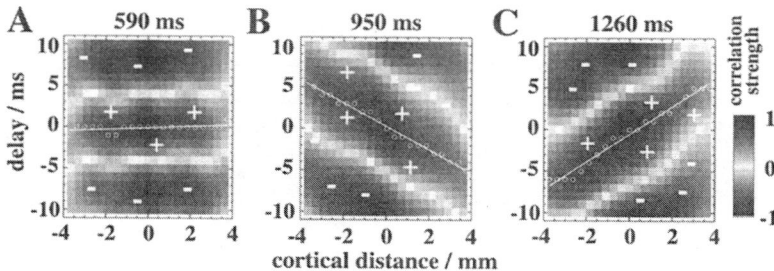

FIGURE 4.7. Snapshots of spatiotemporal phase relations of LFP fast oscillations simultaneously recorded at 7 equally spaced electrodes (0.75 mm interelectrode distance). Correlations of single responses are resolved in space and time (bandpassed at $30 - 70$ Hz) with a sliding window of 20 ms duration (window overlap: 15 ms, normalized with respect to spatial shift). Each correlation results in a two dimensional function of delay and cortical distance. A linear regression fit and its gradient of the FCOs correlation maxima relative to the time axis gives a criterion for linear phase relation and phase velocity with respect to the cortical recording positions. So we could measure the dynamics of single FCO spindles along the recording positions as a function of time. The linear regression fit in A shows a synchronous FCO-epoch over space. 300 ms later (B) the FCOs had changed their common phase with lead to left and lag to right positions. After additional 300 ms the FCOs' phase lead and lag are reversed. Such phase fronts typically occur during sustained visual stimulation (modified from [79])

Fig. 7 is an example of time resolved correlation diagrams from a single response to a single stimulus presentation. The linear regression fit in Fig. 4.7 A shows a wave front over space and time. About 300 ms later (B) the wave front has changed its phase with lead to left and lag to right positions which reversed with respect to lead and lag after additional 300 ms. Such continuous waves in single responses typically occur over the entire width of recording positions during stimulation (here 4 mm; and they were largely absent without). Phase variance increased about linearly with cortical distance (Fig. 4.8). This finding explains the observed decrease of FCO-coherence with cortical distance (Fig. 4.5). In addition, it is a clear indication that average zero-phase correlation is probably not due to common input from a single oscillator circuit (cortical or subcortical). It is rather a hint at continous lateral weak coupling among neigboring neural populations leading to the observed increasing widths of phase distributions with cortical distance. While the averaging methods of signal coherence revealed cortical distances of 3-5 mm over which coherence declines to near zero (Fig. 4.5), the spatial decline of coherence using single wave fronts leads to larger distances (8-10 mm). The latter value can be estimated from Fig. 4.8 if we assume constructive FCO superposition over about a third of the oscillation period at frequencies of 40-45 Hz.

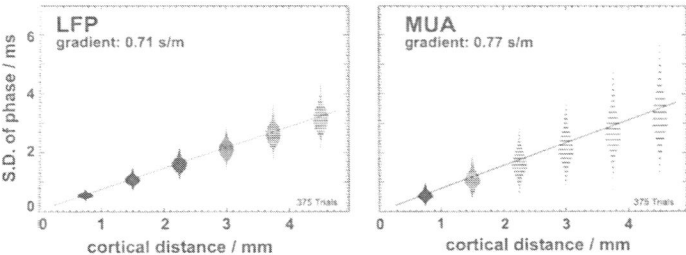

FIGURE 4.8. Phase distribution of FCOs in neighboring recording positions increase in width with cortical distance (see also Fig. 4.7; modified from [79])

Phase Continuity instead of Synchronization may Code Object Continuity. The single-response analysis of spatio-temporal FCO-dynamics revealed that the spatial decline of coherence and the spatially constant average correlation delay of zero are probably due to increasing symmetrical phase jitter with cortical distance (as in Fig. 4.8; [79]). As snapshots of FCO phase relations display phase continuity over cortical distances larger than the 3-5 mm captured by coherence analyses we were forced to modify the former "synchronization hypothesis" by stating that continuity coding of object representations in visual cortex may be supported by phase continuity of FCOs (or more generally, by any other type of phase correlated signal, including stimulus-locked, nonrhythmical). In other words, while feature grouping may be coded locally by near-synchronous population activities the global feature linking may be performed by a continuous overlap of near-synchronized regions whose relative phases are loosely coupled with its neighbor regions so that continuous waves with smoothly changing phases may define a larger scene segment. However, recordings with larger arrays of microelectrodes are required to prove this hypothesis.

4.6.4 Scene segmentation at consecutive levels of processing

Relations among AFs and cRFs. The size of cRFs in mammalian visual systems increase systematically from retina to higher cortical areas, probably due to increasing convergence of feedforward connections [1, 2]. Speculations about the functional role of increasing cRF-size include the stepwise processing of scene segments, the associative inclusion of relevant feature context, the formation of position and other types of invariance, and the classification of objects and their specific representation. Possibly all these functions and some additional unknown play a role. Here we will restrict to aspects of feature grouping and scene segmentation.

Starting point is the hypothesis that under Hebbian learning the AF size at one level of processing determines the cRF size at the next level [22, 80]. For example, the cortical range of coherent activities in V1 (projected to visual space by a superposition of the cRFs of the synchronized V1 neurons) lead to strong and convergent connections in V2 neurons and thereby determine their larger cRFs (Fig. 4.9). For simple cells in V1 this means, for example, that neighboring neurons activated by the same contour will synchronize their activity over a restricted range and send them (at zero-delay average correlation) to the next level of processing where they will activate their target neuron at a higher probability than nonsynchronized inputs. Thus, a local scene segment represented by V1 neurons with neighboring cRFs is associated (integrated) at the next level and drives there a single cell or a local population with similar (overlapping and larger) cRFs.

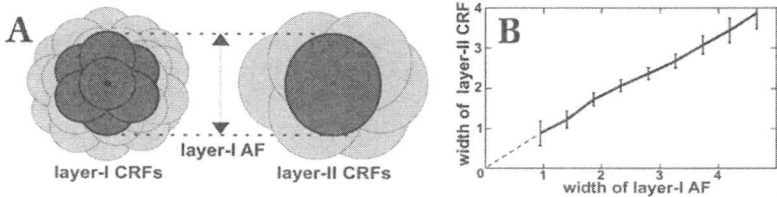

FIGURE 4.9. Dependence of the size of classical receptive fields (cRFs) at higher level of visual processing (e.g., V2) on the size of the association fields (AFs) at the previous level of processing (e.g., V1). Simulation results with a network of spiking neurons with distance depending delays and a Hebbian learning rule (Figure modified from [80])

This hypothesis was tested in a simulation of a network consisting of pulse coding model neurons arranged in two retinotopical layers [80, 111]. Layer-I neurons have overlapping concentric Gaussian cRF profiles and a lateral coupling via linking synapses whose spatial strength distribution resembles a concentric Gaussian profile broader than the corresponding cRFs. Additionally, layer-I is completely feedforward connected to layer-II via excitatory feeding synapses. Each axonal connection exhibits constant signal transmission velocity and therefore increasing delay with distance. Layer-I neurons are driven by trains of action potentials with Poisson interval distribution and constant mean rate. During simulation, the randomly initialized synaptic strengths between layer-I and II were modified by a Hebbian learning rule [80].

This simulation reveals three interesting properties: 1) The output spikes of layer-I neurons synchronize at zero average delay over a restricted range with declining coherence paralleled by spatially decaying coupling strength. The size of this synchronization patch defines the layer-I AF-size (fig. 4.8). 2) After learning, layer-II neurons form retinotopically organized and spatially restricted cRFs which can be fitted to a Gaussian profile (SD = cRF-size). 3) Layer-II cRF-size is linearly related with layer-I AF-size (Fig. 4.8 B; in this example with a slope of 0.8) due to interactions among input correlations, transmission delays, and the temporal course of the learning window.

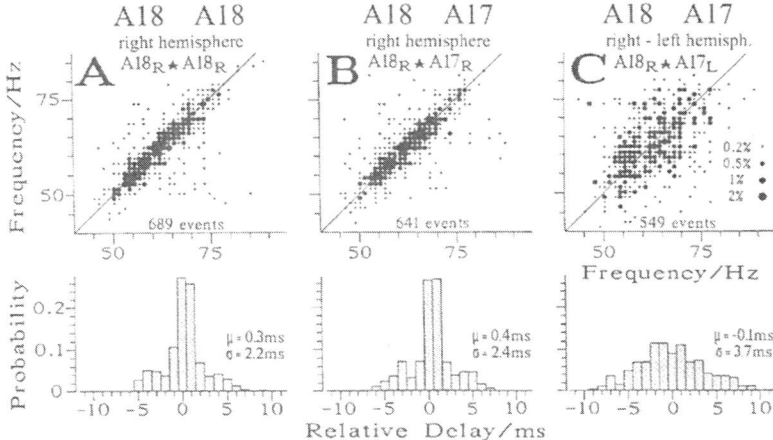

FIGURE 4.10. Distribution of oscillation frequencies (upper) and average phase delay (lower panels) in simultaneous pair recordings from a cat stimulated by a visual texture. A: area 18-18 (V2) recording at 2 mm distance and adjacent receptife fields (cRFs), B: area 17-18 (V1-V2) recording with overlapping cRFs, C: area 17-18 recording from different hemispheres of the visual cortex with partially overlapping cRFs (Figure modified from [69])

In conclusion, this simulation shows that the spatial range of nearly synchronized neurons (AF-size) at one level of processing can determine the cRF-size at the next level under conditions of Hebbian learning and biologically plausible modelling over a large range of parameters. Applied to FCOs and their cortical range of coherence in cat and monkey (3-5 mm in V1 [68]) reveals good correspondence in V1 AF- and V2 cRF-size. Thus, these results support the new hypothesis that the increasing cRF width in the visual hierarchy of processing is related to context processing on the basis of fast synchronization. This principle may well hold also for other sensory systems.

4.7   Additional Properties of FCOs

4.7.1   Frequency and amplitude of FCOs are highly variable

The oscillation frequencies and amplitudes of FCOs are probably not coding sensory features because they are highly variable. Variability of the FCOs' dominant frequency can be 10-20 Hz even if stimuli are constantly applied and do not evoke transient responses. Such variations in frequency are commonly found in all visual areas analyzed so far with respect to FCOs, including preparations at light anesthesia and attentive wakefulness ([4, 5,

33, 81] reviews in [9, 54]). Generally the frequency within a single oscillation spindle (typical duration 80-200 ms) is more constant than in successive oscillatory events. However, some systematic changes in average oscillation frequency are observed when changing the type of stimulation (Sec. 4.7.2).

4.7.2   Visual stimulation influences average oscillation frequency of FCOs

Even though the variability of FCO frequencies and amplitudes is high during stationary visual stimulation, they show some systematic dependency of their average values on stimulation. In general, with increasing *stimulation strength*, and hence, increasing cortical activation, increasing FCO amplitudes are accompanied by decreasing oscillation frequencies in population activities (LFP & MUA [82]). With stimuli activating neurons in V1 and V2 slightly above their maintained rate, incoherent broad band stochastic signals dominate, typically with some short and low amplitude FCOs of high frequency (> 70 Hz). These oscillatory epochs seldomly occur coherently across more than 1 mm cortex. At intermediate levels of activation, spatial domains of synchronized FCOs become on average larger, their population amplitudes increase while their frequencies decrease (50-60 Hz). Strong sustained activation results in high amplitudes of population activity at low frequencies (35-45 Hz) and the cortical domains of coherent FCOs became larger, about 5-7 mm in V1 and V2.

The *velocity and spatial extent* of a visual stimulus also influenced the average oscillation frequency of FCOs [4, 82, 84]. When visual stimuli move slower or are larger in size, FCOs in V1 and V2 of awake monkeys had on average lower oscillation frequencies and higher amplitudes (fig. 4.11). Increase of FCO frequency on stimulus velocity can be explained by a model of spiking neurons by reduced durations of excitation and inhibition periods in local circuits with increased velocity of a stimulus of constant width [83].

It has to be noted in this context that FCOs only occur at high amplitudes and probabilities with stimuli evoking sustained cortical activation, i.e., that do not evoke brisk transients in cortical activity. With a structured visual scene this means that retinal displacements and velocities should be slow, as with natural vision during ocular fixation.

We can explain the lower frequencies observed with larger stimuli by the longer average activation delays within the larger populations. However, lower frequencies may also be due to the increased activation levels with more extended stimuli leading to increased inhibition periods in the local populations [83, 85]. The influence of stimulus velocity on FCOs might have corresponding reasons. Slow velocities allow the formation of larger assemblies engaging in lower frequencies than faster movements [85].

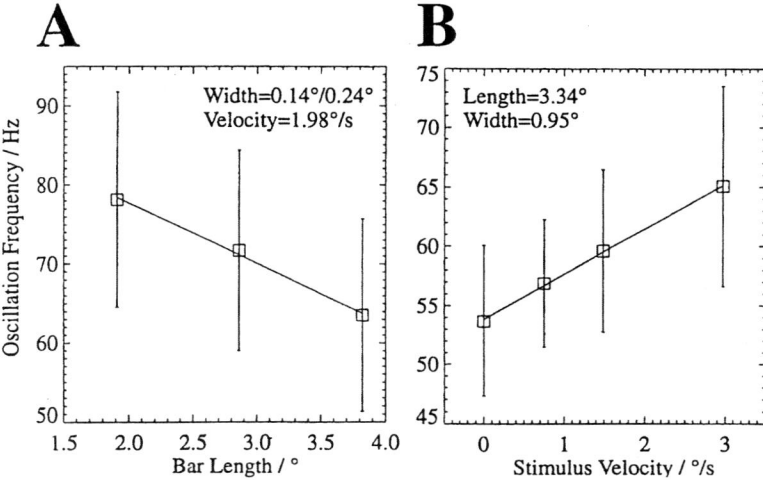

FIGURE 4.11. Changes in oscillation frequencies depended on stimulus size and velocity in the primary visual cortex of an awake monkey. The average oscillation frequencies shifted to lower values accompanied by higher amplitudes, A: if the length of a moving stimulus (light bar) was elongated), or if its width was broadened; B: if its velocity was decreased. Frequency changes with size were restricted to the spatial range in which significant values of coherence were observed in striate cortex (3-5 mm); (modified from Woelbern et al. 1994). (Figure modified from [84]).

4.7.3   FCOs and temporal segmentation

FCOs are chopped cortical activities of short local population bursts followed by inhibitory periods. Chopped cortical activities may provide short snapshots of image representations and thus may be understood as temporal segmentation of the continuous stream of visual information. Temporal segmentation may be necessary for clear percepts, because retinal images move permanently. Even under conditions of rest and ocular fixation microsaccades and drifts of the eyes are always present [86]. Cortical neurones, on the other hand, can persist in their responses over several hun-

dreds of milliseconds even to very short visual stimuli of 10-20 ms duration
(e.g. [87]). However, if a visual object with sharp contours evokes cortical
activations persisting while the retinal image of that figure shifts across
the retina the contour's perception might be smeared, because it activates
many other neurones with neighboring cRFs. Smearing might be prevented
if the continuously varying stream of signals is "chopped" into short "image
frames" that can perceptually be picket out or that are shifted in steps for
superposition.

Three modes with temporal interruptions may play a role in temporal
segmentation: 1) Single-event chopping in the representation range of a
visual object evoked by its sudden retinal displacement; 2) single-event
chopping of the entire scene by ocular saccades of regular and micro amplitudes; and 3) repetitive interruptions by synchronized FCOs. While (1)
causes segmentation of the jerking object against the remaining scene and
has perceptually a strong pop-out effect, (2) is not able to segment single
objects because large parts of a scene move at the same angular velocities
(but purging of the previous representation is possible by it). Finally, with
FCO chopping (3) optimal "frame rates" should be adapted to the rate of
retinal image displacement (velocity) and the retinal size of the object's
image. Faster framing would be desirable with higher movement velocities
and smaller sizes (according to the sampling theorem of signal theory e.g.
[88] which is certainly not fulfilled in visual cortex (compare Fig. 4.11).

If image-chopping by FCOs is perceptually relevant the temporal interval
of a single snapshot should correlate with the performance of observers for
the detection of spatial displacements and the maximal displacement for
observing sharp contours in moving retinal images. For this we can make
a simple calculation. In a typical recording from V1 of an awake monkey
a stimulus of $2°$ visual angle moving at $1°/s$ induces FCOs with dominant
frequencies of about 60 Hz. This corresponds to a retinal image displacement of one arc minute visual angle during a single activation-inhibition
cycle (17ms). An even higher precision in spatio-temporal representation
exists during periods of ocular fixation. With typical ocular drifts of $0.1°/s$
and FCO frequencies of 40 Hz [33, 85] image displacements are 9 arc seconds for a single oscillation cycle (25 ms "frame rate"). This is well in the
range of the highest resolution for line displacements, called hyperacuity
[89].

4.8   Stimulus-Locked Scene Segmentation

Phase-coupled FCOs are induced at high amplitudes, generally during
epochs of slow drift rates of retinal images, but typically during ocular
fixation. In this state cortical cells are sustainedly activated by thalamic
inputs. In contrast, fast transient components are primarily evoked by stim-

ulus epochs of fast changes in luminance, velocity or direction, typically for sudden object or eye movements. For the fast segmentation of a fast changing scene FCOs may be inappropriate as they occur with longer and more variable latencies in the visual cortex (50-100 ms [46]) than needed for fast segmentation tasks. However, transient stimuli do elicit transient phase locked responses in striate cortex appearing at considerably shorter latency (down to 30 ms) than stimulus-induced FCOs (> 50 ms) in V1. This suggests that the transient responses occurring synchronized over the representation of transient scene segments may take over segmentation coding from oscillations. If segmentation in such situations is dominated by stimulus-locked activity, such responses should suppress ongoing oscillations induced by previous more stable scenes in order to avoid confusion in fresh segmentations. This has indeed been found recently [30].

4.8.1 Suppression of FCOs by fast stimulus-locked activations

In visual cortical areas V1 and V2 of cats, ongoing oscillations are gradually reduced in amplitude and finally fully suppressed with increasing amplitudes of fast transient stimulus movements [30] (Fig. 4.12). Comparable measurements were recently made in the awake monkey [90]. Rapid changes of the retinal image due to μ-saccades or sudden changes in an object's contrast or position can immediately disturb or even suppress FCOs. The latter are, as in the cat, generated only during states of stable or slow moving retinal images.

The occurrence of strong stimulus-locked components in cortical responses is, per se, not a sufficient explanation for the observed partial or full suppression of FCOs. In principle, stimulus-locked responses and oscillations might superimpose and coexist independently without major interactions. Instead, we found strong suppression of FCOs by stimulus-locked responses (Fig. 4.12). What are probable explanations of this suppression?

(1) *Inhibition of oscillation circuits by transient afferent activities.* This possibility is supported by the finding of fast stimulus-locked components, being preferentially transmitted via the magnocellular (transient type) afferents, evoking particularly effective fast and long lasting inhibitory responses in the visual cortex (e.g. [91]).

(2) *Perturbation by pushing the oscillatory circuits out of phase.* From theories [92] and simulations of loosely coupled oscillators [21, 56, 57, 93, 94] it is known that oscillations can only be maintained, if strong out-of-phase perturbations are prevented. However, such out-of-phase disturbances were delivered to the cortex in the experiment of Kruse and Eckhorn [30] by the stimulus-locked afferent input occurring at random phases of FCOs. With intermediate levels of stimulus-locked activity, oscillations are only partially suppressed. This can also be explained by perturbations of oscillatory processes by random out-of-phase inputs. Theoretically one would expect a broadening and reduction of the dominant frequency peak of oscillations in the power

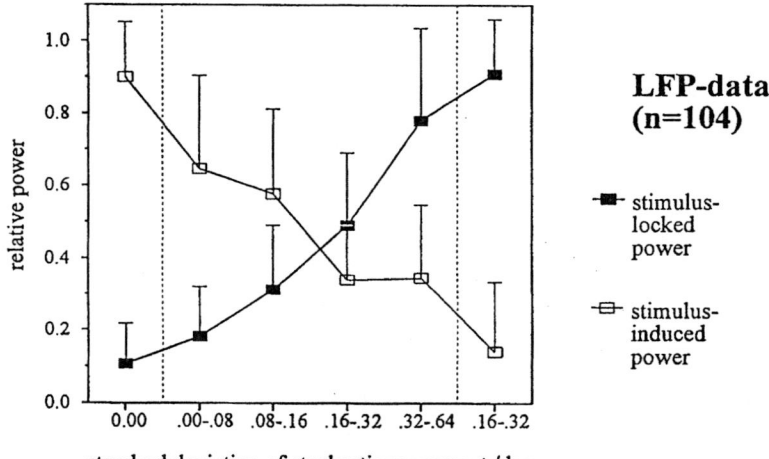

standard deviation of stochastic movement /deg

FIGURE 4.12. Perturbation of FCOs by stimulus-locked signals. Note stimulus-locked responses reduce the amplitudes of population FCOs. Local field potential recordings from cat visual cortex. Stimulus: grating, moving with a random component superimposed on a constant slow velocity ramp movement. Open symbols: average normalized response power of FCOs; filled symbols: normalized power of the stimulus-locked components. Amplitudes (standard deviation) of the random stimulus movement are indicated at the abscissa. Each value is the average of N = 104 single curves. Note that the leftmost values were obtained by applying only the slow constant velocity movement, and rightmost values were measured by exclusive stimulation with stochastic (jerky) movement during which the (average) spatial position of the grating was optimized on the cRFs for strongest stimulus-response coupling (modified from [30]).

spectrum due to an increased phase jitter and a reduced amplitude of local oscillatory processes. This has, indeed, been observed [30, 81].

(3) *Suppression of oscillations by transient reductions of membrane resistance due to strong stimulus locked activations.* If strong phase-locked input activates a considerable number of excitatory and inhibitory synapses of neurones contributing to the oscillations nearly simultaneously, this may transiently lead to massive reductions in membrane resistances of dendrites and somata [78, 95]. Consequently, the temporal and spatial decay constants become shorter and hence, extracellularly recorded MUA and LFPs will decrease in amplitude.

4.8.2 Time courses of stimulus-locked and stimulus-induced FCO-activity

Stimulus-locked responses are evoked in monkey striate cortex typically during a short epoch in response to the presentation (switch on) of a visual stimulus (or comparably, after execution of an ocular saccade). Fig. 4.13 A shows an increase of stimulus-locked LFP power in a time window (50-100 ms duration) centered around 100 ms after stimulus onset with distinct peaks at about 20 and 40 Hz and a rather broad-band increase (50−100 Hz). Even though power increase in the 50 − 100 Hz range is broad-band and of much lower amplitude than the 20 and 40 Hz components, its stimulus related modulation shows a peak near 80 Hz.

Stimulus-induced FCOs are typically not locked in their phases to stimulus events and therefore average out in stimulus-locked averaging. The appearance of stimulus-induced FCO-power is also related to stimulus onset (Fig. 4.13 B) but it increases its amplitude at a higher delay (plus about 50 ms) and less steeply compared to the stimulus-locked components so that the peak power (at about 40 Hz) is reached around 230 ms poststimulus onset. Despite this peak the FCO-power was broad-band containing relevant components up to 100 Hz. Stimulus-modulation (Fig. 4.13 bottom panels) of induced FCO-power starts around 80-100 ms post stimulus onset with a broad-band-characteristic and changes around 230 ms delay to a narrow-band oscillatory state with a sharp peak around 40 Hz.

Stimulus-induced versus stimulus-locked. The strongest modulation of stimulus-induced FCO-power appears about 100 ms later than the maximal modulation of the stimulus-locked FCO-power. However, both types of signals are temporally overlapping, indicative of a continuous transition from strong stimulus phase coupling to looser coupling and finally to activity not phase coupled to the stimulus onset at all. Fig. 4.13 also shows (lower panels A, B) that the stimulus-locked FCO-power is much stronger modulated by the stimulus (max. factor 7) than the stimulus-induced components. Both the identity of frequencies and temporal contiguity suggest that stimulus-locked and stimulus-induced oscillations are caused by the same type of process.

Low frequency LFP components of high power (about $60\mu V^2$) that are independent in their amplitudes and phases to stimulus onset are continuously present (Fig. 4.13 B upper panels), even though the monkey was awake. The stimulus-locked component at low frequency has only about half that power, is very short in duration, and narrowly centered around 10 Hz.

Internally generated FCOs and stimulus-locked activity may support feature binding in different visual situations. In conclusion, coherent FCOs are generated as long as the afferent visual activation remains sustained while coherent single event responses occur with transient retinal stimuli. As both components can appear phase coupled they may both play a role in scene segmentation. For natural vision we can argue that during phases of slowly changing retinal images (like ocular fixation and smooth pursuit) FCOs may support spatial feature grouping and prevent perceptual "smearing" by interrupting the flow of visual information repetitively. However, when a visual object suddenly changes its position, ongoing oscillations are immediately interrupted by a single excitation-inhibition cycle evoked by the object's displacement (these cycles are generally longer than those of

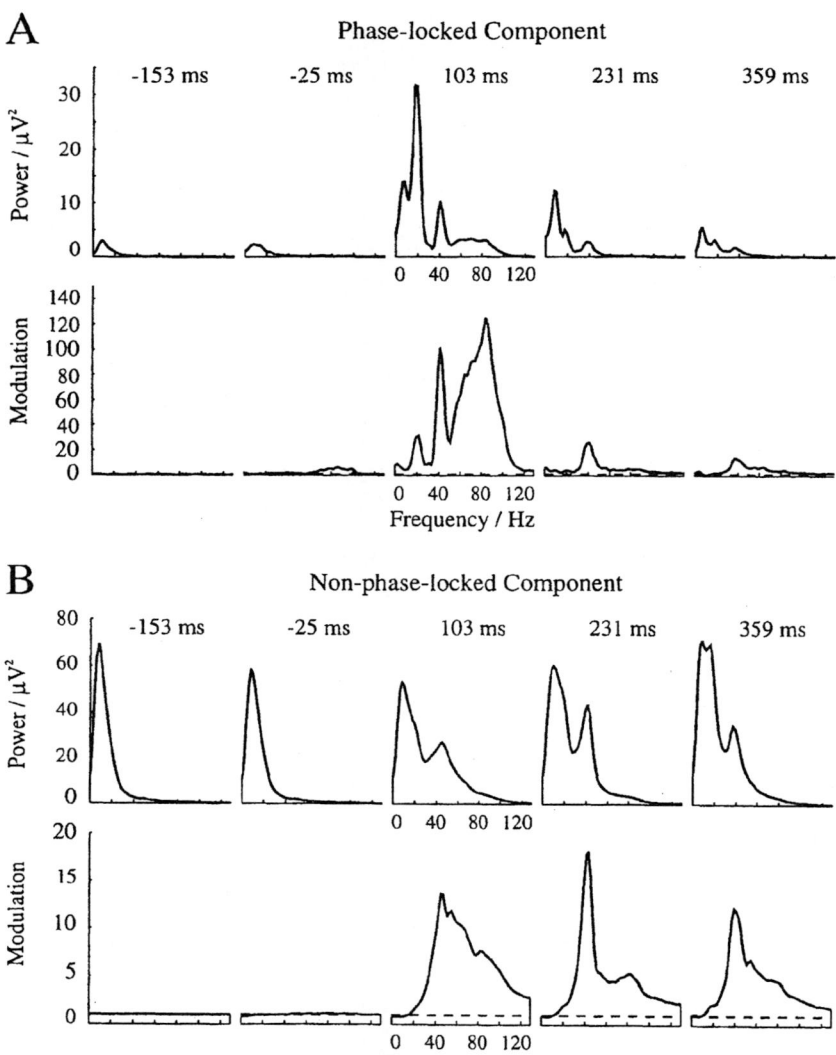

FIGURE 4.13. Spectra of V1 LFPs in successive time windows after stimulus onset show stimulus-locked enhancements in three distinct frequency ranges: $10 - 30$ Hz, $30 - 50$ Hz, and $50 - 100$ Hz. A: Phase-locked spectral components (shift-predictor), and B: power spectra with shift-predictor subtracted. Spectra in the bottom rows were divided by the mean prestimulus values. Data from 3 sessions with recordings from 6 or 7 electrodes simultaneously, total of 895 trials. Note the different scales in A and B (Fig. modified from [96]).

fast oscillations). Hence, the stimulus-locked cycle might act as reset signal and evokes a new, coarse segmentation of the object against the remaining scene.

4.9   Early Labeling of Visual Objects by FCO- or Rate-Coherence?

Preattentive scene segmentation is a prerequisite of labeling, detection and recognition of visual objects (according to criteria of Gestalt psychology [3, 43]). To my knowledge no experimental data have been published that are indicative of labeling scene segments uniquely as visual objects (in contrast to nonobjects and background) already in lower level cortical areas (V1 to V5). However, an indirect hint of early labeling is available by V1 neurons that represent object surfaces exhibiting coherent spike rate enhancement in comparison to a background that is composed of the same type of features (e.g. texture, color, movement, ocular disparity). This object-related rate enhancement emerges at considerably higher latencies (plus 100 ms) than the initial responses to stimulus onset [97, 108] and the appearance of FCOs. Following our working hypothesis we asked in a recent experiment whether phase coupling among FCOs also shows object specific behavior in addition to its already demonstrated potential in region labeling, and if so, whether this has similar or different implications for object labelling [63].

For this MUA and LFP were extracellularly recorded in V1 of a rhesus monkey performing a fixation task. Visual stimulation was by a stationary sinusoidal luminance grating, in which the object was defined by a shifted rectangular part of the grating (2°x4°; i.e., object and background were composed of the same features). One of the edges perpendicularly intersected the linearly aligned receptive fields (cRFs) of the 7 recording positions, or was aligned with all cRFs. The object was shown at two positions mirrored at this edge. We analyzed MUA and LFP amplitude and their time-resolved cross-correlation and coherence.

Object related modulation was extracted by calculating differences between responses to stimulation with and without the object. *Modulation specific components for the object's surface* were extracted by calculating differences between responses to object presentation in the two mirrored positions, where activity specific for the object's contour could be subtracted. Using a single set of data this experiment revealed contour and region specific effects of FCOs, which confirmed all types of results reported above and some additional new ones [63].

Since the contour of the object had sharp local luminance contrast, early stimulus-locked activation (50-120 ms after stimulus onset), and hence spike rate coherence, was strongest along the contour's representation. Object specific rate enhancement emerged at about 140 ms and disappeared at about 260 ms under our conditions. However, reduced FCO-coherence across the object's contour continued until the object disappeared (after about 1 s). Moreover, object specific rate enhancement only occurred within closed contours and decreased with increasing object size (confirming [97]), whereas decrease in FCO-coherence demanded a region boundary, irrespective of the closure of contours and the object's size.

The experiment by Gail and coworkers [63] allows comparison of several aspects of scene segmentation and object coding from a single set of data. Aspects of contour and region coding can be explained by transient stimulus-locked rate coherence at short delay and FCO-coherence across regions and along contours at slightly larger delays. In addition, region separation across the contour was indicated by markedly reduced coherence compared to the coherence within each of the neighboring regions.

In conclusion, FCOs were clearly correlated with aspects of scene segmentation but they did not yet reveal signs of object labeling. In contrast, late rate enhancement, also present in this experiment, emerged only with closed contours as is typical for visual objects. However, long-delay increase of spike rate is probably not related to preattentive mechanisms but is more likely due to top down influences (including mechanisms of attention [98]). These potential contributions of V1 to scene segmentation are obviously not sufficient for object coding and object specification, particularly not for complex scenes containing multiple objects. Probably additional visual structures and possibly other mechanisms are involved.

4.10   Appendix

"Synchronization" in the present context

Synchronization and Postsynaptic Integration Time. In the present context the term *synchronization* is used in relation to the temporal properties of neurones, and in particular to postsynaptic *integration time*. In a first approximation the integration time can be estimated as the half-height duration of an average excitatory postsynaptic potential. If presynaptic action potentials occur on the synaptic inputs of a given neuron within its integration time, we will call them *synchronized*. Integration times can differ in different neurones, in a single neuron at different synapses, and they can change over time. The latter effects depend on postsynaptic membrane properties and on the temporal patterns of input spikes [78, 95, 99]. In cortical neurones integration times can span a broad range of 2 to 100 milliseconds. In the more peripheral visual cortex areas (V1-V5) of awake animals dominant integration times are estimated to 5-15 ms.

Modes of Synchronization: Stimulus-locked and Stimulus-induced. With respect to the processing of behavioral output a synchronization interval of 5-15 ms is rather short and we therefore use the term "fast" for it. Accordingly, neural signals which are correlated positively in this range are called "fast synchronized". In this sense, rhythmic signals with half cycle duration of about 5-15 ms (100 − 30 Hz) are termed synchronized fast cortical oscillations (FCOs).

While the term *"stimulus-locked"* is used for cortical responses being phase-locked to stimulus events (evaluated by stimulus-locked averaging across responses to identical stimulus repetitions) *"stimulus-induced"* responses do not show phase-lock of their components to stimulus events (i.e., they die out in stimulus-locked averages). FCOs are typically stimulus-induced signals.

Activation by Synchronized Signals. The spike encoder of a neuron, including its threshold and refractory dynamics, favors synchronized compared to unsynchronized inputs because the former produce higher and more steeply depolarized membrane excursions at equal numbers of input spikes. Output spikes will be generated at a higher probability if a neuron's excitatory inputs are activated together within its integration time. If input spike patterns are precisely coinciding, a spike encoder with a short refractory period and a high resting threshold can extract this pattern and send it to other targets. In contrast, temporally dispersed input spike patterns, including statistically independent ones with the same number of spikes, will result in lower momentary excursions of membrane potentials at the spike encoder and hence, in lower probabilities of spike firing [13, 101, 102, 103].

Sensitive Detection of Signal Coherence in Local Population Activities (MUA and LFP). Our empirical evidence on the occurrence of stimulus related fast cortical oscillations (FCOs) is based on extracellular recordings. To detect states of weak synchronization it is experimentally advantageous to record local population activities (multiple unit spike activity (MUA) and local slow wave field potentials (LFP; 1 − 150 Hz) via the same microelectrodes instead of restricting to single unit spike trains. The reason for this is that MUA and LFP comprise already the synchronized components of local populations (e.g.[104]). In particular, LFP is a local weighted average of the dendrosomatic postsynaptic signals, mainly reflecting the synchronized components at the inputs of the population within about 0.5 mm of the electrode tip [105]. MUA, on the other hand, comprises in its amplitudes the simultaneity of spikes occurring at the outputs of a local population within about 0.05 mm of an electrode tip [106, 107]. Due to superposition in the extracellular fluid, MUA and LFP amplitudes are the higher the more precise the contributing neural signals are synchronized. This is the main reason why MUA and LFP are more sensitive probes for the investigation of synchronizing effects in neural populations than single unit spike patterns. It has to be noted that synchronized fast oscillations occur as a population signal, generally weakly coupled to the spike trains of single neurones.

Synchronization Quantified by Cross-correlation, Cross-spectrum and

Coherence Function. In the work refereed in the present paper, the degree of synchronization among two neural signals was determined by calculating their *cross-correlation* function [88]. In nonrhythmic signals, generally the area under the cross-correlogram's main peak was taken as the measure of synchrony [13]. In the case of rhythmic signals, cross-correlation was calculated in the spectral domain. This revealed the *cross-spectrum* or the *coherence function* (normalized to the product of the entire powers of the two components, or normalized to the product of the two components for each single frequency, respectively [33, 88]). The coherence can therefore measure the degree of synchronization at each frequency component of the signals selectively.

Acknowledgement

Thanks to present and previous members of the Marburg Group of Neuro-Physics (credited in the figure captions), particularly to Prof. R. Bauer for expert help in conducting the experiments, and U. Thomas and W. Gerber for their excellent technical support. Financial support by the German Research Council is also greatly acknowledged (Projects DFG Ec53/7 and Ro529/12 to R.E.).

4.11 References

[1] J.G. Nicholls, A.R. Martin, B.G. Wallace: *From Neuron to Brain* (Sinauer, Sunderland, Massachusetts 1994).

[2] E.R. Kandel, J.H. Schwartz: *Principles of Neural Science* (Elsevier, New York Amsterdam, Oxford 1985).

[3] M. Wertheimer: Untersuchungen zur Lehre von der Gestalt: II. Psychologische Forschung 4,301 (1923).

[4] R. Eckhorn, R. Bauer, W. Jordan, M. Brosch, W. Kruse, M. Munk, H.J. Reitboeck: *Biol. Cybern.* 60, 121 (1988).

[5] C.M. Gray, P. König, A.K. Engel, W. Singer: *Nature* 338, 334 (1989).

[6] H. Neven, and A. Aertsen: *Biol. Cybern.* 67, 309 (1992).

[7] H. Barlow: in *Brain Theory Biological Basis and Computational Principles*, edited by A. Aertsen, and V. Braitenberg (Elsevier, North Holland 1996) p. 261.

[8] R. Eckhorn: in *Progress in Brain Research* 107. *The Self-Organizing Brain: From Growth Cones to Functional Networks*, edited by J. van Pelt, M.A. Corner, H.B.M. Uylings, F.H. Lopes da Silva (Elsevier, Amsterdam New York, 1994) p. 405

[9] C.M. Gray: *J. Comput. Neurosci.* 1, 11 (1994).

[10] A. Aertsen, M.A. Arndt: Current Opinion in Neurobiology 3, 586 (1993)

[11] A.M. Aertsen, G.L. Gerstein, M.K. Habib, G. Palm: *J. Neurophysiol.* 61, 900 (1989).

[12] K.H. Boven, and A. Aertsen: in *Parallel Processing in Neural Systems and Computers*, edited by R. Eckmiller, G. Hartmann, and G. Hauske (Elsevier, North Holland 1990), p. 53.

[13] M. Abeles: *Corticonics. Neural circuits of the cerebral cortex.* Cambridge Univ. Press, Cambridge, New York, Melbourne, Sydney. (1991).

[14] L. Martignon, H. von Hasseln, S. Gruen, A. Aertsen, and G. Palm: *Biol. Cybernetics* 73, 69 (1995).

[15] E. Vaadia, I. Haalman, M. Abeles, H. Bergman, Y. Prut, H. Slovin, and A. Aertsen. *Nature* 373, 515 (1995).

[16] A. Riehle, S. Gruen, M. Diesmann, and A. Aertsen: *Science* 278, 1950 (1997).

[17] R. Kempter, W. Gerstner, J.L. van Hemmen, H. Wagner: *Neur. Comput.* 10, 1987 (1998).

[18] O. Bernander, C. Koch, M. Usher: *Neural Comput.* 6, 622 (1994).

[19] H. Markram, and M. Tsodyks: *Nature* 382, 807 (1996)

[20] P. König, A.K. Engel, W. Singer: *Proc. Natl. Acad. Sci. USA* 92, 290 (1995b).

[21] R. Eckhorn, H.J. Reitboeck, M. Arndt, P. Dicke: *Neur. Comput.* 2, 293-306 (1990).

[22] R. Eckhorn: in *Proc Artificial Intelligence, Dynamic Perception*, edited by S. Prosch, H. Ritter (Infix, Sankt Augustin 1998) p.127.

[23] C.Y. Li, W. Li: *Vision Res.* 34, 2337 (1994).

[24] U. Polat, D. Sagi: *Vision Res.* 33, 993-999 (1993).

[25] D.J. Field, A. Hayes, and R.F. Hess: *Vision. Res.* 33, 173 (1993).

[26] Y. Fregnac, V. Bringuier, F. Chavane: *J. Physiol.* (Paris) 90, 367 (1996).

[27] E. Bienenstock, R. Doursat: in *Representation of Vision*, edited by A. Gorea et al. (Cambridge Univ. Press 1991) p. 47.

[28] W.A. Phillips, W. Singer: *Behavioral Brain Sci.* 20, 657 (1997).

[29] L. Spillmann, W.H. Ehrenstein: in *Comprehensive Human Physiology*, Vol 1, edited by R. Greger, and U. Windhorst (Springer, Berlin, Heidelberg 1996) p. 861.

[30] W. Kruse, R. Eckhorn: *Proc. Natl. Acad. Sci. USA* 93, 6112 (1996).

[31] E. Juergens, R. Eckhorn: *Biol. Cybern.* 76, 217 (1997).

[32] R. Eckhorn, A. Obermueller: *Exp. Brain Res.* 95,177 (1993).

[33] A. Frien, R. Eckhorn, R. Bauer, T. Woelbern, H. Kehr: *NeuroReport* 5, 2273 (1994).

[34] A.P. Georgopoulos: *Quantit. Biol.* 55, 849 (1990).

[35] E. Zohary, P. Hillman, and S. Hochstein: *Biol. Cybernetics* 62, 475 (1990).

[36] M.N. Shadlen, and W.T. Newsome: *Current Opinion Neurobiol.* 4, 569 (1994).

[37] A. Frien, R. Eckhorn, R. Bauer, T. Woelbern, A. Gabriel: *Europ. J. Neurosci.* 12, 1453 (2000).

[38] C. Spengler: Masters Thesis, Philipps-University, supervised by R. Eckhorn (1996).

[39] H. Sompolinsky, and R. Shapley:*Current Opinion Neurobiol.* 7, 514 (1997).

[40] C.M. Gray, D.A. McCormick: *Science* 274, 109 (1996).

[41] A.M. Sillito, H.E. Jones, G.L. Gerstein, and D.C. West, :*Nature* 369, 479 (1994).

[42] A. Frien, R. Eckhorn: *Europ. J. Neurosci.* 12, 1466 (2000).

[43] I. Rock, and S. Palmer: *Sci. Amer.* 263, 48 (1990).

[44] A.F. Kramer, A. Jacobson:*Percept. Psychophys.* 50, 267 (1991).

[45] U. Polat, and A.M. Norcia, A.M.: *Vision Res.* 36, 2099 (1996).

[46] C.M. Gray, A.K. Engel, P. König, W. Singer: *Europ. J. Neurosci.* 2,607 (1990).

[47] C.M. Gray, G. Viana Di Prisco: *J. Neurosci.* 17, 3239 (1997).

[48] A.K. Engel, P. König, C.M. Gray, W. Singer: *Europ. J. Neurosci.* 2, 588 (1990).

[49] A.K. Engel, P. König, A.K. Kreiter, W. Singer: *Science* 252, 1177 (1991ε).

[50] A.K. Engel, A.K. Kreiter, P. König, W. Singer: *Proc. Natl. Acad. Sci. USA* 88, 6048 (1991b).

[51] A.K. Kreiter, and W. Singer: *Europ. J. Neurosci.* 4, 369 (1992).

[52] P. König, A.K. Engel, P.R. Roelfsema, W. Singer: *Neural Comp.* 7, 469 (1995a).

[53] M.H.J. Munk, P.R. Roelfsema, P. König, A.K. Engel, W. Singer: *Science* 272, 271 (1996).

[54] A.K. Kreiter, W. Singer: *J. Neurophysiol.* 16, 2381 (1996).

[55] M. Brosch, R. Bauer, R. Eckhorn: *Cerebral Cortex* 7, 70 (1997).

[56] H. Sompolinsky, D. Golomb, D. Kleinfeld: *Proc. Natl. Acad. Sci. USA* 87, 7200 (1990).

[57] T.B. Schillen, P. König: *Biol. Cybern.* 70, 397 (1994).

[58] R. Ritz, W. Gerstner, U. Fuentes, J.L. van Hemmen: *Biol. Cybernetics* 71, 349 (1994a).

[59] R. Ritz, W. Gerstner, J.L. van Hemmen: in *Models of Neural Networks* II, edited by E. Domany, J.L. van Hemmen, K. Schulten (Springer Verlag, New York 1994b) p. 177.

[60] U. Schott: Masters Thesis, Philipps University, Dept Neurophysics, Marburg, supervised by R. Eckhorn (1995).

[61] M. Stoecker, H.J. Reitboeck, R. Eckhorn: *Neurocomputing* 11, 123 (1996).

[62] A. Guettler, R. Eckhorn, E. Juergens, A. Frien: in Göttingen Neurobiology Report edited by N. Elsner, H. Wässle (Thieme, Stuttgart, New York 1997) p. 551.

[63] A. Gail, H.J. Brinksmeyer, R. Eckhorn: *Cerebral Cortex* 10, 840 (2000).

[64] W. Jordan: Dissertation, Philipps-University Marburg supervised by R. Eckhorn (1989).

[65] R. Bauer, M. Brosch, R. Eckhorn: *Brain Res.* 669, 291 (1995).

[66] M. Brosch, R. Bauer, R. Eckhorn: *Europ. J. Neurosci.* 7, 86(1995).

[67] A. Frien, R. Eckhorn, H.J. Reitboeck: *Soc. Neurosci. Abstr.* 22, 255.5 (1996).

[68] E. Juergens, R. Eckhorn, A. Frien, T. Woelbern: *in Brain and Evolution*, edited by N. Elsner, and H.-U. Schnitzler (Thieme, Berlin, New York 1996) p. 418.

[69] R. Eckhorn, T. Schanze, M. Brosch, W. Salem, R. Bauer: in *Induced Rhythms in the Brain*. Brain Dynamics Series, edited by E. Basar, T.H. Bullock (Birkhäuser, Boston, Basel, Berlin 1992) p. 47.

[70] H.A. Swadlow: *Brain Res. Reviews* 6, 1 (1983).

[71] J.I. Nelson, P.A. Salin, M.H.J. Munk, M. Arzi, J. Bullier: *Visual Neurosci.* 9, 21 (1992).

[72] L.G. Nowak, M.H.J. Munk, P. Girard, J. Bullier: *Vis. Neurosci.* 12, 371 (1995b).

[73] J.I. Nelson, P.A. Salin, M.H.J. Munk, M. Arzi, J. Bullier: *Visual Neurosci.* 9, 21 (1992).

[74] L.G. Nowak, M.H.J. Munk, J.I. Nelson, A.C. James, J. Bullier:*J. Neurophysiol.* 74, 2379 (1995a).

[75] M. Steriade: *Current Opinion Neurobiol.* 3, 619 (1993).

[76] M. Steriade, F. Amzica, D. Contreras: *J. Neurosci.* 16, 392 (1996).

[77] V. Braitenberg, A. Schüz: *Anatomy of the Cortex. Statistics and Geometry* (Springer, Berlin 1991).

[78] H. Agmon-Snir, I. Segev: *J. Neurophysiol.* 70 (1993).

[79] A. Gabriel, R. Eckhorn: *Göttingen Neurobiology Report* 1999, Thieme, Stuttgart, New York, p 489 (1999).

[80] M. Saam, R. Eckhorn: in *New Neuroethology on the Move*, edited by N. Elsner, R. Wehner (Thieme 1998) p.767.

[81] R. Eckhorn, T. Schanze: in *Self-Organization, Emerging Properties and Learning*, edited by A. Babloyantz (Plenum Press, New York 1991) p. 63.

[82] R. Eckhorn, A. Frien: in *Brain Processes, Theories and Models: An International Conference in Honor of W.S. McCulloch 25 Years After His Death*, edited by R. Moreno-Diaz and J. Mira-Mira (MIT-Press, Cambridge MA 1995) p. 381

[83] T. Wennekers, M. Erb, G. Palm, R. Eckhorn: *Europ. J. Neurosci.* Suppl. 7, 66.03 (1994).

[84] Woelbern T, Frien A, Eckhorn R, Bauer R, Kehr H: in *Proceedings of the 22nd Göttingen Neurobiology Meeting*, edited by N. Elsner, H. Breer (Thieme, Stuttgart New York 1994) p. 518.

[85] R. Eckhorn: *IEEE Neural Networks* 10, 464 (1999).

[86] M. Alpern: in *Handbook of Sensory Physiology* Vol VII – 4, Visual Psychophysics, edited by D. Jameson, L.M. Hurvich (1972).

[87] E.T. Rolls, M.U. Tovee: *Proc. Roy. Soc. Lond.* B 257, 9 (1994).

[88] L. Padulo, M.A. Arbib: *System Theory* (Saunders, New York 1974).

[89] G. Westheimer: *Progress Sens. Physiol.* 1, 1 (1981).

[90] A. Guettler, R. Eckhorn, A. Frien, T. Woelbern: in *Brain and Evolution*, edited by N. Elsner, and H.-U. Schnitzler (Thieme, Berlin, New York 1996) p. 421.

[91] R.J. Douglas, K.A.C. Martin: *J. Physiol.* 440, 735 (1991).

[92] Y. Kuramoto: *Physica D* 50, 15 (1991).

[93] H.G. Schuster, P. Wagner: *Biol. Cybernetics* 64, 72 (1990).

[94] W. Gerstner, R. Ritz, J.L. van Hemmen: *Biol. Cybernetics* 68, 363 (1993).

[95] M. Nelson:*Neural Computation* 6, 242 (1994).

[96] E. Juergens, A. Guettler, R. Eckhorn: *Exp. Brain Res.* 129, 247 (1999).

[97] V.A.F. Lamme: *J. Neurosci.* 15, 1605 (1995).

[98] C. van der Togt, V.A.F. Lamme, H. Spekreijse: *Europ. J. Neurosci.*, 10, 1490 (1998).

[99] M. Häusser, A. Roth: *J. Neurosci.* 17, 7606 (1997).

[100] K. Fox, N. Daw: (1992) *Neural Computation* 4, 59 (1992).

[101] P. König, A.K. Engel, W. Singer: *Trends Neurosci.* 19, 130 (1996).

[102] M. Volgushev, M. Chistiakova, W. Singer: *Neurosci.* 83, 15 (1998).

[103] E.D. Lumer, G.M. Edelman, and G. Tononi: *Cerebral Cortex* 7, 228 (1997).

[104] R. Eckhorn: in *Information Processing in the Cortex, Experiments and Theory*, edited by A. Aertsen, V. Braitenberg (Springer-Verlag, Berlin, Heidelberg New York 1992) p. 385.

[105] U. Mitzdorf: *Int. J. Neurosci.* 33, 33 (1987).

[106] A.D. Legatt, J. Arezzo, H.G. Vaughan, Jr.: *J. Neurosci. Meth.* 2, 203 (1980).

[107] C.M. Gray, P.E. Maldonado, M. Wilson, B. McNaughton: *J. Neurosci. Meth.* 63, 43 (1995).

[108] K. Zipser, V.A.F. Lamme: *J. Neurosci.* 16, 7376 (1996).

[109] T. Wennekers, G. Palm G: in *Time and the Brain*, edited by R. Miller, Gordon & Breach, Lausanne, p. 202 (2000).

[110] R. Eckhorn, A. Frien, R. Bauer, T. Woelbern, H. Kehr: *NeuroReport* 4, 243 (1993).

[111] M. Saam, R. Eckhorn: *Biol. Cybern.* 83, L1 (2000).

[112] B. Jagadeesh, C.M. Gray, D. Ferster: *Science* 257, 552 (1992).

[113] M.H.J. Munk, L.G. Nowak, G. Chouvet, J.I. Nelson, J. Bullier: *Eur. J. Neurosci. Suppl.* 5, 21 (1992).

[114] V.N. Murthy, E.E. Fetz:*Neural Computation* 6, 1111 (1994).

5

Figure-Ground Segregation and Brightness Perception at Illusory Contours: A Neuronal Model

E. Peterhans, R. van der Zwan, B. Heider, and F. Heitger

ABSTRACT It has been shown in animals and man that illusory contours are represented at an early stage of visual processing. Animal studies revealed that neurons which signaled illusory contours usually also responded to contrast borders (bars, edges), and that the orientations of these contours are represented in similar cortical maps. In humans, illusory contour representations have been found at a comparable level of processing. Further, evidence of perception suggests that illusory contours often coincide with occluding contours and that mechanisms segregating figure and ground at such contours are also implemented at an early stage of processing. We studied this question in the visual cortex of the alert monkey by recording the responses of single neurons in stimulus conditions which defined illusory contours and the associated step in depth on the basis of occlusion cues (light and dark line-ends, or corners). In area V2, we found neurons sensitive to the figure-ground direction that human observers perceive at such contours. Most neurons showed this sensitivity independent of the contrast polarity that the stimuli induced at the contour, the remainder preferred a certain combination of figure-ground direction and contrast polarity. We explain these results in terms of a computational model using end-stopped operators for the detection of occlusion cues. In computer simulations we show that this model reproduces the figure-ground direction and the contrast polarity that human observers perceive at illusory (occluding) contours.

5.1 Introduction

Visual perception requires a stable segregation of figure and ground, especially in vision of cluttered scenes where near objects can occlude those further away. The definition of object borders (contours) from the retinal image of such scenes is often difficult because they can be defined by a number of cues, most often by discontinuities of brightness, color, texture, motion, or binocular disparity. Furthermore, these definitions change depending on scene illumination and body movements of the observer. The visual system includes several mechanisms to stabilize perception in such situations. For example, the perception of contours of waxing and waning luminance contrast are stabilized by filling in illusory contours at sites of fading contrast (see [28, 46, 53]). Illusory contours often coincide with occluding borders [4], and evidence of perception suggests that they are inferred from the contrast pattern produced by spatial occlusion. Such patterns include interposition or occlusion cues which terminate at the occluding borders (line-ends, corners, and different types of junctions). Examples of artificial figures which illustrate perception are shown in Fig. 5.1. The left figure pair induces the perception of a dark (A), or bright (B), triangle which appears to occlude six striped objects in the background. The right figure pair, instead, produces the impression of a triangular window cut out of a dark (A), or bright (B), surface with the striped objects seen in the background. It is the spatial alignment of the occlusion cues (line-ends) which determines the orientation of the illusory contours. The direction of these cues indicates the location (left or right) of the occluding surface relative to the contour, and their contrast polarity the overall step in luminance perceived between the occluding surface and the background.

Perceptual studies suggest that figure-ground segregation at such contours occurs early in visual processing (striate or prestriate cortex), at least before the form of objects and their general attributes are defined [48, 39, 38]. However, little is known about the neuronal mechanisms involved. Cortical representations of illusory contours per se have been found both in animal and man. Hirsch et al.[24] reported representations of illusory contours in human visual cortex (area V2) using functional magnetic resonance imaging (fMRI) technique, and Fftyche and Zeki [8] confirmed these results using positron emission tomography (PET) technique. Methods of single cell physiology revealed corresponding representations in cat and monkey cortex [57, 49, 43, 56, 17]. Further, Sheth et al. [52] revealed regular orientation maps of such contours in cat visual cortex, just as has been reported for solid contrast borders [16]. They also showed that the proportion of neurons sensitive to illusory contour stimuli increased from striate to prestriate cortex, a finding presumably related to the different proportions of simple and complex cells found in these cortical areas [34].

However, neural mechanisms defining the depth order that human observers perceive at such contours are virtually unknown, as are mechanisms explaining the brightness illusion associated with the occluding (illusory) surfaces (see Fig. 5.1). Baumann et al. [2] studied these questions in the visual cortex of the awake monkey and found neuronal representations of the depth order at illusory contours mainly in prestriate cortex. Further,

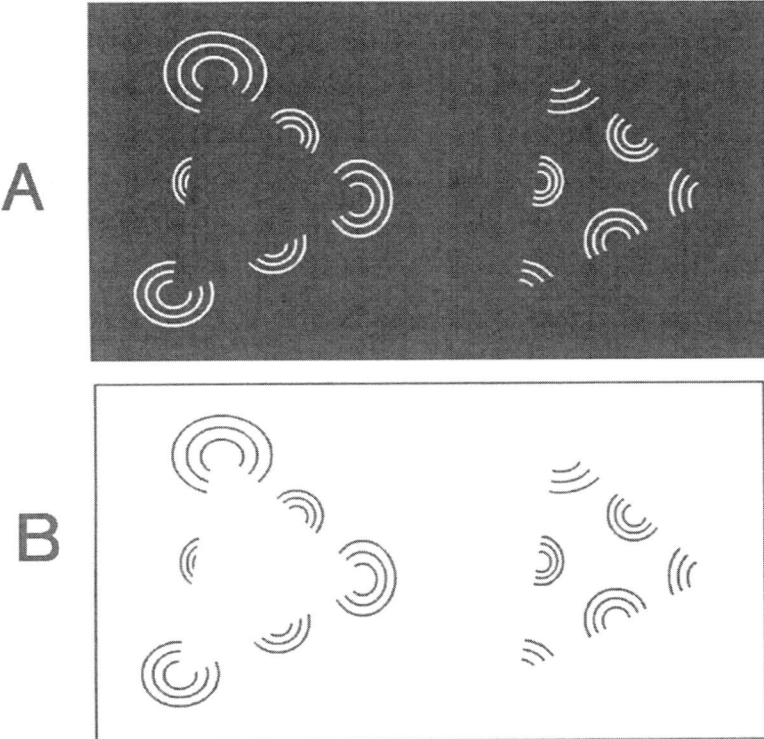

FIGURE 5.1. Artificial figures which mimic situations of spatial occlusion. (**A**) Light line-ends (occlusion cues) induce the perception of a dark triangle which appears to occlude six striped disks in the background (left panel), or a window cut out of a dark surface with the disks partially visible in the background (right panel). (**B**) Same figures but opposite contrast polarity. Figures modified with permission from [36].

they reported that this representation also included neurons which indicated the apparent step in luminance associated with such contours. Here,
we review these findings and provide an explanation in terms of a model
which uses end-stopped operators for the detection of occlusion cues.

5.2   Methods

The physiological experiments were done in awake rhesus monkeys (*Macaca
mulatta*) that were trained on a visual fixation task which reinforced foveal
viewing. When the animals reached a performance rate of $\geq 85\%$, they were
prepared for single cell recording. In successive operations, a head-holder
and two recording chambers (one over each hemisphere) were mounted to
the skull under general anesthesia and aseptic conditions. We studied the
responses of single neurons while the animal fixated its gaze on the visual
target. The receptive field of each neuron was first defined by determining
the response field for a bar or an edge defined by luminance contrast. In the
figures of the present review, these response fields are indicated by ellipses.
Subsequently, we studied as many neurons as possible with occluding contour stimuli as shown in Figs. 5.3 and 5.4 (A-D). Typically, the contours
were moved back and forth over the response field of the neuron under
study. (Note that the position of the contours in Figs. 5.3 and 5.4 indicate
the beginning of the forward sweep.) In human observers, these stimuli induced the perception of an opaque, dark (A, C) or light (B, D) rectangle
which appeared to occlude a line-grating of opposite contrast polarity. In
some neurons we used static stimuli with the contours kept stationary in
the center of the receptive field. All contours were presented at the neurons'
preferred orientation. For further details of our physiological methods see
[2].
    The computational model consisted of a series of operators which mim-
icked as close as possible the functions of different types of cortical neuron. It included operators for the detection of contrast borders (S- and
C-operators), occlusion cues (ES1- and ES2-operators), and illusory contours (Grouping 1 and 2). (For a detailed description of our computational
implementations see [22, 23]).

5.3   Results

5.3.1   Neurophysiology

Neuronal sensitivity to illusory contours

A prerequisite of cortical representations of figure-ground directions that
human observers perceive at illusory contours are neurons which are se-

lective for the orientation of such contours and for one of two possible figure-ground directions. Neurons sensitive to illusory contour stimuli were first reported about 15 years ago in cat and monkey visual cortex, and details about their physiological properties and anatomical location were reported in a number of studies thereafter (see Introduction). Hence, only basic properties shall be reviewed here. Neurons which show sensitivity to illusory contour stimuli have been shown to be selective for the orientation of illusory contours as for solid contrast borders (light and dark bars or edges). This suggests that these neurons contribute to a mechanism which initiates the generalization of contours in a perceptual sense – object borders are perceived independently of the stimulus parameters by which they are defined in the retinal image. Such a mechanism is important in everyday life when figure-ground relations change continuously due to changes of illumination, or self motion of the observer. Typical examples of the responses of such a neuron are shown in Fig. 5.2. It was similarly selective for the orientation of a light bar (solid line) and a contour between abutting gratings (dashed line), and the responses to solid contrast borders were slightly stronger than the responses to the illusory contour stimulus. This result and other properties of these neurons, for example the required number and spacing of lines, show a good correlation with human perception (see [54]). However, with regard to figure-ground direction, the abutting-grating stimulus is ambivalent – both the left or the right half of the stimulus can be perceived as foreground or background, and with similar probability. Hence, it is unknown whether such neurons can also be selective for the perceived figure-ground direction at such contours. A study which investigated this question became recently available [2].

Neuronal sensitivity to occluding contours

In this study, Baumann et al. [2] recorded the responses of a total of 182 orientation selective neurons in areas V1 (N = 66) and V2 (N = 116) in stimulus conditions which induced the perception of occluding (illusory) contours in human observers (see Figs. 5.3 and 5.4 for examples). All neurons included in this study preferred long contours; none had an end-stopped receptive field. In the following, we summarize these results in the light of a neuronal model detecting figure-ground direction and contrast polarity at occluding (illusory) contours. For a full account of the physiological properties of these neurons and for a comparison between cortical areas V1 and V2, see [2].

Neurons not sensitive to occluding contours. In both cortical areas these authors found neurons which were only weakly activated by occluding contour stimuli, or failed to respond to such stimuli (V1, 18/66 or 27%; V2, 10/116 or 8%). Others, were unselective giving similar responses to all four types of occluding contour stimuli (V1, 22/66 or 33%; V2, 51/116 or 44%). The remainder were selective for the overall contrast polarity at such contours (V1, 13/66 or 20%; V2, 23/116 or 20%). They gave similar responses to

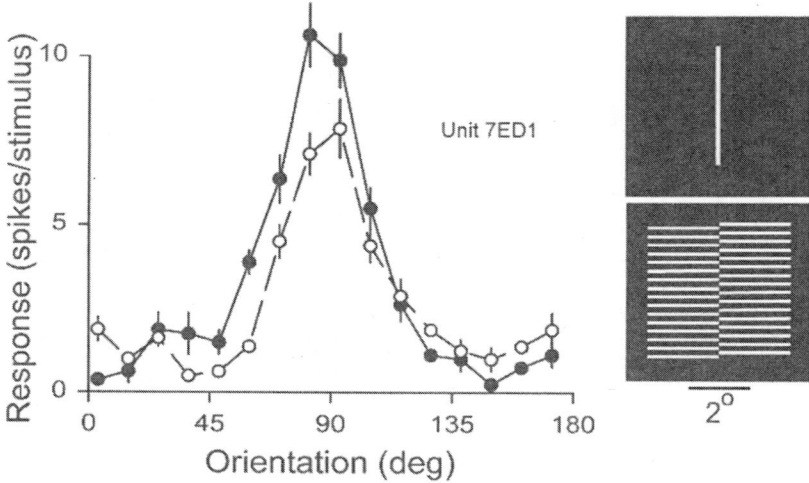

FIGURE 5.2. Neuronal sensitivity to illusory contours. Responses of a neuron which showed similar orientation selectivity for a contour between abutting gratings and for a solid contrast border (light bar). As often found in these neurons, the bar responses (filled circles) were slightly stronger than the responses to the illusory contour stimulus (open circles). For further properties of these neurons, see [56].

occluding contour stimuli and to solid contrast borders of the same contrast polarity, either light/dark, or dark/light.

Neurons sensitive to figure-ground direction at occluding contours. However, a distinct group of neurons recorded in area V2 (25/116 or 22%) actually did show sensitivity to occluding contours. These neurons were selective for the figure-ground direction that human observers perceive at such contours and showed this selectivity independent of contrast polarity. This type of neuron was not found in area V1. The responses of such a neuron are shown in Fig. 5.3. It preferred oblique orientations as indicated in the stimulus inset by the orientation of the bar response field (ellipse). Of the occluding contour stimuli it preferred the lower stimulus pair (C, D) which had the occluding surface to the right and above the contour. It gave much weaker responses to the upper stimulus pair (A, B) which had the opposite figure-ground arrangement. This selectivity for one of two possible figure-ground directions was independent of the overall contrast polarity at these contours: The preferred contours (C and D) actually had opposite contrast polarity (C, light/dark; D, dark/light). This result was confirmed by the responses to the corresponding solid edges (same stimulus configurations, but with a uniform surface instead of a line-grating). All edges evoked similar responses (E-H). Note also that the responses to occluding contours (C, D) were much stronger than the corresponding responses to solid edges

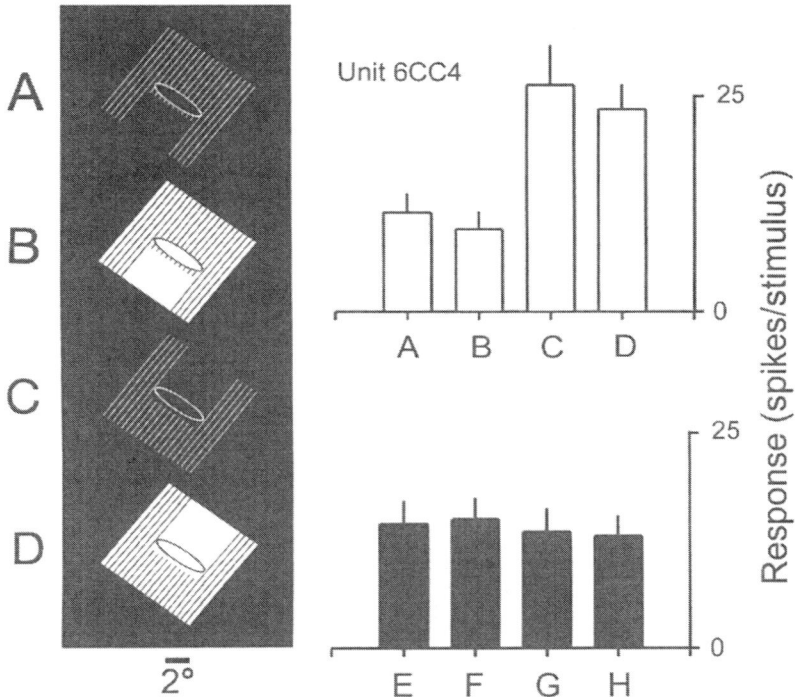

FIGURE 5.3. Neuronal sensitivity to figure-ground direction independent of contrast polarity. Stimulus insets show two pairs of occluding contour stimuli (A, B and C, D), presented at the neuron's preferred orientation. The histograms on the right show the responses to these stimuli (A-D, open bars), and to the corresponding solid edges (E-H, filled bars). One can see that the neuron preferred contours with the occluding surface above and to the right (C, D), and that it gave much weaker responses to contours with the opposite figure-ground arrangement (A, B). This selectivity was independent of contrast polarity: Note that the preferred contours (C, D) had opposite contrast polarity and that all solid edges evoked similar responses (E-H). The contours were moved back and forth over the neuron's response field as mapped with a light bar (ellipse). (For further details see [2]).

(G, H), although the mean luminance contrast at the occluding contours was only about half of that of the solid edges. This suggests that these neurons receive two kinds of input, one producing responses to contrast borders, the other signaling illusory (occluding) borders (see below).

Neurons sensitive to a certain combination of figure-ground direction and contrast polarity. Besides the type of neuron described above, which showed sensitivity to figure-ground direction independent of contrast polarity, we found neurons which preferred a certain combination of figure-ground direction and contrast polarity. These neurons were rarely found, but recorded in both cortical areas (V1, 5/66 or 8%; V2, 5/116 or 4%). Fig. 5.4 shows an example of the responses of such a neuron. One can see that it responded only to one of the four types of occluding contour stimuli (C). It did not respond to the other three types of contour which either had a different figure-ground direction (B), contrast polarity (D), or both (A). The neuron's sensitivity to contrast polarity was confirmed by the responses to solid edges (E-H); dark/light edges (F, G) evoked much stronger responses than light/dark edges (E, H).

In the following, we call the two types of neuron described above "neurons sensitive to figure-ground direction," knowing that the majority of these neurons showed this sensitivity independent of contrast polarity, and that some preferred a certain combination of figure-ground direction and contrast polarity.

5.3.2   Computational model

Model overview

The properties of cortical neurons showing sensitivity to illusory contour stimuli can be explained in terms of a model which invokes two inputs for these neurons: one generating the responses to solid contrast borders (bars, edges), the other producing the described sensitivity to illusory contour stimuli [45, 43, 56]. Here, we show that a logical extension of this model also explains the selectivity of cortical neurons for certain figure-ground directions and contrast polarities at such contours. Fig. 5.5 shows the scheme of the extended model. Each stage consists of a set of operators which mimic the function of a certain type of cortical neuron. The operators of the first two stages are denoted by S and C, respectively. They are intended to simulate the function of "simple" (S) and "complex" (C) type cells of the primary visual cortex [25, 27] for detecting one-dimensional luminance variations in images (bars, edges). The next higher level includes $ES1$- and $ES2$-operators which are intended to mimic the function of cortical "end-stopped cells" with inhibitory zones either at one (single-stopped) or both ends (double-stopped) of their receptive field [26, 27]. Their function is the detection of two-dimensional luminance variations such as line-ends and corners. In the next higher stage, the signals of these ES-operators are combined by grouping mechanisms (Grouping 1 and 2),

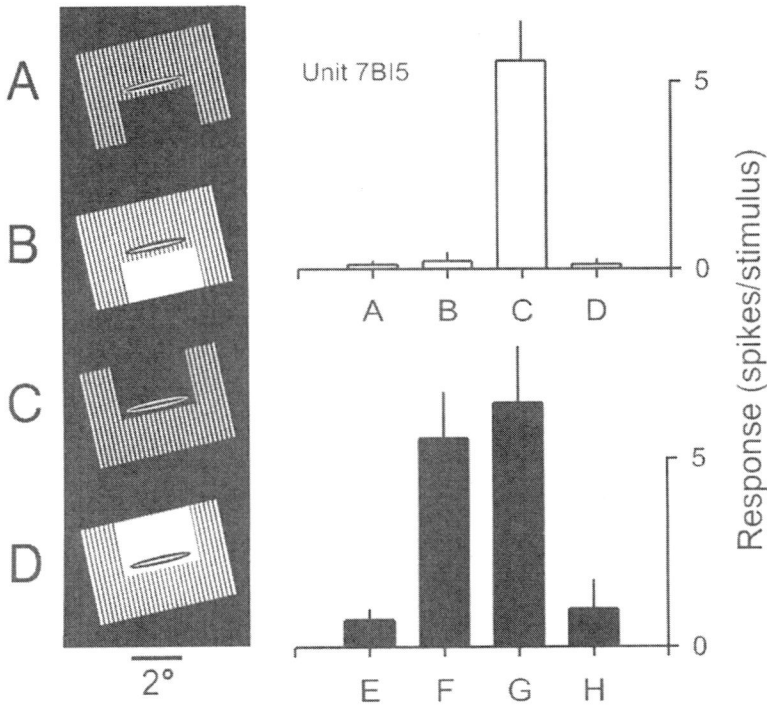

FIGURE 5.4. Neuronal sensitivity to a certain combination of figure-ground direction and contrast polarity. Of the occluding contour stimuli shown in the inset on the left (A-D), this neuron preferred stimulus (C) which had the occluding surface to the left and above the contour and a dark/light contrast polarity. The remaining stimuli either had the "wrong" contrast polarity (D), the "wrong" figure-ground direction (B), or both (A). Note that the sensitivity to contrast polarity was confirmed by the responses to the corresponding solid edges (E-H). Dark/light edges (F, G) evoked much stronger responses than light/dark edges (E, H). Conventions as Fig. 5.3.

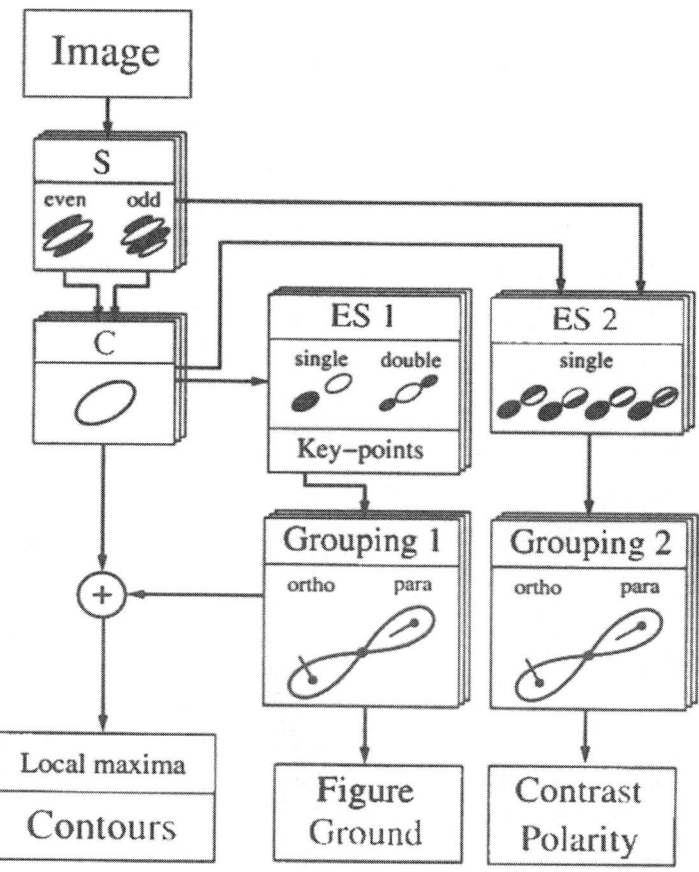

FIGURE 5.5. Model overview. Images were convolved with sets of different operators which mimicked functions of cortical neurons. S-operators (Gabor-like
functions) mimicked the function of cortical simple cells, C-operators (modulus
of odd- and even-symmetrical S-operators) the function of cortical complex cells.
The maxima of the activity of the C-operator stage were used to reconstruct the
contrast borders of the image. ES1- and ES2-operators mimicked the function of
cortical end-stopped cells with complex and simple type excitatory fields, respectively. Single-stopped and double-stopped operators were defined by the number
of inhibitory end-zones (filled ellipses). The combined activity of the single- and
double-stopped ES1-operators was used to detect the "key-points" of the image
(line-ends, corners). ES1 single-stopped operators were further used to reconstruct the illusory contours and the perceived figure-ground direction at these
contours (Grouping 1 process, see text). In addition, ES2 single-stopped operators were used to reconstruct the luminance step perceived at illusory (occluding)
contours (Grouping 2 process, see text).

first for representations of illusory contours, and second for identifying the perceived figure-ground direction and contrast polarity at such contours.

Representations of contrast borders

S- and C-operators were used for representations of solid contrast borders. S-operators represented orientation selective odd- and even-symmetrical function pairs, similar to two-dimensional Gabor functions [31, 5, 19], but with the difference that they had polar separable Fourier spectra (i.e., similar orientation response functions for lines and edges), and that both even- and odd-symmetrical operators integrated to zero in the space domain (frequency sweep, see below).

S-operators were defined on the basis of odd- and even-symmetrical Gabor functions which included a gradual decrease in frequency with increasing distance from the center of the Gaussian envelope. The profile of odd-symmetrical functions was denoted by

$$G_{\text{odd}}(x) = e^{-\frac{x^2}{2\sigma^2}} \cdot \sin[2\pi\nu_0 x\xi(x)] \tag{5.1}$$

and that of even symmetrical functions as

$$G_{\text{even}}(x) = e^{-\frac{x^2}{2\sigma^2}} \cdot \cos[2\pi\nu_0 x\xi(x)] \tag{5.2}$$

where σ determined the width of the Gaussian envelope, ν_0 the frequency at the origin, and $\xi(x)$ the frequency sweep. The frequency sweep was defined by

$$\xi(x) = ke^{-\lambda(x/\sigma)^2} + (1 - k) \tag{5.3}$$

where k was the relative amplitude of the frequency sweep ($0 < k < 1$; we used $k = 0.5$), and λ was chosen so that

$$\int_{-\infty}^{\infty} G_{\text{even}} = 0. \tag{5.4}$$

For a more detailed definition of parameters and for examples of the performance of these functions in line- and edge detection, see [22].

From these basic functions, we constructed the two-dimensional S-operators by defining the radial variation in the Fourier plane using the complex valued spectra of these functions $\mathbf{R}(r)$ and an angular variation of the power of a cosine function:

$$\mathbf{F}(r, \varphi) = \mathbf{R}(r) \cdot \cos^{2m}(\varphi_0 - \varphi) \tag{5.5}$$

where φ_0 determined the orientation of the operator and m the sharpness of the orientation tuning (for further properties, see [22]). The convolutions with images were based on this two-dimensional Fourier representation, actually carried out by multiplying operator and image spectra and transforming the result back to the space domain. The sampling was discrete using 6 orientation channels spaced by 30 degrees with a width at half height of $\pm 23.5°(m = 4)$ and a spectral bandwidth of 1.5 octaves.

C-operators were defined as the modulus (square-root of sum of squares) of the outputs of pairs of odd- and even-symmetrical S-operators at a given orientation, similar to a "local energy" representation [1, 37].

$$C_i = \sqrt{O^2_{i,\text{odd}} + O^2_{i,\text{even}}} \tag{5.6}$$

where $O_{i,odd}$ and $O_{i,even}$ represented convolutions of the image with the two-dimensional S-operators (odd- and even-symmetrical type, respectively) over 6 orientation channels $\varphi_i (i = 1, \ldots, 6)$. The square root was used to adjust the contrast response function of C-operators to that of S-operators. For a detailed discussion concerning the construction of S- and C-operators, see [22].

Representations of occlusion cues

ES-operators were designed for the detection of occlusion cues (line-ends, corners). The function of these operators were akin to the properties of cortical end-stopped cells which can have either "complex" or "simple" type excitatory fields (see below). Correspondingly, we defined two types of ES-operators: The first type, $ES1$, had a C-operator type excitatory field and was therefore activated by all types of occlusion cues (light and dark line-ends and corners). The second type, $ES2$, had an S-operator type excitatory field and was most strongly activated by one of four types of occlusion cues (light or dark line-end, or corner), depending on the type of S-operator involved (see Fig. 5.6B).

$ES1$-operators were defined for each orientation channel by taking the first (single-stopped) and second (double-stopped) derivatives of C-operators along the line of the operator's preferred orientation (see Fig. 5.6A). Practically, $ES1$-operators were constructed by taking the rectified differences of two (single-stopped operator), or three (double-stopped operator) spatially separated C-operators. For horizontal orientations, single-stopped operators were defined as

$$ES1_S(x,y) = \{C(x-d,y) - C(x+d,y)\}^+ \tag{5.7}$$

and double operators as

$$ES1_D(x,y) = \left\{C(x,y) - \frac{1}{2}\left[C(x-2d,y) + C(x+2d,y)\right]\right\}^+ \tag{5.8}$$

where C denoted the C-operator, d a constant and $\{\}^+$ clipping of negative values. Thus, each $ES1$-operator consisted of two (Fig. 5.6A) or three

(Fig. 5.6B) "subunits" which had the same overall size and orientation preference. Since these operators produced spurious activity at nonoptimal orientations (i.e., positive signals did not cancel entirely the negative signals), we added an "inhibitory surround" which suppressed these responses successfully (for details, see [22]). For convolutions with images we also used six orientation channels at each sampling point – that is, 6 double-stopped and 12 single-stopped operators. The simulations revealed that the centers of two-dimensional luminance variations of an image ("keypoints") were most accurately localized by the local maxima of the summed activity of single- and double-stopped operators. Single-stopped operators played a further, special role in the detection of occluding contours: Because of their asymmetric fields, they were only activated by terminations (line-ends, corners) pointing in one of two possible directions of a given orientation channel. For example, operator a of Fig. 5.6A can be activated by terminations pointing to the left, but not by terminations pointing to the right. Hence, single-stopped operators provide information about the orientation and the direction of occlusion cues, and thus indicate the location (left or right) of an occluding surface relative to the contour (Grouping 1 process, see below).

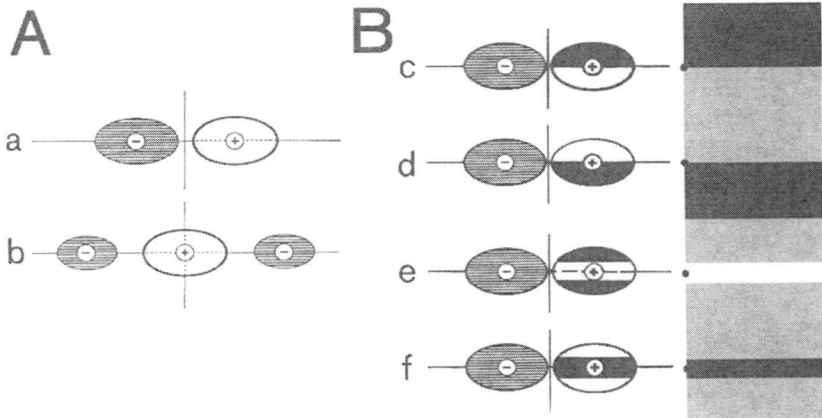

FIGURE 5.6. Types of ES-operators. (**A**) $ES1$-operators. The single-stopped version (a) consisted of a C-operator type "excitatory" (open ellipse) and "inhibitory" field (hatched ellipse), the double-stopped operator (b) of one C-operator type excitatory (open ellipse) and two inhibitory fields (hatched ellipses). (**B**) $ES2$-operators. Single-stopped operators are shown only. Each operator had an S-type excitatory and a C-type inhibitory field (hatched ellipses). Four different $ES2$-operators were used: two with odd-symmetrical (c, d), two with even symmetrical fields (e, f). Image features which produced maximum activity in these operators are shown on the right. For comparison, $ES1$-operators were similarly activated by all four types of stimulus.

$ES2$-operators were constructed as described above for $ES1$-operators, the only difference being the "excitatory" field which was derived by S- instead of C-operators. Four types of $ES2$-operators were defined (see Fig. 5.6B), one pair with odd-symmetrical fields (c, d), and another pair with even-symmetrical fields (e, f). For horizontal orientations, single-stopped operators were defined as

$$ES2_S(x,y) = \{S(x-d,y) - C(x+d,y)\}^+ \tag{5.9}$$

and double-stopped operators as

$$ES2_D(x,y) = \left\{ S(x,y) - \frac{1}{2}\left[C(x-2d,y) + C(x+2d,y)\right] \right\}^+ \tag{5.10}$$

where S represented one of the four types of S-operators defined by Eqns. 5.1 and 5.2. We also added inhibitory surrounds to $ES2$-operators as described for $ES1$-operators. Image features, which produced the strongest activation in these operators, are shown in the stimulus inset of Fig. 5.6B: Operators with odd-symmetrical fields were strongest activated by terminating edges (corners), operators with even-symmetrical fields by terminating lines (line-ends). $ES1$-operators, for comparison, were similarly activated by all four types of stimulus. Thus, $ES2$-operators provided information about the direction of terminations, their type and contrast polarity. In the Grouping 2 process, these operators were used to define the apparent step in luminance perceived at illusory contours (see Fig. 5.1). Since the localization of "key-points" and their direction was first determined from the activity of $ES1$-operators, $ES2$-operator activity was only considered at locations of maximum $ES1$-responses.

Evidence for neuronal sensitivity to occlusion cues

Hubel and Wiesel [26, 27] discovered that neurons with end-stopped receptive fields include inhibitory zones either at one ("single-stopped cells") or both ("double-stopped cells") ends of the excitatory part of their receptive field. Hence, these neurons prefer short stimuli of a certain (optimal) length and respond only weakly or not at all to long stimuli which extend beyond the excitatory part of the receptive field. In addition, these authors showed that these neurons also respond to long stimuli which terminate in their receptive field (line-ends, corners). They may therefore contribute to the detection of occlusion cues. In our laboratory, we have studied end-stopped cells in this light in the visual cortex of the alert monkey [44]. We compared responses elicited by bars and edges of optimal length with those evoked by line-ends and corners and found in a sample of about 100 neurons only two that failed to respond to terminating stimuli. About a third of these neurons even gave stronger responses to terminating stimuli than to bars and edges of optimal length [55].

In addition, we determined the symmetry of the receptive fields of end-stopped cells quantitatively from the responses to long stimuli which spared either one or the other inhibitory zone. By definition, we called a neuron single-stopped, when the maximum response to one stimulus end was twice as strong or stronger than the maximum response to the other end. By this criterion, about half of the neurons studied (45/98 or 46%) were single-stopped, the remainder (53/98 or 54%) were double-stopped [55]. In a separate analysis of areas V1 and V2 we found similar proportions of single- and double-stopped cells in these two cortical areas.

It has been shown in neurons of cat and monkey visual cortex that the excitatory part of the receptive fields of end-stopped cells can be either of the "simple" or "complex" type (see Discussion). Thus, depending on their excitatory field, end-stopped cells may, or may not be sensitive to certain types (line-ends, corners) and contrast polarities of occlusion cues. Neurons with classical complex type fields are expected to be unselective for stimulus type and contrast polarity, whereas neurons with simple type fields are expected to prefer certain stimulus types. We have studied the stimulus selectivity of end-stopped cells in areas V1 and V2 with short stimuli of optimal length (light and dark bars, edges of opposite contrast polarity) and with terminating stimuli (light and dark line-ends, corners). The results suggest similar selectivities for short stimuli and for terminating stimuli. Of the 119 end-stopped cells studied, we found 42% (50/119) which gave similar responses to all stimulus types, 29% (35/119) preferred pairs of stimuli (bar and edge), and 29% (34/119) were highly selective for one stimulus type [20]. Fig. 5.7 shows examples of the responses of highly selective neurons preferring either a light or a dark bar-end (units 7BD8 and 4CC5), or corners of opposite contrast polarity (units 4CC6 and 7DB3). These results indicate that end-stopped cells with simple or complex type excitatory fields can be sensitive to occlusion cues and thus contribute to mechanisms segregating figure and ground at contours.

Representations of occluding contours

The reconstruction of occluding (illusory) contours was based on a grouping mechanism which used the output of single-stopped ES1-operators [21, 23]. As previously described, these operators also convey information about the direction of occlusion cues, and thus indicate the location (left or right) of the occluding surface relative to the contour. The basic element of the grouping mechanism was a club-shaped field with a certain preferred orientation as illustrated in Fig. 5.8A. This field was denoted by

$$F(r,\theta) = \begin{cases} e^{-\frac{r^2}{2\sigma^2}} \cdot \cos^{2n}(\theta) & \text{if } -\frac{\pi}{2} \leq \theta \leq \frac{\pi}{2} \\ 0 & \text{otherwise,} \end{cases} \quad (5.11)$$

where r and θ were polar coordinates, and σ and n determined the spatial support of the Gaussian and angular width, respectively.

Grouping operators of a particular orientation channel were represented by pairs of such elementary fields which pooled the activity of ES-operators

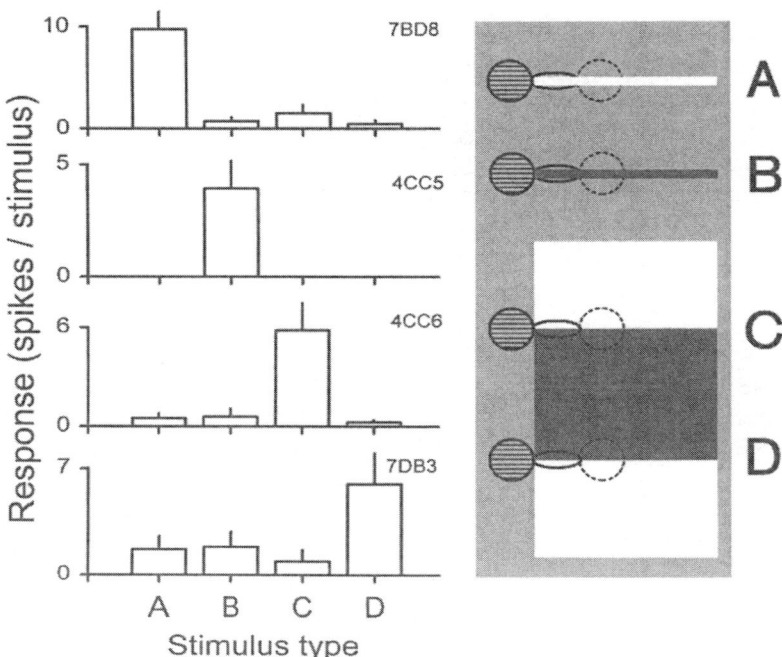

FIGURE 5.7. Neuronal selectivity for certain types of occlusion cues. Histograms of the responses of four end-stopped cells to four different stimulus types: A, light line-ends; B, dark line-ends; C and D, corners of opposite contrast polarity. All neurons were clearly end-stopped giving strong responses to short stimuli of optimal length, and only weak responses or none to long stimuli centered in the receptive field. One neuron had a single-stopped receptive field (unit 4CC5), the remainder had double-stopped fields.

(weighted with distance). Six orientations channels, separated by $30°$, were used at each sampling point. The grouping response G of a particular orientation channel was defined by

$$G = [(F^{\rightarrow} * Q) . (F^{\leftarrow} * Q)]^{\frac{1}{2}} \qquad (5.12)$$

where $F^{\rightarrow}$ and $F^{\leftarrow}$ denoted elementary fields of opposite directions. Q represented the weighted ES-signals within the denoted field. The multiplication of opposite field activities ensured that grouping activity was confined to regions between "key-points." Isolated "key-points" were not included. There were two modes of grouping of ES-signals, depending on the type of occlusion cue (line-end, corner). Briefly, *ortho* grouping (Fig. 5.8B, left panel) pooled the activity of ES-operators indicating components orthogonal to a given orientation channel, and *para* grouping (Fig. 5.8B, right panel) those indicating parallel components. Furthermore, reconstructions

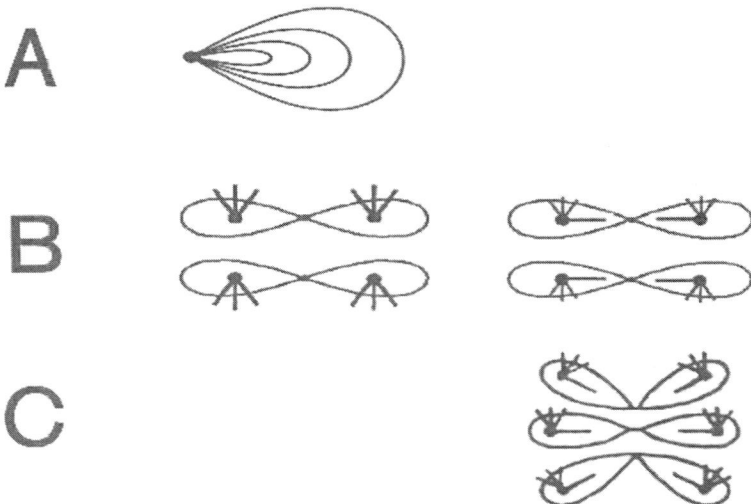

FIGURE 5.8. Basic elements of the grouping mechanism. (**A**) Elementary group-ing field. Lines indicate the weighting of ES-signals. (**B**) Scheme of *ortho* (left panel) and *para* (right panel) grouping shown for a horizontal orientation channel. Small lines indicate the signals of ES-operators oriented either orthogonal or par-allel to the occluding contour. Oblique lines indicate that ES-operators oriented $\pm 30°$ from orthogonal also contributed to the process, though with less weight. Two spatially separate pairs of grouping fields indicate that the ES-signals of the two possible figure-ground directions (up and down in this case) were combined separately. The sum of the pooled activity of the two directions was taken as a measure for contour strength, the difference as an indication for the perceived figure-ground direction. (**C**) In order to optimize the reconstruction of curved contours, the activity of grouping fields oriented $\pm 30°$ from optimal was also included, though with less weight.

of curved contours were facilitated by including *ES*-operators oriented $\pm 30^o$ from orthogonal, though with less weight. Similarly, the activity of grouping fields oriented $\pm 30^o$ deg from a given orientation channel also contributed to the final contour signal (Fig. 5.8C). Finally, we computed separate grouping responses ($G \uparrow$ and $G \downarrow$) for the two possible directions of terminations at a given orientation channel:

$$G \uparrow \;\; = \;\; \left[\left(F^{\rightarrow} * Q^{\uparrow \rightarrow} \right) . \left(F^{\leftarrow} * Q^{\uparrow \leftarrow} \right) \right]^{\frac{1}{2}} \tag{5.13}$$

$$G \downarrow \;\; = \;\; \left[\left(F^{\rightarrow} * Q^{\downarrow \rightarrow} \right) . \left(F^{\leftarrow} * Q^{\downarrow \leftarrow} \right) \right]^{\frac{1}{2}} . \tag{5.14}$$

The vertical arrows in the superscript of Q indicate the two directions of terminations at a particular contour, i.e., upward and downward for a horizontal orientation channel (see Fig. 5.8B). The sum of the grouping responses in the two directions ($G \uparrow$ and $G \downarrow$) provided a local measure for "strength of contour," the difference indicated the perceived figure-ground direction. This logic of combinations produced high grouping selectivity comparable to that observed in perception: *Ortho* grouping generated boundary activity in regions between terminations, *para* grouping between corners of certain spatial arrangements (for further details on grouping selectivity see [23]).

The model overview of Fig. 5.5 suggests two grouping processes working in parallel, one for the reconstruction of the occluding (illusory) contours and the perceived figure-ground directions (Grouping 1), and another for the reconstruction of the contrast polarity perceived at such contours (Grouping 2). The two processes rely on different inputs: The Grouping 1 process depends on *ES*1-operators, the Grouping 2 process on *ES*2-operators. In the Grouping 2 process we assigned four grouping operators to each orientation channel. Each one consisted of a pair of elementary grouping fields which collected information from either side of the sampling point. Two operators were necessary to detect all possible types of occlusion cues, one for the detection of light line-ends and corners, and another for the detection of dark line-ends and corners. These operators were doubled for separate sampling of the two figure-ground directions (see [42]).

Computer simulations

We tested the performance of the computational model using two-dimensional, binary images which induced the perception of illusory contours and the perception of occluding contours and surfaces in human observers. Fig. 5.9 shows the result for the abutting grating stimulus which is ambivalent with regard to the perceived figure-ground direction at the illusory contour. The image (A) was analyzed using six evenly spaced orientation channels at each sampling point. The result (B) shows a trace of the maxima of the activity of the *C*-operator stage (solid, horizontal lines) and the

Grouping 1 process (illusory contours). It shows that the performance of the model was similar at straight and at curved contours.

Fig. 5.10 shows the results regarding figure-ground direction and contrast polarity using the example of the Kanizsa triangle [28]. Two images (A) of opposite contrast polarity were used. In human perception the two figures induce identical forms and figure-ground relations (the triangles appear to occlude three disc-shaped objects in the background). The only difference is the apparent brightness of the occluding surface (triangle) which appears brighter than the background in the left image and darker in the right image.

As for the abutting grating stimulus, the images were analyzed discretely using six orientation channels at each sampling point. The maxima of the activity produced by the C-operator stage and the Grouping 1 process are shown in (B). The result represents a reconstruction of the perceived contours of the image. The figure-ground relations as reproduced by the Grouping 1 process, are shown in (C). The short lines orthogonal to the illusory (occluding) contours indicate the direction of the background, their lengths the strength of the signal. For a more detailed description of these results and for a discussion about the figure closure which occurred at the corners of the triangles, see [23]. The reconstruction of the contrast polarity perceived at the illusory (occluding) contours was determined by the Grouping 2 process and is shown in (D). Here, the short orthogonal lines indicate the net contrast polarity detected at each sampling point. One can see that the result suggests the perception of a bright triangle for the left, and a dark triangle for right image. In summary, the results show that the model simulates the critical aspects of perception – that is, the illusory (occluding) contours, the depth order perceived at such contours, and the apparent step in luminance perceived at these contours which also indicates the perceived brightness enhancement of the occluding surface.

5.4   Discussion

This chapter provides experimental evidence for a neuronal mechanism of figure-ground segregation and contrast induction at illusory (occluding) contours in the visual cortex of the awake monkey. This mechanism is based on the responses of cortical neurons (end-stopped cells) which detect occlusion cues (line-ends, corners). In computer simulations we show that the proposed mechanism reproduces perception – that is, the illusory (occluding) contours, the perceived figure-ground directions, and the apparent contrast polarity perceived at such contours. This information may be used at later stages of processing for defining the depth order and the perceived brightness of surfaces associated with such contours.

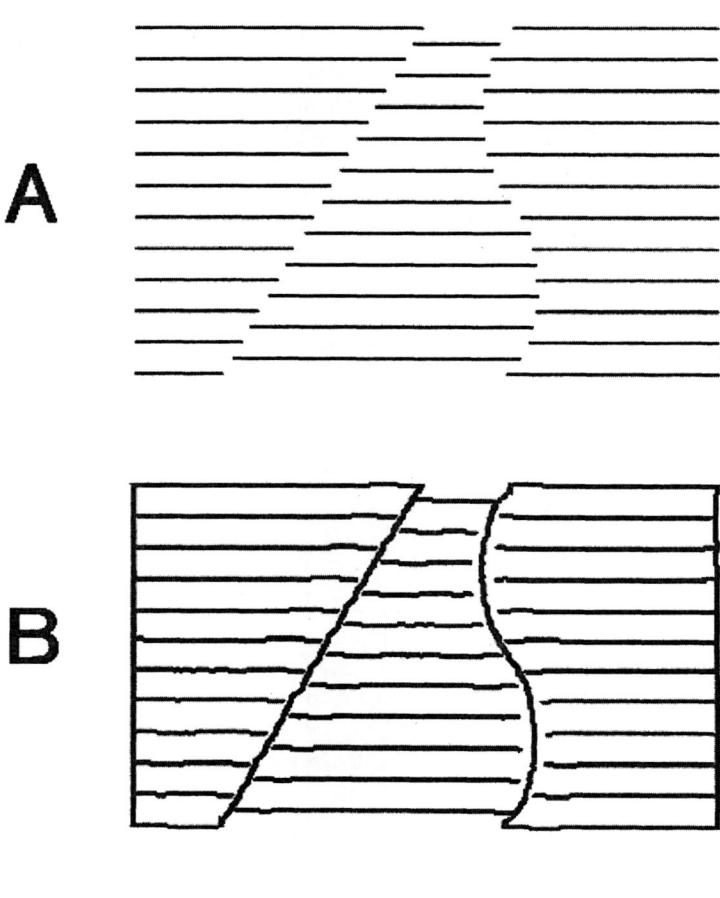

FIGURE 5.9. Computer simulations of illusory contours between abutting gratings. (**A**) shows the stimulus, (**B**) the result of the C-operator stage and of the Grouping 1 process, both represented by the maxima of the activity produced at each sampling point. The horizontal bar indicates the size of the elementary grouping field (i.e., 2σ of equation 5.11 at the 5% sensitivity level).

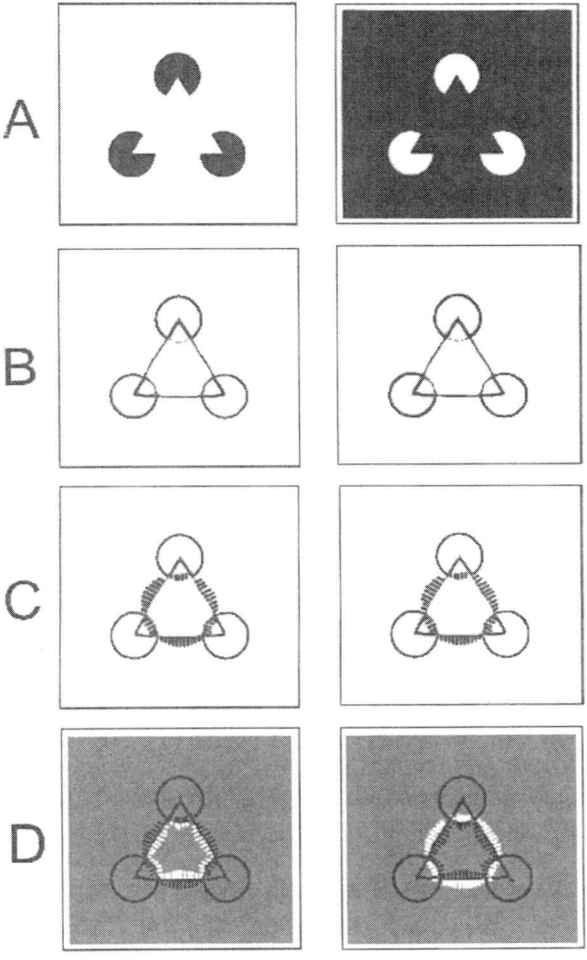

FIGURE 5.10. Computer simulation showing the result of different stages of the model. (**A**) Kanizsa triangles of opposite contrast polarity were used as images. (**B**) Reconstruction of real and illusory contours represented by the local maxima of the sum of the activity of the C-operator stage and the Grouping 1 process. (**C**) Reconstruction of the perceived figure-ground direction at contours as indicated by the Grouping 1 process (see text). (**D**) Reconstruction of the perceived contrast polarity at contours as indicated by the Grouping 2 process (see text).

5.4.1   Occlusion cues

Symmetry of end-inhibition

Since the original studies of Hubel and Wiesel [26, 27], little attention has been given to the questions of field symmetry and stimulus selectivity of end-stopped cells. Orban et al. [40] found evidence for asymmetrical end-inhibition in neurons of the cat visual cortex from records of unilateral length-response functions. The example of their Fig. 9 shows that the response reduction as a function of bar length was different for the two halves of the receptive field, thus indicating a weak inhibitory zone at one end of the field and a strong one at the other. Our results of monkey visual cortex suggest that end-stopped cells can show all degrees of field symmetry ranging from complete asymmetry to complete symmetry [55]. Since knowledge about the cortical circuitry generating end-stopped cells is sparse, possible connections and interactions within inhibitory zones and between excitatory and inhibitory zones are virtually unknown. Hubel and Wiesel [26] proposed two inputs for single-stopped cells, an excitatory and an inhibitory one with abutting receptive fields. For double-stopped cells they gave two alternatives, either one excitatory and two inhibitory inputs, or a dual input as for single-stopped cells, but with overlapping fields of different size. Bolz et al. [3] studied the generation of end-stopped receptive fields in double-stopped cells and found support for the second hypothesis.

Excitatory fields of end-stopped cells

Hubel and Wiesel [26, 27] proposed complex type excitatory fields for end-stopped cells. However, simple type end-stopped cells have been reported since, both in cat and monkey visual cortex [7, 51, 50, 14, 29]. Our results in the awake monkey also suggest both types of excitatory fields [20]. From our results we cannot provide a classical definition of simple- and complex type fields since the fixational eye movements of the animal blurred the spatial relationship of these responses to some extent, especially in neurons with narrow receptive fields (for a discussion of the effect of eye movements in our experiments see [2]). However, our aim was to determine possible functions of these neurons in the visual process rather than the spatial organization of their receptive fields. Thus, we can only infer possible field types from their stimulus selectivity, based on the theoretical scheme of Fig. 5.6. Highly selective responses, such as those of unit 4CC5 of Fig. 5.7, cannot be explained on the basis of this scheme. The excitatory fields of such neurons seem to be akin to those of the S1-type simple cells of Schiller et al. [51] and include nonlinearities which cannot be explained in terms of Gabor functions alone.

Computational models

In our scheme we defined single- and double-stopped ES-operators ($ES1$ and $ES2$) for the detection and localization of contour terminations. Lourens [35] developed this model further, especially with regard to corner detection, and showed performance at corners of different angles, scales, and orientations. Other models of end-stopped cells have been used to analyze curvilinear patterns following the original suggestion of Hubel and Wiesel [26] that end-stopped cells may serve to measure curvature. Fukushima [12] developed single- and double-stopped operators with complex type excitatory fields for this purpose, and Dobbins et al. [6] used double-stopped operators with simple type excitatory fields. For a more detailed discussion of biologically inspired models of end-stopped cells and for reviews of computational approaches see also Heitger et al. [22] and Lourens [35].

5.4.2   Occluding contours and surfaces

Figure-ground segregation at illusory contours

The basic elements of the model of this chapter are single-stopped operators which detect different directions of occlusion cues. The Grouping 1 process uses these signals to infer the illusory (occluding) contours and to reconstruct figure-ground directions. Alternative models of illusory contour formation and reconstructions of situations of spatial occlusion have been discussed in detail by Lesher [32], Grossberg [18], and Heitger et al.[23]. Here, we concentrate on specific models which involve end-stopped operators to reconstruct depth order and/or contrast polarity at illusory (occluding) contours. For example, Finkel and Edelman [9] used single-stopped operators and a feedback mechanism for detecting the direction of foreground and background at contours. Finkel and Sajda [10] used end-stopped operators in a complex scheme of successive representations of contour terminations, junctions, and curvatures which also revealed the depth order of the objects of a visual scene. Grossberg and coworkers used double-stopped cells with complex type excitatory fields in a cooperative (push-pull) network of orthogonal orientations for detecting ends of lines ("end-cuts") and for initiating boundary completion (see [18]). For depth segregation, however, they invoked a mechanism based on stereoscopic cues which depended on the mutual interaction of end-stopped and bipole (grouping) cells sensitive to binocular disparity.

Brightness induction at illusory contours

The physiological results of this chapter fit with evidence of perception which suggests that illusory contours and the depth order at such contours can be perceived independent of the contrast polarity of the inducers (see for example [47]): The majority of V2 neurons that were sensitive to figure-ground direction at contours showed this sensitivity independent of the contrast polarity of the inducers. We explain this result by the Grouping 1 process. However, many figures also induce the perception of an apparent step in luminance at illusory contours, and thus an apparent brightening (or darkening) of the occluding surface relative to ground. This phenomenon was also found at the neuronal level: Some neurons were selective for a certain combination of figure-ground direction and contrast polarity at illusory contours. We explain these responses by the Grouping 2 process. Because these neurons preferred the same contrast polarity for real and illusory borders, they could provide input to neuronal mechanisms "filling-in" real and illusory surfaces. Frisby and Clatworthy [11] were the first to propose a physiologically based model of the brightness illusion associated with illusory surfaces. They suggested that the phenomenon was initiated by retinal ganglion cells which induced "brightness buttons" at line-ends due to their unbalanced center-surround receptive fields (see also [30]). At later stages of processing, these signals could then be "spread" between contours and thus be used to "fill-in" illusory (occluding) surfaces [13, 41]. This two-stage model was used in many subsequent studies on brightness induction at illusory contours (for overviews see [33, 53]). Furthermore, Gove et al. [15] proposed lateral geniculate nucleus neurons with concentric receptive fields as inducers of "brightness buttons" at line-ends, and proposed that this brightness enhancement was induced by a corticogeniculate feedback circuit. For subsequent surface representations these authors also invoked the "spreading" of these signals between previously defined contours. In contrast, our results suggest that illusory (occluding) contours as well as the figure-ground segregation and the brightness induction at such contours, are generated by the same, feed forward mechanism. At further stages of processing, these signals may be used in representations of occluding surfaces.

5.5 Acknowledgment

This research was supported by SNF-ESPRIT grant #6019, HFSP grant #RG-31/93, and SNF grant #5002-44891.

5.6 REFERENCES

[1] Adelson EH, Bergen JR (1985) Spatio-temporal energy models for the perception of motion. *J Opt Soc Am* A 2:284-299.

[2] Baumann R, van der Zwan R, Peterhans E (1997) Figure-ground segregation at contours: a neural mechanism in the visual cortex of the alert monkey. *Eur J Neurosci* 9:1290-1303.

[3] Bolz J, Gilbert CD, Wiesel TN (1989) Pharmacological analysis of cortical circuitry. *Trends Neurosci* 12:292-296.

[4] Coren S (1972) Subjective contours and apparent depth. *Psychol Review* 79:359-367.

[5] Daugman JG (1983) Six formal properties of two-dimensional anisotropic visual filters: Structural principles and frequency/orientation selectivity. *IEEE Transactions on Systems, Man and Cybernetics* 13:882-887.

[6] Dobbins A, Zucker SW, Cynader MS (1987) Endstopped neurons in the visual cortex as a substrate for calculating curvature. *Nature (Lond)* 329:438-441.

[7] Dreher B (1972) Hypercomplex cells in the cat's striate cortex. *Invest Ophthalmol* 11:355-356.

[8] Ffytche DH, Zeki S (1996) Brain activity related to the perception of illusory contours. *Neuroimage* 3:104-108.

[9] Finkel LH, Edelman GM (1989) Integration of distributed cortical systems by reentry: a computer simulation of interactive functionally segregated visual areas. *J Neurosci* 9:3188-3208.

[10] Finkel LH, Sajda P (1992) Object discrimination based on depth-from-occlusion. *Neural Computation* 4:901-921.

[11] Frisby JP, Clatworthy JL (1975) Illusory contours: curious cases of simultaneous brightness contrast? *Perception* 4:349-357.

[12] Fukushima K (1970) A feature extractor for curvilinear patterns: A design suggested by the mammalian visual system. *Kybernetik* 7:153-160.

[13] Gerrits HJM, Vendrik AJH (1970) Simultaneous contrast filling-in process and information processing in man's visual system. *Exp Brain Res* 11:411-430.

[14] Gilbert CD (1977) Laminar differences in receptive field properties of cells in cat primary visual cortex. *J Physiol* 268:391-421.

[15] Gove A, Grossberg S, Mingolla E (1995) Brightness perception illusory contours and corticogeniculate feedback. *Visual Neurosci* 12:1027-1052.

[16] Grinvald A, Lieke E, Frostig RD, Gilbert CD, Wiesel TN (1986) Functional architecture of cortex revealed by optical imaging of intrinsic signals. *Nature* 324:361-364.

[17] Grosof DH, Shapley RM, Hawken MJ (1993) Macaque V1 neurons can signal "illusory" contours. *Nature* 365:550-552.

[18] Grossberg S (1997) Cortical dynamics of three-dimensional figure-ground perception of two-dimensional pictures. *Psychological Review* 3:618-658.

[19] Hawken MJ, Parker AJ (1987) Spatial properties of neurons in monkey striate cortex. *Proc R Soc Lond* B 231:251-288.

[20] Heider B, Meskenaite V, Peterhans E (2000) Anatomy and physiology of a neural mechanism defining depth order and contrast polarity at illusory contours. *Eur J Neurosci* 12:4117-4130.

[21] Heitger F, von der Heydt R (1993) A computational model of neural contour processing: Figure-ground segregation and illusory contours. In: *Proc 4th Int Conf Computer Vision*, IEEE Computer Society Press, Berlin, pp. 32-40.

[22] Heitger F, Rosenthaler L, von der Heydt R, Peterhans E, Kübler O (1992) Simulation of neural contour mechanisms: From simple to end-stopped cells. *Vision Res* 32:963-981.

[23] Heitger F, von der Heydt R, Peterhans E, Rosenthaler L, Kübler O (1998) Simulation of neural contour mechanisms: Representing anomalous contours. *Image and Vision Computing* 16:407-421.

[24] Hirsch J, De La Paz RL, Relkin NR, Victor J, Kim K, Li T, Borden P, Rubin N, Shapley R (1995) Illusory contours activate specific regions in human visual cortex: evidence from functional magnetic resonance imaging. *Proc Nat Acad Sci USA* 92:6469-6473.

[25] Hubel DH, Wiesel TN (1962) Receptive fields, binocular interaction and functional architecture in the cat's visual cortex. *J Physiol* (Lond) 160:106-154.

[26] Hubel DH, Wiesel TN (1965) Receptive fields and functional architecture in two nonstriate visual areas (18 and 19) of the cat. *J Neurophysiol* 28:229-289.

[27] Hubel DH, Wiesel TN (1968) Receptive fields and functional architecture of monkey striate cortex. *J Physiol* (Lond) 195:215-243.

[28] Kanizsa G (1979) *Organization in Vision. Essays on Gestalt Perception.* Praeger, New York

[29] Kato H, Bishop PO, Orban GA (1978) Hypercomplex and simple/complex cell classification in cat striate cortex. *J Neurophysiol* 41:1071-1095.

[30] Kennedy JM (1979) Subjective contours, contrast, and assimilation. In: Nodine CF, Fisher DF (eds) *Perception and Pictorial Representation*. Praeger, New York, pp. 167-195.

[31] Kulikowski JJ, Marčelja S, Bishop PO (1982) Theory of spatial position and spatial frequency relations in the receptive fields of simple cells in the visual cortex. *Biol Cybern* 43:187-198.

[32] Lesher, GW (1995) Illusory contours: Toward a neurally based perceptual theory. *Psychonomic Bulletin and Review*, 2:279-321.

[33] Lesher, GW, Mingolla, E (1993) The role of edges and line-ends in illusory contour formation. *Vision Res*, 33:2253-2270.

[34] Leventhal, AG, Zhou, Y (1994) Cat visual cortical cells are sensitive to the orientation and direction of "illusory" contours. Soc for *Neurosci Abstract*, 20:1053.

[35] Lourens, T (1998) A biologically plausible model for corner-based object recognition from color images. PhD thesis, University of Groningen, The Netherlands

[36] Minguzzi GF (1987) Anomalous figures and the tendency to continuation. In: Petry S, Meyer GE (eds) *The Perception of Illusory Contours*. Springer, Berlin, pp. 71-75.

[37] Morrone MC, Burr DC (1988) Feature detection in human vision: A phase-dependent energy model. *Proc R Soc Lond B* 235:221-245.

[38] Nakayama K, Shimojo S (1990) Da Vinci stereopsis: Depth and subjective occluding contours from unpaired image points. *Vision Res* 30:1811-1825.

[39] Nakayama K Shimojo S, Silverman G H (1989) Stereoscopic depth: its relation to image segmentation, grouping, and the recognition of occluded objects. *Perception* 18:55-68.

[40] Orban GA, Kato H, Bishop PO (1979) Dimension and properties of end-zone inhibitory areas in receptive fields of hypercomplex cells in cat striate cortex. *J Neurophysiol* 42:833-849.

[41] Paradiso MA, Nakayama K (1991) Brightness perception and filling-in. *Vision Res* 31:1221-1236.

[42] Peterhans E, Heitger F (2001) Simulation of neuronal responses defining depth order and contrast polarity at illusory contours in monkey area V2. *J Comput Neurosci* (in press).

[43] Peterhans E, von der Heydt R (1989) Mechanisms of contour perception in monkey visual cortex. II. Contours bridging gaps. *J Neurosci* 9:1749-1763.

[44] Peterhans E, von der Heydt R (1991) Elements of form perception in monkey prestriate cortex. In: Gorea A, Frégnac Y, Kapoulis Z, Findlay J (eds) *Representations of Vision: Trends and Tacit Assumptions.* Cambridge University Press, Cambridge, pp. 111-124.

[45] Peterhans E, von der Heydt R, Baumgartner G (1986) Neuronal responses to illusory contour stimuli reveal stages of visual cortical processing. In: Pettigrew JD, Sanderson KJ, Levick WR (eds) *Visual Neuroscience.* Cambridge University Press, Cambridge, pp. 343-351.

[46] Petry S, Meyer GE (1987) *The Perception of Illusory Contours.* Springer, Berlin

[47] Prazdny K (1983) Illusory contours are not caused by simultaneous brightness contrast. *Perception and Psychophysics* 34:403-404.

[48] Ramachandran VS, Anstis S (1986) Figure-ground segregation modulates apparent motion. *Vision Res* 26:1969-1975.

[49] Redies C, Crook J M, Creutzfeldt OD (1986) Neuronal responses to borders with and without luminance gradients in cat visual cortex and dorsal lateral geniculate nucleus. *Exp Brain Res* 61:469-481.

[50] Rose D (1977) Responses of single units in cat visual cortex to moving bars of light as a function of bar length. *J Physiol* 271:1-23.

[51] Schiller PH, Finlay BL, Volman SF (1976) Quantitative studies of single-cell properties in monkey striate cortex. I. Spatiotemporal organization of receptive fields. *J Neurophysiol* 39:1288-1319.

[52] Sheth BR, Sharma J, Rao SC, Sur M (1996) Orientation maps of subjective contours in visual cortex. *Science* 274:2110-2115.

[53] Spillmann L, Dresp B (1995) Phenomena of illusory form: can we bridge the gap between levels of explanation? *Perception* 24:1333-1364.

[54] Soriano M, Spillman L, Bach M (1996) The abutting grating illusion. *Vision Res* 36:109-116.

[55] van der Zwan R, Baumann R, Peterhans E (1995) End-stopped cells in the visual cortex of the alert monkey. *Perception* 24:43.

[56] von der Heydt R, Peterhans E (1989) Mechanisms of contour perception in monkey visual cortex. I. Lines of pattern discontinuity. *J Neurosci* 9:1731-1748.

[57] von der Heydt R, Peterhans E, Baumgartner G (1984) Illusory contours and cortical neuron responses. *Science* 224:1260-1262.

6

Controlling the Focus of Visual Selective Attention

Ernst Niebur, Laurent Itti, and Christof Koch

ABSTRACT Selecting only a subset of the available sensory information before further detailed processing is crucial for efficient perception. In the visual modality, this selection is frequently implemented by suppressing information outside a spatially circumscribed region of the visual field, the so-called 'focus of attention." The model for the control of the focus of attention in primates presented here is based on a "Saliency Map" which is a topographic representation of the instantaneous saliency of the visual scene.

6.1 Introduction

Access to information about an organism's environment is essential for its survival. As a consequence, highly sensitive and efficient sensory organs were developed during evolution. Nearly as essential as acquiring information is sorting through it and deciding which part of it requires more detailed analysis and which is irrelevant. For animals with a highly developed sensorium (which includes all higher phyla), it is impractical to process all sensory input at all times. The problem is the mismatch between the need for sensors which provide information about the outside world required *at some time*, and the information processing capacity of the brain which is not capable to treat the information provided by all sensors *simultaneously*. The solution adopted by biology is to provide animals with a multitude of powerful sensors capable of providing detailed information in different modalities, to *select* a small portion of the available information, and to discard all the other information. This process is usually called "selective attention".

The triage of the incoming sensory information into parts to be discarded and parts to be retained can go along different lines, i.e. along different feature dimensions. Selective filters can be employed which selectively enhance one or more distinct visual "channels" across the visual field, such as to facilitate perception of all stimuli of a specific color, spatial frequency, or direction of motion. One of the most important of these filters uses spatial location as the selection criterion, by suppressing all stimuli outside a spatially circumscribed region of the visual field relative to stimuli inside this region. This makes sense functionally because the location of stimuli is only weakly correlated with their other properties. In other words, most objects can appear with nearly equal probabilities anywhere in the visual field. This allows the system to reduce the complexity of the object recognition problem significantly since it can then operate in a space constructed as the *sum* of the feature space and the locality space, rather than the outer *product* of these states. Indeed, there is psychophysical evidence that space does play a special role among features [56, 82, 77]; but see [11] for a different view. The most direct evidence for a spatially defined focus of attention was recently found by imaging methods [3]. We refer the reader to ref. [59] for a more general discussion of selective attention from a computational point of view.

A space-based mechanism is also supported by the observed anatomical and physiological structure of the primate visual system. This system uses anatomically distinct pathways for encoding spatial information of objects in the environment and the specific features of these objects. The locations of visual stimuli are represented in the "dorsal" (or "where") pathway, while detailed feature processing and object recognition are localized in the "ventral" (or "what") pathway. In earlier work, we have presented neuronal models for attentional control in the ventral pathway [60, 57]. Here, we focus on the selection process in the dorsal pathway.

The purpose of this chapter is to present a neuronally based model of visual selection, paying attention to the underlying neurophysiology and anatomy. Two more technically oriented reports about aspects of this work have been published previously [58, 37]. The following section 6.2 describes the model we have developed and implemented. Section 6.3 presents the results obtained on different classes of stimuli, ranging from simple synthetic stimuli akin to those used in prototypical psychophysical experiments to images of natural scenes. Finally, section 6.4 discusses the results, establishes the relation to previous work, and concludes with an outlook to future developments. A C++ installation of the model and numerous examples of predictions of the model can be retrieved from http://www.klab.caltech.edu/~itti/attention.

6.2   A Computational Model of The Dorsal Pathway

6.2.1   Model Assumptions

Our model is limited to the bottom-up control of attention, i.e. to the control of selective attention by the properties of the visual stimulus. The present model is based on the following hypotheses:

- Selection is based on iconic (appearance-based) scene representations rather than on categorical decisions.

- Visual input is represented in subcortical and early cortical structures in feature maps. A crucial step in the construction of these representations consists of center-surround computations in every feature, to avoid excessive redundancy in the processing.

- Information from these feature maps is combined in an additional feature map which represents the local saliency.

- *Per definitionem,* the instantaneous maximum of this saliency map is the most conspicuous location at a given time. The focus of attention has to be pointed to this location.

- The saliency map is endowed with internal dynamics which allow the perceptive system to scan the visual input such that its different parts are visited by the focus of attention in the order of decreasing saliency.

6.2.2   General architecture

Fig. 6.1 shows an overview of the model *Where* pathway and selective attention mechanism. Input is provided in the form of digitized images from a variety of sensors, such as NTSC cameras or image scanners. Low-level vision features (red, green, blue, and yellow color channels; orientation; and brightness) are extracted from the original color image, at several spatial scales, using linear filtering. The different spatial scales are created using Gaussian pyramids [1], which consist of progressively low-pass filtering and subsampling the input image. Pyramids have a depth of 9 scales, providing horizontal and vertical image reduction factors ranging from 1:1 (level 0; the original input image) to 1:256 (level 8) in consecutive powers of two. Each feature is computed in a center-surround structure akin to visual receptive fields. Center-surround operations are implemented as differences between a fine and a coarse scale for a given feature: The center of the receptive field corresponds to the value of a pixel at level $n \in \{2, 3, 4\}$ in the pyramid, and the surround to the corresponding pixel at level $n + \delta$, with $\delta \in \{3, 4\}$. We hence compute six feature maps for each type of feature. The feature maps are then normalized so as to allow the direct comparison between a priori not comparable modalities, as well as to enhance the feature maps in which a small number of highly significant stimuli are found. Within each modality, the features are linearly combined across scales, yielding three *conspicuity maps* for color, intensity, and orientation. These are normalized, linearly combined and fed to a network of integrate-and-fire neurons representing the *saliency map*. The saliency map input, as well as the three conspicuity maps, is a single image lying at level 4 (reduction factor 1:16 horizontally and vertically).

Short descriptions of the different feature maps are presented in the next section (6.2.3). We then (section 6.2.4) address the question of the integration of the input in the saliency map, a topographically organized map which codes for the instantaneous conspicuity of the different parts of the visual field, and present the dynamical mechanism controlling the focus of attention.

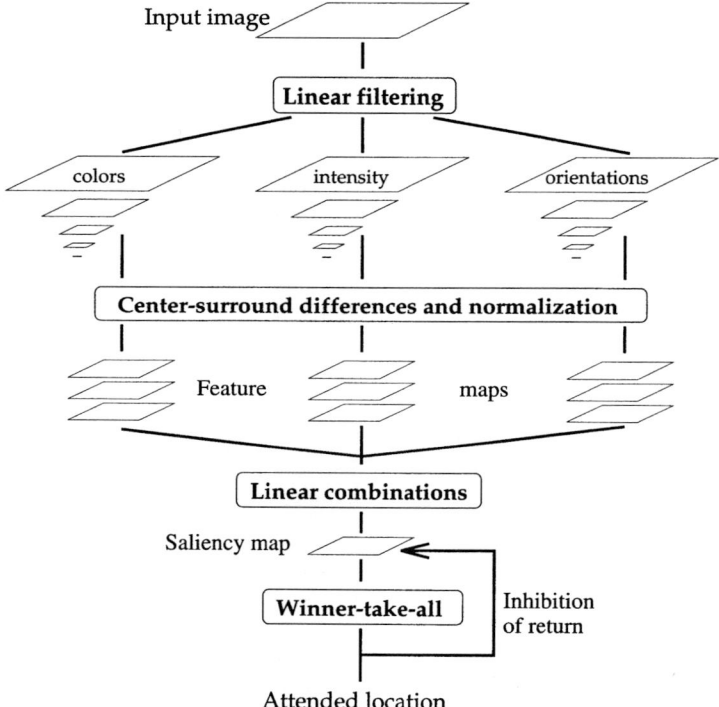

FIGURE 6.1. Overview of the model *Where* pathway. Features are computed as center-surround differences between several fine and coarse scales. The feature maps are normalized, combined, and integrated in the saliency map which provides input to an array of integrate-and-fire neurons with global inhibition. This array has the functionality of a winner-take-all network and provides the output to the ventral pathway as well as feedback to the saliency map through an inhibition of return (IOR) mechanism.

6.2.3   Input features

Input is represented in a hierarchical scheme based on basis functions which are similar to those realized in the primate visual system. For the representation of shape in our model in the form of oriented edges, it has been shown that similar functions can be learned from natural scenes [66, 10]. For the chromatic information, it was shown by principal components analysis that the most efficient transformation of incoming data, either from a spectrally flat Gaussian noise process [4] or from natural scenes [50] is a three-stage mechanism. The three principal components are, in the order of decreasing signal energy, (1) intensity, (2) red minus green, and (3) blue minus yellow. These features will be discussed in the following two sections.

Intensity

Intensity information is obtained from the chromatic information of the original color input image. With R, G, and B being the red, green and blue channels, respectively, the intensity I is obtained as $I = (R + G + B)/3$. A more accurate computation could be implemented if only images from a unique known source were to be used (e.g. always the same camera). A Gaussian pyramid is created from the intensity image. The entries in the intensity feature maps are given by the modulus of the contrast, i.e., $|I_{center} - I_{surround}|$. This corresponds roughly to the sum of two single-opponent cells of opposite phase, i.e. bright-center — dark-surround and vice-versa.

Chromatic input

Psychophysical results [44] show that color information is available for preattentive selection. We therefore implemented inputs to the saliency map which depend on the chromatic contrast of the input image. The red, green and blue components of the original color image are first normalized by the intensity image computed in the previous step, in order to clearly decorrelate intensity and hue information. Three Gaussian pyramids are then created for these three isoluminant color components. At any given scale, yellow is computed as $(R + G)/2$. A quantity corresponding to the double-opponency cells in primary visual cortex is finally computed by center-surround differences across scales. For instance, for the red-green filter, we first compute at each pixel the value of (red-green) at the scale of the center. From this, we then subtract (green-red) of the surround. Finally, we take the absolute value of the result.

Shape: Orientated edges

Local orientation is obtained at all scales through the creation of oriented Gabor pyramids from the intensity image. In such pyramids, orientation-selective Gabor filters of increasing spread are successively applied to down-scaled versions of the original intensity image. The oriented Gabor pyramids were implemented in a computationally efficient way by using overcomplete steerable pyramids [26]. Four different orientations are used (0, 45, 90 and 135 degrees). Center-surround operations are performed, within each oriented pyramid, yielding orientation feature maps. These maps encode, as a group, how different the average local orientation is between the center and surround scales. It is possible in our implementation to use an arbitrary number of orientations, but we noticed that using more oriented filters than the four abovementioned did not alter the performances of the model drastically.

6.2.4   The saliency map

Implementation of the saliency map

In section 6.1, we discussed that "decoupling" of feature dimensions is computationally advantageous, and that one implementation of such decoupling is sequential scanning of different parts of the visual field. This leads to the metaphor of the "focus of attention" and requires a spatially defined selection scheme which controls where the focus of attention is deployed at any given time. In 1985, Koch and Ullman [41] suggested an efficient way for the coordination of this control mechanism is a spatially organized feature map, which codes for the "saliency" of every location in the visual field.

The task of the saliency map is the computation of the salience at every location in the visual field and the subsequent selection of the most salient areas or objects. At any time, only one such area is selected. The output of the saliency map consists of a spike train from neurons corresponding to this selected area in the topographic map which projects to the ventral (What) pathway.

Fusion of information

Once all relevant features have been computed in the various feature maps, they have to be combined to yield the salience, i.e. a scalar quantity. The main difficulty in this task is that features from different modalities have different dynamic ranges and consequently are not directly comparable. Normalizing all feature maps to the same dynamic range is not a satisfactory solution because it artifactually amplifies those maps in which only a low, noisy response was originally present. Moreover, evidence has been found in monkey primary visual cortex [12] for the presence of nonlinear amplification stages, whose gains vary with the responses of neighboring cells. This supports the idea that different feature maps should not be normalized to a fixed range, but rather according to their content. Also, psychophysical studies performed on humans have shown that strong competition between spatially close stimuli yields mutual lateral inhibition of areas in which a large number of conspicuous responses are present [6].

We currently use a simple normalization scheme, consisting of promoting those feature maps in which a small number of strong peaks of activity are present, while suppressing feature maps eliciting comparable peak responses at numerous locations over the whole visual scene. The first step is to normalize all the feature maps to the same dynamic range, in order to eliminate across-modality amplitude differences due to dissimilar feature extraction mechanisms. In biological systems, this step may reflect rapidly adapting fine tuning of weighting factors assigned to the various feature maps. Then, for each map, the global maximum M of the activity is found, and an average $\overline{m}$ of all the other local maxima in the map is computed. Finally, the map is globally multiplied by $(M - \overline{m})^2$. The direct consequence of this normalization scheme is to strongly amplify those feature maps in which the most conspicuous location (global maximum) elicits a much stronger response than, on average, the other conspicuous locations (secondary maxima). Such contents-based amplification of the feature maps might be implemented in biological systems (in a local neighborhood rather than for the whole visual field) by lateral inhibition mechanisms, suppressing large areas of rather uniform peak activities while enhancing strong isolated peaks.

After normalization, feature maps are combined into three separate channels for intensity, color, and orientation. These combinations are performed across scales and within each channel to yield three conspicuity maps at the spatial scale of the saliency map. The reduction of each feature map to the scale of the saliency map is obtained by using auxiliary Gaussian pyramids. The intensity conspicuity map is obtained simply by adding up the six normalized (using the normalization method described previously) and downscaled intensity feature maps. The color conspicuity map is obtained by adding up the six red-green and six blue-yellow normalized and downscaled feature maps. Four intermediate orientation maps are first obtained by adding up the six normalized and downscaled center-surround maps for a given orientation (0, 45, 90 or 135 degrees). Each of these four maps is then normalized by the same method as used previously. The four orientation maps are finally added up, to form the orientation conspicuity map. The three conspicuity maps for intensity, color and orientation each undergo one final normalization, and are then added together, to form the final input to the saliency map. A total of 50 maps is hence computed by our system (42 feature maps, 4 intermediate orientation maps, 3 conspicuity maps, and the final input to the saliency map neural network). The motivation for the creation of three separate channels and their individual dedicated normalization is the hypothesis that similar features compete strongly for salience, while different modalities contribute independently to the saliency map.

The biophysical assumption underlying the linear combination of normalized feature maps is that of linear summation of incoming synaptic potentials in the dendritic tree. This feedforward procedure for the fusion of information has the advantage of being very fast compared to relaxation methods suggested in a similar context [52].

Internal dynamics and trajectory generation

By definition, the activity in a given location of the saliency map represents the relative conspicuity of the corresponding location in the visual field. At any given time, the maximum of this map is therefore the most salient stimulus to which the focus of attention should be directed next, to allow more detailed inspection by the "what" pathway with its powerful object recognition capabilities not available to the "where" pathway. To find the most salient location, we have to determine the maximum of the saliency map.

This maximum is selected by application of a winner-take-all (WTA) mechanism. Different mechanisms have been suggested for the implementation of neural winner-take-all networks [41, 90]. In our model, we used a two-dimensional layer of integrate-and-fire neurons with strong global inhibition in which the inhibitory population is reliably activated by any neuron in the layer.[1] Therefore, when the first of these cells fires, it will inhibit all cells (including itself), and the neuron with the strongest input will generate a sequence of action potentials. All other neurons are quiescent.

For a static image, the system described so far would continuously attend the most conspicuous stimulus. This is neither observed in biological vision nor desirable from a functional point of view; instead, after inspection of any point, there is usually no reason to dwell on it any longer and the nextmost salient point should be attended. An additional temporal effect comes into play after the focus of attention has moved away from a momentaneously attended location. There is strong psychophysical evidence that the visual system tries to avoid shifting back the focus of attention to a location which it has just visited. One of the most direct observations of this behavior is in terms of reaction times which are longer when the subject is requested to return attention to a location which had just[2] been attended [69, 87, 17]. Kwak and Egeth [39] showed that this "inhibition of return" has a strong spatially defined component, i.e., the return is inhibited to the *location* of the last attended item. Although location is not the only feature on which inhibition of return is based [45], this behavior is in agreement with previously mentioned data [56, 82, 77] which showed that spatial location has a somewhat special role among object features.

We achieve this behavior by introducing feedback from the winner-take-all array to the saliency map. When a spike occurs in the WTA network, the integrators in the saliency map receive additional input with the spatial structure of an inverted Mexican hat, i.e. a difference of Gaussians (Fig. 6.2). The inhibitory center (with a standard deviation of half the ra-

---

[1] A more realistic implementation would consist of populations of neurons. For simplicity, we model such populations by a single neuron with very strong synapses.

[2] A lengthening of the reaction time was observed if the time between the cue and the stimulus exceeded about 300 ms and was less than about 1.5 s

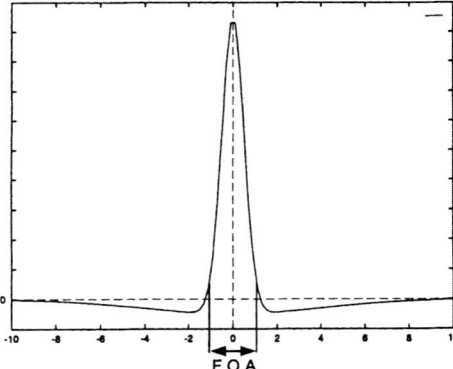

FIGURE 6.2. Radial section of the inhibitory feedback triggered around the last attended location. The central lobe provides strong inhibition which causes the focus of attention (F.O.A.) to jump towards another location. The left and right excitatory lobes (whose sign is, of course, opposite to that of the central inhibitory lobe) slightly enhance the salience of the immediate surround of the currently attended location, favoring short jumps of the focus of attention over long jumps.

dius of the focus of attention) is at the location of the winner. Cells in the saliency map located there receive maximal inhibition and this location, which used to be the global maximum of saliency, now becomes a (local or global) minimum of the saliency map. As a consequence, attention switches to the nextmost conspicuous location (Fig. 6.3). The function of the positive lobes (half width of four times the radius of the focus of attention) of the inverted Mexican hat is to favor locality in the displacements of the focus of attention: If two locations are of nearly equal conspicuity and one of them is close to the previously attended location and the other is far away, attention will jump to the closer location rather than to the distant one.

The inhibitory feedback of the saliency map is consistent with the aforementioned inhibition of return phenomenon observed psychophysically. Not all observed properties of the inhibition of return phenomenon can be explained, however, in the present version of our model. For instance, we assume that the location at which inhibition of return occurs is computed relative to the global stimulus environment. It has been observed, in fact, that this location can be relative to other stimuli [80]. If more than one frame of reference is available (e.g., the global environment and a nearby object), the coordinates are combined in a nontrivial, not fully understood way [29]. From a functional point of view, it seems desirable that inhibition of return should act in a world-based frame of reference, rather than in a retina-based coordinate system as our model does. This requires that a model of attentional selection which includes the inhibition-of-return location takes into account the different coordinate systems and their interactions. However, recent psychophysical results [36] indicate that visual search strategies make use of surprisingly little information across movements of the focus of attention, thus obviating the need to keep track of

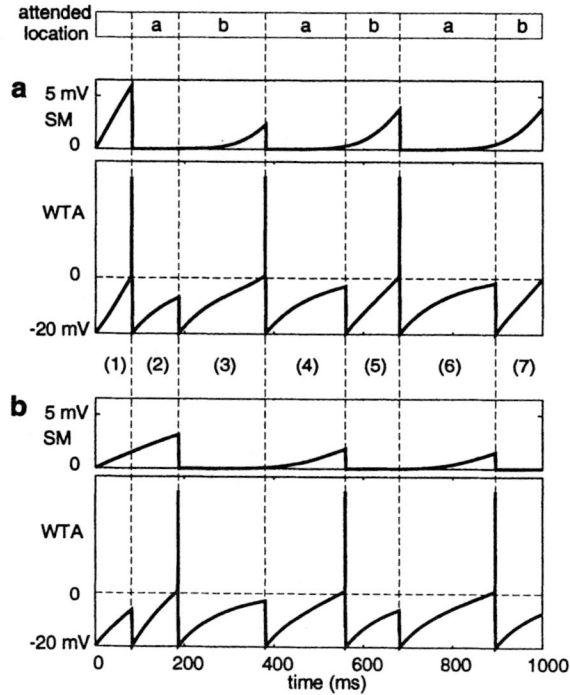

FIGURE 6.3. Dynamical evolution of the potential of some simulated neurons in the saliency map (SM) and in the winner-take-all (WTA) networks. The input contains one salient location (a), and another input of half the saliency (b); the potentials of the corresponding neurons in the SM and WTA are shown as a function of time. During period (1), the potential of both SM neurons (a) and (b) increases as a result of the input. The potential in the WTA neurons, which receive inputs from the corresponding SM neurons but have much faster time constants, increases faster. The WTA neurons evolve independently of each other as long as they are not firing. At about 80ms, WTA neuron (a) reaches threshold and fires. A cascade of events follows: First, the focus of attention is shifted to (a); second, both WTA neurons are reset; third, inhibition of return (IOR) is triggered, and inhibits SM neuron (a) with a strength proportional to that neuron's potential (i.e., more salient locations receive more IOR, so that all attended locations will recover from IOR in approximately the same time). In period (2), the potential of WTA neuron (a) rises at a much slower rate, because SM neuron (a) is strongly inhibited by IOR. WTA neuron (b) hence reaches threshold first. (3)—(7): In this example with only two active locations, the system alternatively attends to (a) and (b). Note how the IOR decays over time, allowing for each location to be attended several times. Also note how the amount of IOR is proportional to the SM potential when IOR is triggered (e.g., SM neuron (a) receives more IOR at the end of period (1) than at the end of period (3)). Finally, note how the SM neurons do not have an opportunity to reach threshold (at 20 mV) and to fire (their threshold is ignored in the model). Since our input images are noisy, we did not explicitly incorporate noise into the neurons' dynamics.

different coordinate systems. If substantiated, these results clearly support our model.

6.3   Simulation Results

We have studied the behavior of the system with input from two large classes of simulated visual stimuli. The first are visual scenes constructed analogously to the stimuli typically presented in psychophysical studies of visual search (discussed in Section 6.3.1). The second class of input are color images of natural and artificial environments. Results from the second class are discussed in Section 6.3.2.

6.3.1   Synthetic stimuli

One of the simplest possible tasks is the detection of a bright spots on a dark backgrounds, or dark spots in bright backgrounds. This task is solved reliably by our model, and the focus of attention immediately jumps to such stimuli. If there is more than one such stimulus, the system scans them one-by-one, in the order of decreasing contrast from the background. The same is true for stimuli that have a color or orientation different from that of the background.

More challenging tasks are defined by the typical stimuli employed in visual search tasks, which is one of the most active fields of human psychophysics. A typical experiment consists in a speeded alternative forced-choice task in which the presence of a certain item in the presented display has to be either confirmed or denied. Salient visual features "guide" attention to the stimuli possessing these features, and these stimuli are selected efficiently. In particular, it is known that stimuli which differ from nearby stimuli in one or more feature dimensions can be easily found in visual search, typically in a time which is nearly independent of the number of other items ("distractors") in the visual scene. In contrast, search times for targets which differ from distractors by a combination of features (a so-called "conjunctive task") are typically proportional to the number of distractors.

In 1980, Treisman and Gelade [81] published an elegant explanation of this fact. The underlying computational principle is that the detection of a few elementary features is most economically done by massively parallel processes early in the visual hierarchy. Because the number of possible combination of features increases very rapidly with the number of features to be combined, there are parallel, preattentive maps only for the elementary features. "Conjunctions" of the features can only be processed by a central attentional authority, which gives rise to the sequential attentional process. Although this "Feature Integration Theory" explained many properties of visual search, it was shown later that it is invalid in its simplest form [20, 85, 53, 68].

We reproduced one of the original experiments used by Treisman and Gelade in order to relate the performances of our model on a "simulated psychophysical experiment" to human psychophysics. Artificial stimulation images were generated by the computer. They were of three classes: (1) one red target (rectangular bar) among green distractors (also rectangular bars) with the same orientation; (2) one red target among red distractors with orthogonal orientation; and (3) one red target among green distractors with the same orientation and red distractors with orthogonal orientation. In order not to artifactually favor any particular direction, the orientation of the target was chosen randomly for every image generated. Also, in order not to obtain ceiling performance in the first two tasks (100% pop-out), we added strong orientation noise to the stimuli (between -17 and +17 degrees with uniform probability) and strong speckle noise to the images (each pixel had a 15% probability, drawn from a uniform distribution, to become a maximally bright red, green, blue, cyan, purple, yellow, or white). The positioning of the stimuli along a uniform grid was also randomized (by up to ± 40% of the spacing between stimuli, in the horizontal and vertical directions), to eliminate any possible influence of our discrete image representations (pixels) on the system. Twenty images were computed for a total number of stimuli per image varying between 4 and 36, yielding the evaluation of a total of 540 images.

Results are presented in Fig. 6.4. Clear pop-out was obtained for the first two tasks (color only and orientation only), independently of the number of distractors in the images. Slightly worse performances are found when the number of distractors is very small, which seems sensible since in these cases the distractors are nearly as conspicuous as the target itself. Evaluation of these types of stimulation images without introducing any of the distracting noises exposed above yielded 100% pop-out (target found as the first attended location) in all images. The conjunctive search task yielded a linear increase, with the number of distractors, in the number of false detections prior to the correct detection of the target. This result is in good accordance with human psychophysics. Notice that the large error bars in our results indicate our model usually finds the target either quickly (in most cases) or after scanning a high number of locations. The target was always found, and the system was never trapped into a recurrent cycle passing through only a limited number of locations.

6.3.2   Natural images

One of the most severe tests an artificial vision system can undergo is the evaluation of its performance when applied to natural images. We have therefore studied the behavior of our model attentional system using such images as input and we describe some of the results in this section. A substantial difficulty is, however, that it is not straightforward to establish objective criteria for the performance of the system. Unfortunately, nearly

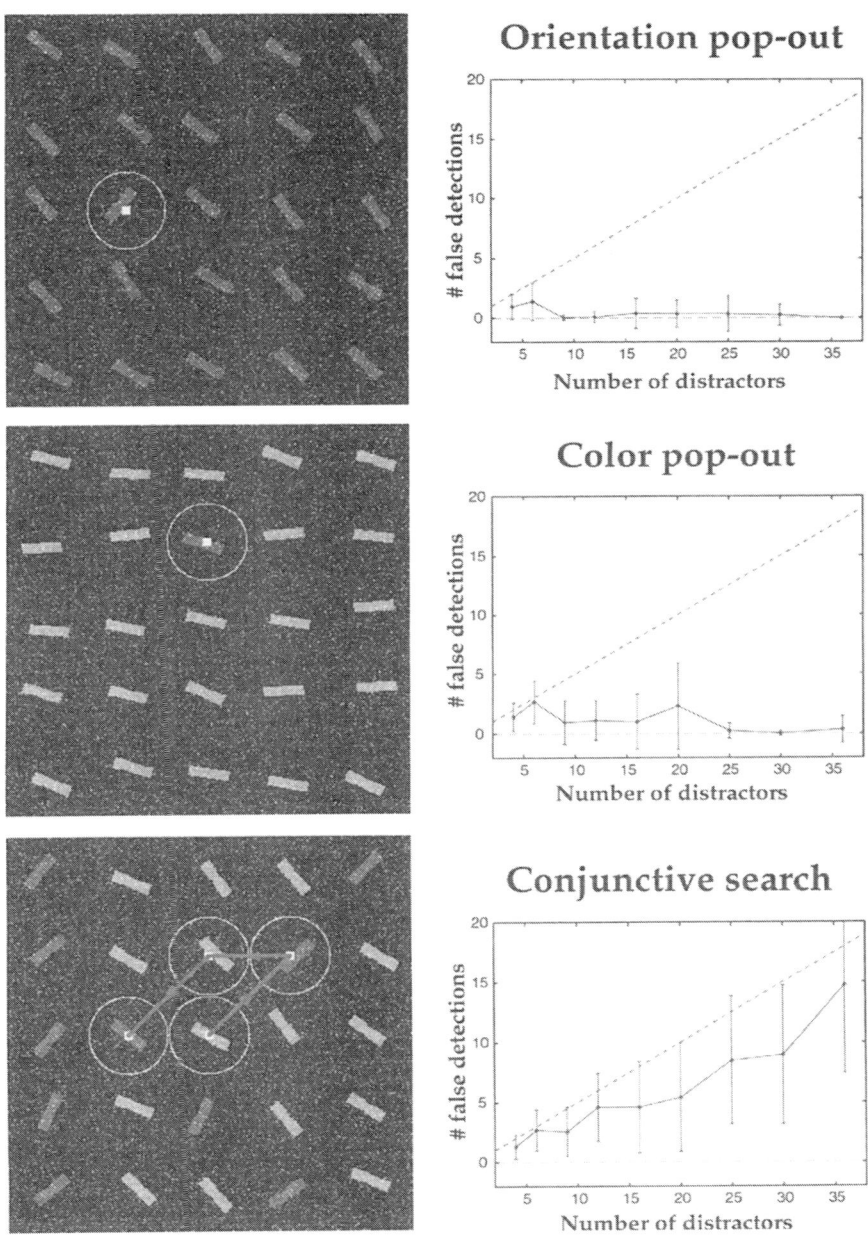

FIGURE 6.4. Performance of our current visual attention model on the pop-out and conjunctive tasks pioneered by A. Treisman. The typical search slopes of human observers in feature and conjunction search, respectively, are successfully reproduced by the model.

all quantitative psychophysical data on attentional control have been obtained based on synthetic stimuli similar to those discussed in section 6.3.1. Consequently, we are largely limited to plausibility arguments, i.e., we have to judge the performance on natural images by making arguments about the probable functional significance and usefulness of the strategy the system uses. The scan paths of overt attention (eye movements) are much better known than those of covert attention [42, 64, 89]. It is unclear, however, to what extent these scan paths are similar to the motion of covert attention since the requirements and limitations (e.g. spatial and temporal resolutions) of the two systems are most likely quite different.

We tested our model on a wide variety of real images, ranging from natural outdoor scenes to artistic paintings. All images were in color. They contained significant amounts of noise, strong local variations in illumination, shadows and reflections, large numbers of "objects" often partially obstructed, and strong textures. We present in Figs. 6.5 and 6.6 some trajectories obtained without introducing any particular modification or tuning in the model. The examples shown (as well as some others we studied and which are not illustrated here) indicate that the system scans the image in an order which makes functional sense in most behavioral situations.

One particularly interesting application is to use the model as a target detector in complex natural scenes. Although our algorithm will focus on salient objects irrespectively of their nature, human-made objects usually are fairly salient in natural environments and are quickly detected by the system. Examples of such applications which we have investigated include the detection of traffic signs on low-resolution color video frames (Fig. 6.5), or the detection of very small military vehicles (Fig. 6.7) in very large digitized photographs (6144 × 4096 pixels) [79]. Further details about these applications may be found on our World-Wide-Web site, at http://www.klab.caltech.edu/~itti/attention/.

6.4   Discussion

6.4.1   Psychophysical and physiological basis of the model

We present in this chapter a prototype for a system mimicking the control of visual selective attention. The model identifies the most salient points one by one in a visual scene and scans the scene autonomously in the order of decreasing saliency. This allows the control of a subsequently activated processor which is specialized for detailed object recognition.

The model is formulated in terms of interacting neuronal populations. The elements of the model are integrate-and-fire neurons, a crude but not unreasonable approximation for many neurons in the nervous system. Our model is compatible with the known anatomy and physiology of the primate visual system, and the way its different parts communicate by signals which are neurally plausible. In particular, there is evidence for a representation of visual information in terms of feature maps, for the generation of the feature

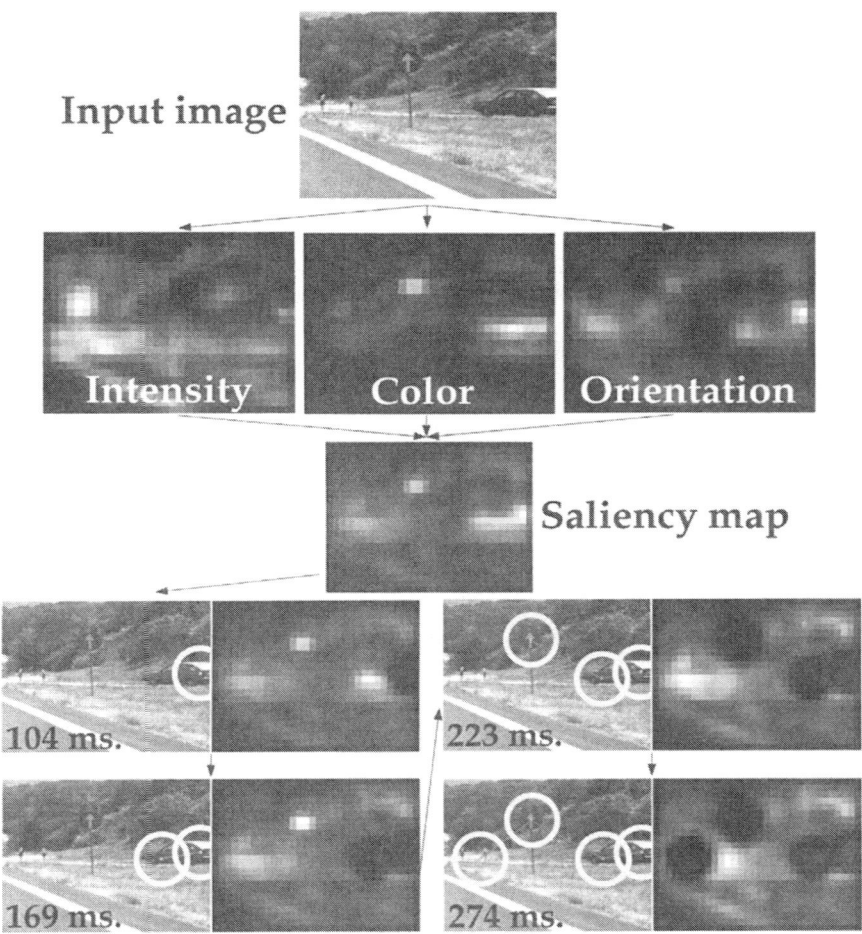

FIGURE 6.5. Example of the global working of our model. Feature maps extracted from the input image at several scales are combined into the saliency map. A winner-take-all neural network then successively selects, in order of decreasing saliency, the attended locations. Once a location has been attended, it is transiently suppressed by the inhibition of return mechanism. Note how the inhibited locations recover over time (e.g., the first attended location has regained some activity at 274 ms).

FIGURE 6.6. More examples of the application of the model to real-world images. Shown are only the trajectories of the focus of attention.

maps in terms of center-surround operations etc; for a recent example of physiological support, see [55].

In addition to its biologically plausible structure and to the realistic input the model operates on, it also provides output which can be used immediately for information selection. The output of the model consists of elevated activity at the location representing the instantaneously attended location in visual space in a topographic representation. This is exactly the type of input required for attentional selection in the ventral pathway. In previous work [57], we developed a model which can use the output from the present model to implement the attentional selection process for object recognition. The same is true for a variation of that model [60] except that it would require that the output of the "where" pathway have a periodic (repeating in time) structure, which could be generated either by intrinsically oscillating neurons or by network effects. The present model demonstrates a neurally plausible substrate that can generate the appropriate control signals which select visual targets from their background.

FIGURE 6.7. Example of detection of a military vehicle in a highresolution (6144 × 4096 pixels) color photograph. This image is part of a database of 44 images for which human search times have been measured [79]. After scaling of the model's time such that it made three attentional shifts per second on average, and an addition of 1.5 sec. to account for human motor reaction time, the model found the target faster than humans in 75% of the images.

Sequential scanning of the visual scene requires movement of the focus of attention, analogously to eye movements.[3] Several independent sets of data indicate that the time required for one shift of attention is on the order of $30 - 50 \; ms$ (for a direct measurement, see [76]) which is consistent with the behavior of the model. Longer dwell-times of the focus of attention are observed in conditions in which a stimulus is flashed, the subject attends to it, then the next stimulus is flashed and attended etc [18]. It is possible that flashing a stimulus disrupts the attentional process, possibly by its action on the saliency map.

There is stronger evidence in favor of a *functional* saliency map than there is for a *spatially localized* saliency map. In other words, the functionality of a saliency map may be spread over different anatomical areas. Robinson and Petersen [71] reviewed data showing that the pulvinar nuclei of the thalamus play a significant role in the selection of visual targets. However, it seems unlikely that the pulvinar is the only location implicated in this selection process. Other candidate areas are the posterior parietal cortex [5, 46, 51] and the superior colliculus [31]. A possible scenario is that of a saliency map distributed over two or more of these structures.

Psychophysical evidence indicates that the map may not be organized retinotopically but instead is realized with a coordinate system relative to the visual environment of the observer [69] or relative to the observed objects [80]. Recent data provides evidence for a neural substrate of object-centered coordinate systems, at least for overt attention [67] and the functional arguments in favor of coordinate systems varying with the visual scene and task requirements. On the other hand, we have already mentioned results [36] which indicate that surprisingly little information may be conserved across movements of the focus of attention. If these results turn out to be of general validity, our simple retinotopic coordinate system were all that is required. More experimental evidence is required before the question of the coordinate system can be answered authoritatively.

The dynamics of the saliency map are determined in our model by the interplay between the winner-take-all mechanism and the feedback provided at the location of the instantaneous winner. As a result, the system will find a winner, direct attention to its location, and then move on to the next-salient location by suppressing the activity at the previous winner. One of the side effects of these dynamics is that the system will have a tendency to avoid the location of a previous winner, i.e. the previous focus of attention, for some time. The existence of this effect is well-established in the psychophysical literature [69, 48, 29, 80]. Furthermore, suppression of the attended location has been observed experimentally in area 7a of

---

[3]Analogous to the physical limitions which allow only one point of fixation for the eyes, there is strong evidence that the focus of attention is unique, i.e., there is only one focus of attention at any given time. An interesting exception are split-brain patients who may dispose of two foci, presumably one for each hemisphere [25, 43].

rhesus monkeys performing a match-to-sample task [75, 74]. Inhibition of return is a natural consequence of our model.

6.4.2 Limitations of the model

Our model does not explain every aspect of attentional control. We have already mentioned that, for the time being, we focus on stimulus-driven (bottom-up) control and make no attempt to implement top-down attentional influences in any detail. The prototypical top-down attention task is perhaps the one formalized by Posner [70] although its roots go back to, at least, Hermann von Helmholtz [32]. In this task, subjects are instructed to attend to one part of their visual field which is identified to them by information requiring cognitive processing or memory or both, e.g. by an arrow pointing towards the area to be attended, or by a verbal command ("attend to the upper left quadrant.") A robust increase in performance, either measured as decreased reaction time or increased accuracy, is observed for stimuli appearing in the attended area. It is likely that at least some aspects of this effect could be captured in our model by providing additional input to the saliency map

Top-down attention is not limited to space-based selection. We can attend to objects and events using nonspatial criteria. One wellcontrolled experimental paradigm which involves selection based on the memory of non-spatial properties is the delayed-match-to-sample paradigm (e.g, [14]), in which the animal subject has to select one of several stimuli based on a cue stimulus presented some time (typically, a few seconds) earlier and held in working memory. In work complementary to the present approach, Usher and Niebur recently developed a model for attentional selection in this condition [84]. Another approach to object-based attentional selection was recently modeled by Ballard and collaborators [72]. Using a representation in terms of multiple scales of oriented filters and chromatic information – similar to that presented here and in our previous report [58] – they modeled selection strategies for saccadic eye movements ("overt attention") by matching iconic target and scene representations. Clearly, this approach could be used without any change in the framework of the present model.

There are other phenomena that our model does not explain explicitly but could possibly be incorporated relatively easily. One example are "express" attentional shifts, first observed by Mackeben and Nakayama [49]. These authors found that their subjects can generate rapid shifts of attention by using exactly the same paradigm supposed to underlie express saccades [23], i.e. by turning the fixation point off and thus facilitating disengagement of attention. A related hypothesis for the interaction between attention and express saccades was put forward by Fischer and Weber [24]. It has been proposed that express saccades can be explained in a framework similar to that proposed in the present report as "normal" saccades with abnormally short dwell times. The short dwell time was proposed to be caused by very rapid updating of the saliency map and consequently rapid issue of the motor command to execute the next eye movement [72]. A similar mechanism might be at work for covert attentional shifts rather than eye movements; if so, express attentional shifts could be explained in a similar way by the present model.

We do not take into account a wealth of psychophysical results on the finer properties of stimuli and their interactions. An example is the observation of Wolfe [88] showing that the metrics in different feature maps may be different from those chosen in our model. For instance, he found that mirror-symmetric orientations are more similar to each other than the numerical values for the angular separation would indicate. Another example are part-whole interactions between elements of search stimuli [86] or the somewhat special role that color seems to play, insofar as it was the only feature tested by Nothdurft [62] that did not require a local feature contrast for parallel detection. Again, it appears that our model should be general enough to allow such effects to be added.

Our model does not explain grouping (see [33] for a connectionist model which focuses on grouping in visual search). There is some evidence indicating that grouping (at least of texture elements) is performed not preattentively but at a second (attentive) perceptual stage; for instance, Nothdurft found that only salient stimuli group [61]. This would indicate that grouping happens at a stage of processing which is beyond that modeled in the present work.

6.4.3   Relationship to other models

Our model is a member of a much larger class of models based on the dichotomy between parallel, effortless, pre-attentive processing and sequential, effortful, attentive processing [7, 34, 35, 54]. Variations of the saliency map concept have been used in multiple instances and with different terminologies (e.g. the "Master Map" of Treisman [83] or the "Activation Map" of Wolfe and collaborators [85, 88]). Desimone and Duncan [15] recently suggested that no saliency map may be required at all and selective attention is instead a consequence of interactions between feature maps only. While this is certainly a possibility, one argument in favor of a saliency map is that it provides a convenient structure for the fusion of information required to compute a single location from the data from a multitude of feature maps.

There is strong psychophysical evidence indicating that integration of information across visual dimensions takes place in attentional selection [63]. Even when saliency was produced in a feature domain irrelevant for the task, targets were detected as quickly as when saliency was generated completely within the relevant feature dimension. Such integration across dimensions (and nonspecific with respect to the target properties) is exactly what the saliency map introduced in section 6.2.4 is based upon.

A position intermediate between the protagonists and antagonists of a saliency map is taken by Braun and Sagi [8, 9] who suggested that bottom-up and top-down mechanism have different capabilities and selection criteria and that only the bottom-up control relies on a saliency map. Since we focus in this work on bottom-up influences, our model is compatible with their results.

Another model in the mentioned class was developed over the last years by Wolfe and collaborators [85, 88]. In previous work, most notably in the earlier versions of Treisman's influential Feature Integration Theory, the activity in preattentive feature maps is observed by a cognitive process which is only capable of deciding whether a given stimulus is the target or not. In contrast, Wolfe et al. assumed that activities in several feature maps can be combined and "guide" the focus of attention toward the most promising locations. Thus, we have "Guided Search". This aspect is, of course, very similar to our approach and, as a consequence, many predictions made by our model are also made by Guided Search (e.g., triple conjunctions should be easier to find than simple conjunctions; this was confirmed experimentally [85]). There are, however, significant differences to our model. The most important is the level of modeling. While our model is explicitly formulated in terms of neuronal populations and their interactions, Guided Search is a functional model without immediate constraints imposed by a physiological substrate.

Other models were described over the last years in the connectionist literature [2, 22, 30, 33, 38, 47, 73, 78]. In this paradigm, networks are constructed from interacting units (assumed to roughly correspond to neurons or groups of neurons) which are connected in various ways. The basic functions in many connectionist networks are, however, quite different from those of biological neurons (for instance, units may exchange informations about pointers, abstract addresses, etc.). This makes it difficult to compare predictions of connectionist architectures with physiological observations.

In contrast to these connectionist approaches, Olshausen and collaborators [65] developed a model for the neural basis of attention at a similar level of neural plausibility as the one presented in this report. Their model is based on the assumption of "shifter circuits" which switch the synaptic input to neurons in higher areas and thereby select which part of the sensory information is made accessible to higher cortical areas. Since in this model the input to the neurons is gated whereas in our model the activity of the cells themselves is modulated, Desimone [16] called the model in ref [65] "input-gated" and models of the class in this report "cell-gated". Both models make clear predictions which should be experimentally verifiable. For instance, one of the defining properties of the [65] model is the conservation of spatial relations within the focus of attention. In contrast, since *all* stimuli in the focus of attention are tagged in the model in this report, the loss of spatial relationships within the focus of attention is predicted by our model.

6.4.4   Predictions

The mammalian visual system is characterized by a sequence of cortical areas which, despite all their differential specializations, shows a uniform trend: As we go from close to the sensory periphery to more central areas, neuronal responses are characterized by a simultaneous increase in feature specifity and a decrease in spatial specificity. For instance, many cells in area V1 (close to the sensory periphery) respond to *any* elongated structure of a certain orientation (close to the preferred orientation of the cell under study), provided it is in its *small* receptive field. On the other hand, cells in inferotemporal cortex (distant from the periphery) will only respond to *specific* stimuli e.g., a face, but they will do so in a very *large* receptive field.

One reason why this architecture evolved may be found in the combinatorial character of highly specific stimuli: It is simply not possible to provide more than a few cells which are sensitive to such stimuli, and these cells therefore have to be usable in large parts of the visual field. It was realized early on that this architecture leads, however, to a vexing problem: Since the location information seems to be largely lost in higher areas, how do these neurons "know" *where* to attribute the different properties of two or more stimuli present simultaneously in the visual field?

Over the last few years, several groups have proposed that this so-called "binding problem" may be solved by attaching "tags" to the different parts of an object whose neuronal realization is the temporal structure of the spiketrains coding for each object [13, 19, 40, 57, 60]. In the present model, the problem would be solved in a very simple way, or rather, it would not exist in the first place. We suggest that at any given time, only one out of the possibly many simultaneously present stimuli is selected and all other stimuli are suppressed.[4] This seems to be one of the simplest methods to solve the binding problem. Thus, attention controls access to visual awareness and we can only be aware of a single stimulus at a time, compatible with much psychophysical evidence.

Support for our model and, in fact, all models which are based on the existence of an anatomically identifiable saliency map has been provided recently by Friedman-Hill and coworkers [21]. These authors identified a patient with bilateral parietal-occipital lesions who showed exactly the kind of problems to be expected in a patient who is lacking significant parts of a saliency map. In particular, this patient routinely shows evidence of "illusory conjunctions," i.e., he miscombines colors and shapes even under free viewing conditions (i.e., in the absence of high perceptual load which

---

[4]The model presented in this report is only concerned with the selection of the attended stimuli, not the suppression of unattended stimuli. In earlier work [60, 57], we have shown how tags consisting of temporal modulation of spike trains can be used to inhibit neurons responding to nonattended stimuli.

is required to generate illusory conjunctions in normal subjects). In the context of our theory, we would predict that the absence of a saliency map (or important parts of it) leads to the absence of the signal which is used to distinguish the (usually unique) attended object from all others, therefore leading to the miscombination of the features of several objects.

A prediction of our model is that the input to central sensory areas (e.g. the inferotemporal areas) should rapidly change under free viewing conditions in a complex environment. This prediction should be verifiable in electrophysiological experiments, but there are only few studies of cortical neuronal responses under free viewing conditions [27, 28].

Another prediction is that inhibition should be observed at the attended location in structures likely to control the position of the focus of attention, due to the inhibition underlying the scanning mechanism discussed in section 6.2.4. This is a particularly valuable prediction because it seems counterintuitive. Furthermore, in earlier work, *enhanced* activity was observed in neurons representing the attended location [5]. However, more recent recordings in posterior parietal cortex of awake behaving monkey, using more consistent deployment of attention to identified locations in the visual field than it was the case in the earlier work, provided evidence for a substantial suppression of activity of those neurons that represent the attended location [75]. The fate of our model is at this time in the hands of experimentalists.

Acknowledgments

Work at Caltech on this project was supported by NSF, the NSF- funded Center for Neuromorphic Systems Engineering and by ONR. Work at Johns Hopkins University was supported by NSF and by the Alfred P. Sloan Foundation.

6.5 REFERENCES

[1] E.H. Adelson, C.H. Anderson, J.R. Bergen, P.J. Burt, and J.M. Ogden. Pyramid methods in image processing. *RCA Engineer*, Nov-Dec., 1984.

[2] S. Ahmad and S. Omohundro. Efficient visual search: a connectionist solution. *Proc. 13th Ann. Conf. Cog. Sci. Soc.* 1991.

[3] J. A. Brefczynski and E. A. DeYoe. A physiological correlate of the spotlight of visual attention. *Nature Neuroscience*, 2:370–374, 1999.

[4] G. Buchsbaum and A. Gottschalk. Trichomacy, opponent colour coding and optimum colour information transmission in the retina. *Proceedings of the Royal Society of London B*, 220:89, 1983.

[5] M. C. Bushnell, M. E. Goldberg, and D. L. Robinson. Behavioral enhancement of visual responses in monkey cerebral cortex. I. Modulation in posterior parietal cortex related to selective visual attention. *J. Neurophysiol.*, 46:755–772, 1981.

[6] J. Braun. Visual search among items of different salience: removal of visual attention mimics a lesion in extrastriate area V4. *J. Neuroscience*, 14:554–567, 1994.

[7] D. E. Broadbent. *Perception and Communication.* Pergamon, London, 1958.

[8] J. Braun and D. Sagi. Vision outside the focus of attention. *Perception and Psychophysics*, 48:45–58, 1990.

[9] J. Braun and D. Sagi. Texture-based tasks are little affected by second tasks requiring peripheral or central attentive fixation. *Perception*, 20:483–500, 1991.

[10] A. J. Bell and T. J. Sejnowski. The independent components of natural scenes are edge filters. *Vision Research*, 37:3327–3338, 1999.

[11] C. Bundesen. Visual selection of features and objects: is location special? a reinterpretation of Nissen's (1985) findings. *Perception & Psychophysics*, 50:87–89, 1991.

[12] M. Carandini and D.J. Heeger. Summation and division by neurons in primate visual cortex. *Science*, 264:1333–1336, 1994.

[13] F. Crick and C. Koch. Towards a neurobiological theory of consciousness. *Seminars in the Neurosciences*, 2:263–275, 1990.

[14] L. Chelazzi, E.K. Miller, A. Lueschow, and R. Desimone. Dual mechanisms of short-term memory: ventral prefrontal cortex. *Soc. for Neuroscience Abstracts*, 19:975, 1993.

[15] R. Desimone and J. Duncan. Neural mechanisms of selective visual attention. *Ann. Rev. Of Neurosci.*, 18:193–222, 1995.

[16] R. Desimone. Neural circuits for visual attention in the primate brain. In G. Carpenter and S. Grossberg, editors, *Neural Networks for Vision and Image Processing.* MIT Press, Cambridge, 1992.

[17] S. Danziger, A. Kingstone, and J. Snyder. Inhibition of return to successively stimulated locations in a sequential visual search paradigm. *Journal of Experimental Psychology: Human Perception and Performance*, 24(5):1467–75, Oct 1998.

[18] J. Duncan, R. Ward, and K. Shapiro. Direct measurement of attentional dwell time in human vision. *Nature*, 369:313–315, 1994.

[19] A. K. Engel, P. König, Kreiter A.K., T. B. Schillen, and W. Singer. Temporal coding in the visual system: new vistas on integration in the nervous system. *Trends in Neurociences*, 15:218–226, 1992.

[20] H.E. Egeth, R.A. Virzi, and H. Garbart. Searching for conjunctively defined targets. *J. Experimental Psychology*, 10(1):32–39, 1984.

[21] S. R. Friedman-Hill, L. C. Robertson, and A. Treisman. Parietal contributions to visual feature binding: evidence from a patient with bilateral lesions. *Science*, 269:853–855, 1995.

[22] K. Fukushima, S. Miyake, and T. Ito. Neocognitron: A neural network model for a mechanism of visual pattern recognition. *IEEE Transactions on Systems, Man, and Cybernetics*, 13(5):826–834, 1983.

[23] B. Fischer and E. Ramsperger. Human express-saccades: extremely short reaction times of goal directed eye movements. *Experimental Brain Research*, 57:191–195, 1984.

[24] B. Fischer and H. Weber. Express saccades and visual attention. *Behavioral and Brain Sciences*, 16:553–610, 1993.

[25] M.S. Gazzaniga. Independent hemispheric attentional systems mediate visual search in split-brain patients. *Nature*, 342:543–545, 1989.

[26] H. Greenspan, S. Belongie, R. Goodman, P. Perona, S. Rakshit, and C.H. Anderson. Overcomplete steerable pyramid filters and rotation invariance. In *Proceedings of the IEEE Conference on Computer Vision and Pattern Recognition*. IEEE, 1994.

[27] J. L. Gallant, C. E. Connor, H. Drury, and D. C. Van Essen. Neural responses in monkey visual cortex during free viewing of natural scenes – mechanisms of response suppression. *Investigative Ophthalmology and Visual Science*, 36(4):S1052, 1995.

[28] J. L. Gallant, C. E. Connor, and D. C. Van Essen. Neural activity in areas v1, v2 and v4 during free viewing of natural scenes compared to controlled viewing. *Neuroreport*, 9(9):2153–2158, 1998.

[29] B.S. Gibson and H. Egeth. Inhibition of return to object-based and environment-based locations. *Perception & Psychophysics*, 55(3):323–339, 1994.

[30] S. Grossberg, E. Mingolla, and W.D. Ross. A neural theory of attentive visual search: interactions at boundary, surface, spatial and object recognition. *Psychological Review*, 101(3):470–489, 1994.

[31] M. E. Goldberg and R. H. Wurtz. Activity of superior colliculus in behaving monkey II: The effect of attention on neuronal responses. *J. Neurophysiology*, 35(560-574), 1972.

[32] H. von Helmholtz. *Handbuch der physiologischen Optik.* Voss, Leipzig, 1867.

[33] G.W. Humphreys and H.J. Müller. Search via recursive rejection (SERR): A connectionist model of visual search. *Cognitive Psychology*, 25:43–110, 1993.

[34] J. E. Hoffman. Search through a sequentially presented visual display. *Perception & Psychophysics*, 23(1):1–11, 1978.

[35] J. E. Hoffman. A two-stage model of visual search. *Perception & Psychophysics*, 2325(4):319–327, 1979.

[36] T. S. Horowitz and J. M. Wolfe. Visual search has not memory. *Nature*, 394:575–577, 1998.

[37] L. Itti, E. Niebur, and C. Koch. A model of saliency-based fast visual attention for rapid scene analysis. *IEEE Transactions on Pattern Analysis and Machine Intelligence*, 20(11):1254–1259, November 1998.

[38] S. R. Jackson, R. Marrocco, and M. I. Posner. Networks of anatomical areas controlling visuospatial attention. *Neural Networks*, 7(6/7):925–944, 1994.

[39] H. Kwak and H. Egeth. Consequences of allocating attention to locations and to other attributes. *Perception & Psychophysics*, 51(5):455–464, 1992.

[40] P. König, A. K. Engel, S. Löwel, and W. Singer. How precise is neuronal synchronization? *Neural Computation*, 7(3):469–485, 1995.

[41] C. Koch and S. Ullman. Shifts in selective visual attention: towards the underlying neural circuitry. *Human Neurobiol.*, 4:219–227, 1985.

[42] M. F. Land and S. Furneaux. The knowledge base of the oculomotor system. *Philosophical Transactions of the Royal Society of London B*, 352(1358):1231–1239, 1997.

[43] S.J. Luck, S.A. Hillyard, G.R. Mangun, and M.S. Gazzaniga. Independent attentional scanning in the separated hemispheres of split-brain patients. *J. Cog. Neurosci*, 6(1):84–91, 1994.

[44] A. Lüschow and H. C. Nothdurft. Pop-out of orientation but no pop-out of motion at iso-luminance. *Vision Res.*, 33:91–104, 1993.

[45] M. B. Law, J. Pratt, and R. A. Abrams. Color-based inhibition of return. *Perception & Psychophysics*, 57(3):402–408, 1995.

[46] V. B. Mountcastle, R. A. Andersen, and B. C. Motter. The influence of attentive fixation upon the excitability of the light-sensitive neurons of the posterior parietal cortex. *J. Neurosci.*, 1:1218–1232, 1981.

[47] R. Milanese, J.M. Bost, and T. Pun. Visual indexing with an attentive system. In *Lecture Notes in Artificial Intelligence*, pages 415–419. Springer Verlag, Berlin, 1991.

[48] E. A. Maylor and R. Hockey. Inhibitory components of externally controlled covert orienting in visual space. *Journal of Experimental Psychology: Human Perception and Performance*, 11:777–787, 1985.

[49] M. Mackeben and K. Nakayama. Express attentional shifts. *Vision Research*, 33:85–90, 1993.

[50] I. R. Moorhead. Human colour vision and natural images. In *Colour in Information Technology and Information Displays*, number 61, page 21. Institution of Electronic and Radio Engineers, Alderman, Ipswich, 1985.

[51] V. B. Mountcastle. The parietal system and some higher brain functions. *Cerebral Cortex*, 5(5):377–390, 1995.

[52] R. Milanese, T Pun, and H. Wechsler. A non-linear integration process for the selection of visual information. In V. Roberto, editor, *Intelligent Perceptual Systems*, pages 323–336. Springer Verlag, Berlin, 1993.

[53] J.T. Mordkoff, S. Yantis, and H.E. Egeth. Detecting conjunctions of color and form in parallel. *Perception & Psychophysics*, 48(2):157–168, 1990.

[54] U. Neisser. *Cognitive Psychology*. Appleton-Century-Crofts, New York, 1967.

[55] H. C. Nothdurft, J. L. Gallant, and D. C.N Van Essen. Response modulation by texture surround in primate area V1: correlates of "popout" under anesthesia. *Visual Neuroscience*, 16(1):15–34, Jan-Feb 1999.

[56] M.J. Nissen. Accessing features and objects: is location special? In M.I. Posner and O.S.M Marin, editors, *Mechanisms of attention: Attention and Performance XI*, pages 205–219. Hillsdale, NJ, 1985.

[57] E. Niebur and C. Koch. A model for the neuronal implementation of selective visual attention based on temporal correlation among neurons. *Journal of Computational Neuroscience*, 1(1):141–158, 1994.

[58] E. Niebur and C. Koch. Control of selective visual attention: Modeling the "where" pathway. In D. S Touretzky, M. C. Mozer, and M. E. Hasselmo, editors, *Advances in Neural Information Processing Systems*, volume 8, pages 802–808. MIT Press, Cambridge, MA, 1996.

[59] E. Niebur and C. Koch. Computational architectures for attention. In R. Parasuraman, editor, *The Attentive Brain*, chapter 9, pages 163–186. MIT Press, Cambridge, MA, 1998.

[60] E. Niebur, C. Koch, and C. Rosin. An oscillation-based model for the neural basis of attention. *Vision Research*, 33:2789–2802, 1993.

[61] H. C. Nothdurft. Feature analysis and the role of similarity in preattentive vision. *Perception & Psychophysics*, 52:355–375, 1992.

[62] H. C. Nothdurft. The role of features in preattentive vision: comparison of orientation, motion and color cues. *Vision Res.*, 33:1937–1958, 1993.

[63] H. C. Nothdurft. Saliency effects across dimensions in visual search. *Vision Res.*, 33:839–844, 1993.

[64] D. Noton and L. Stark. Scanpaths in eye movements. *Science*, 171:308–311, 1971.

[65] B. Olshausen, C. Andersen, and D. Van Essen. A neural model of visual attention and invariant pattern recognition. *J. Neuroscience*, 13(11):4700–4719, 1993.

[66] B. Olshausen and D. J. Fields. Emergence of simple-cell receptive field properties by learning a sparse code for natural images. *Nature*, 381:607–609, 1996.

[67] C. Olson and S. Gettner. Object-centered direction selectivity in the macaque supplementary eye field. *Science*, 269:985–988, August 1995.

[68] J. Palmer. Set-size effects in visual search: the effect of attention is independent of the stimulus for simple tasks. *Vision Res.*, 34:1703–1721, 1994.

[69] M. I. Posner and Y. Cohen. Components of visual orienting. In H. Bouma and D. G. Bouwhuis, editors, *Attention and Performance X*, pages 531–556. Hilldale, NJ, 1984.

[70] M.I. Posner. Orienting of attention. *Quart. J. Exp. Psychol.*, 32:3–25, 1980.

[71] D. L. Robinson and S. E. Petersen. The pulvinar and visual salience. *Trends in Neurosciences*, 15(4):127–132, 1992.

[72] R. P. N. Rao, G. J. Zelinsky, M. M. Hayhoe, and D. H. Ballard. Eye movements in visual cognition: a computational study. Technical Report 97.1, Department of Computer Science, University of Rochester, March 1997.

[73] P.A. Sandon. An attentional hierarchy. *Behavioral and Brain Sciences*, 12:414–415, 1989.

[74] M. A. Steinmetz and C. Constantinidis. Neurophysiological evidence for a role of posterior parietal cortex in redirecting visual attention. *Cerebral Cortex*, 5:448–456, 1995.

[75] M. A. Steinmetz, C. E. Connor, C. Constantinidis, and J. R. McLaughlin. Covert attention suppresses neuronal responses in area 7A of the posterior parietal cortex. *J. Neurophysiology*, 72:1020–1023, 1994.

[76] J. Saarinen and B. Julesz. The speed of attentional shifts in the visual field. *Proc. Nat. Acad. Sci., USA*, 88:1812–1814, 1991.

[77] S. Shih and G. Sperling. Visual search, visual attention and feature-based stimulus selection. *Investigative Ophthalmology and Visual Science*, 34(4):1288, 1993.

[78] G.W. Strong and B.A. Whitehead. A solution to the tag-assignment problem for neural networks. *Beh. Brain Sci.*, 12:381–433, 1989.

[79] A. Toet, P. Bijl, F. L. Kooi, and J. M. Valenton. *A High-Resolution Image Dataset for Testing Search and Detection Models (TNO-TM-98-A020)*. TNO Human Factors Research Institute, Soesterberg, The Netherlands, 1998.

[80] S. P. Tipper, J. Driver, and B. Weaver. Short report: object-centered inhibition or return of visual attention. *Quarterly Journal of Exp. Psychology*, 43A:289–298, 1991.

[81] A. Treisman and G. Gelade. A feature-integration theory of attention. *Cognitive Psychology*, 12:97–136, 1980.

[82] Y. Tsal and N. Lavie. Location dominance in attending to color and shape. *Journal of Experimental Psychology: Human Perception and Performance*, 19(1):131–139, 1993.

[83] A. Treisman. Features and objects: the fourteenth Bartlett memorial lecture. *Quart. J. Exp. Psychol.*, 40A:201–237, 1988.

[84] M. Usher and E. Niebur. A neural model for parallel, expectation-driven attention for objects. *J. Cognitive Neuroscience*, 8(3):305–321, 1996.

[85] J.M. Wolfe, K.R. Cave, and S.L. Franzel. Guided search: an alternative to the feature integration model for visual search. *J. Exp. Psychology*, 15:419–433, 1989.

[86] J.M. Wolfe, S.R. Friedman-Hill, and A.B. Bilsky. Parallel processing of part-whole information in visual search tasks. *Perception & Psychophysics*, 55:537–550, 1994.

[87] D. G. Watson and G. W. Humphreys. Visual marking: Prioritizing selection for new objects by top-down attentional inhibition of old objects. *Psychological Review*, 104(1):90–122, 1997.

[88] J.M. Wolfe. Guided search 2.0–a revised model of visual search. *Psychonomics Bulletin & Review*, 1(2):202–238, 1994.

[89] A.L. Yarbus. *Eye Movements and Vision*. Plenum Press, New York, 1967.

[90] A. L. Yuille and N. M. Grzywacz. A winner-take-all mechanism based on presynaptic inhibition feedback. *Neural Computation*, 2:334–344, 1989.

7

Activity–Gating Attentional Networks

J. Eggert and J. L. van Hemmen

ABSTRACT In the visual system, "attention" selectively enhances and expedites the processing of a subset of the available stimuli vs. the rest. Attention can be directed to many different feature dimensions, such as location, form, color, texture and direction of movement. In this work, we present a model of attentional processing that makes extensive use of the feedforward, lateral and feedback connections known to exist in the visual cortex. The model uses local modulations of the activity of neuronal ensembles to superpose additional saliency and attentional information on top of the sensory data. The additional signals "gate" the information through the entire network and trigger response competition, resulting in an attentional concentration of the processing resources. At the network level, the model consists of two complementary information counterstreams that process separately sensory and attentional data: A sensory, feedforward stream directly analyses the features available in the stimulus, while an attentional stream provides expectations and global hypotheses about the stimulus. We explain the function of such a network as a hypothesis generating and confirming system. We also explain the architecture, components and dynamics necessary for the implementation of such an activity–gating network. The goal is to arrive at a consistent and unified model of attentional processing in the visual system that explains the different types of attention within a single framework.

7.1 Introduction

7.1.1 Different types of attention

The visual cerebral cortex of primates is composed of many distinct, functionally specialized processing areas. They can be roughly grouped into a hierarchy containing multiple levels that represent increasingly complex information about the visual scene (see e.g. [13, 54]). Although the different areas are often mainly dedicated to process visual information conveyed by feedforward connections from retinal sources, there is a large amount of evidence that neurons in the visual cortex can also be affected by extraretinal or indirect visual influences.

"Attention" in the visual cortex can be characterized as a very special indirect influence. It acts in a *modulatory* way, in the sense that it enhances or suppresses already existing sensory information, and, in addition, this modulation is *selective*, meaning that at any single moment it acts only on a restricted portion of the visual scene [3, 12, 22, 23, 37]. Therefore the name "selective attention".

Selective attention in the visual system is frequently implemented in the spatial domain, using a geometrically well-defined region of the visual field, the so-called "focus of attention" (FOA) [5, 12, 46] (see also the chapter by Niebur et al. in this volume). Models based on the FOA idea serve to explain some of the experimental evidence found about locationally guided attention.

In addition, there has been found abundant experimental evidence that attentional processing is not restricted to a purely spatial domain, but that there exist many different nonspatial, feature-based attentional mechanisms (with "feature-based" meaning here any stimulus dimensions other than exclusively spatial ones). Attention can be directed to form features, color, texture or motion, to name only a few known.

7.1.2   Why attentional processing at all?

Why is attention necessary at all? Information processing in the visual cortex is organized along several hierarchically organized pathways. Along with the information flow in each pathway, a specialization takes place, meaning that the cells get more selective to particular stimulus features (see Fig. 7.1). This goes hand in hand with an increasing generalization capability and a loss of information that is irrelevant to the particular pathway. For example, along the form-processing ventral pathway of primates, cells get more selective to particular forms and acquire increasingly more translation invariance. The increasing selectiveness on one hand and the loss of information on the other constitutes a severe information bottleneck if the final result is to be extracted at the top of the pathways only. For example, consider a stimulus with two letters X and U at different locations. One pathway may extract the form information, arriving at the result that both X and U are present, but not knowing in detail where. A second pathway may extract the exact positions of the objects, without being able to identify their form. One of the problems that arises in such an architecture is the binding of the top information between different processing pathways, i.e., in the example, which form belongs to which position.

In some circumstances, like the case of conflicting stimuli, or in cases which require binding of information across pathways, not only the information at the top of the processing hierarchy is necessary, but a part of the entire tree of information flow along the hierarchy. This may comprise the route information takes in a single pathway, or it may involve two or more pathways, as in the X-U example. The binding problem is then equivalent to a tracking of the tree of information flow through the network. Such a tracking process relies heavily upon local operations, a characteristic that

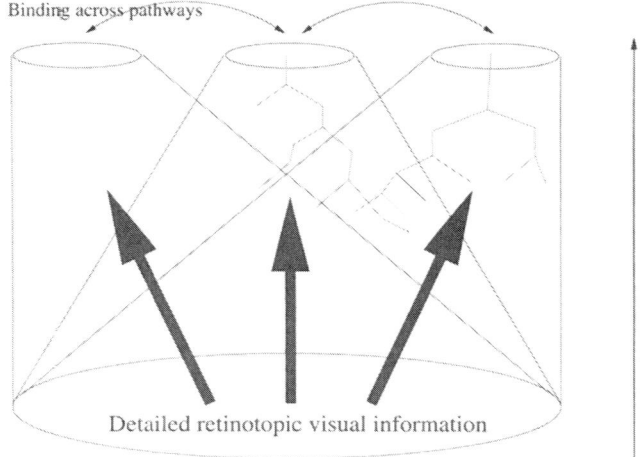

Different processing pathways

Binding across pathways

Detailed retinotopic visual information

Growing specialization,
loss of irrelevant information

FIGURE 7.1. Diagram of the organization of visual processing into several pathways. In each pathway, a specialization and a loss of irrelevant information takes place. This causes a binding problem at the top of the pathways. Nevertheless, the necessary binding information is still contained in the common area of high-resolution at the bottom of the pathways. Attention may be used to solve the binding problem by reaccessing the low-level information and making it explicit at the top of the pathways. We propose a network architecture to accomplish this task. Also shown: Partial trees of the information flow through the network. Solving the binding problem is equivalent to tracking the trees.

will be utilized in a later section when we develop a metaphor to illustrate the main points of our model.

We propose that the role of attention in the visual system is to solve this tracking problem, both along and across the processing pathways. This cannot be done for too many stimuli simultaneously, because the different trees of information flow interfere with each other. To remain tractable, the interference requires attention to act selectively concentrating resources on a subset of the available information, so that only the most important trees are tracked (in other words, attentional resources are limited). By tracking the information trees through the network, attentional processing allows to circumvent the information bottleneck mentioned above, binding and (re)accessing information that was lost when moving upwards in the processing hierarchies.

If the information contained in the entire network would always be *accessible*[1], there would be no such thing as a bottleneck or a binding problem, because at the high-resolution, low-level areas the binding information is present. Complete explicit access is not possible in a finite system, so that attention becomes necessary. In this work, we postulate that for *some specific tasks*, the access of low level information is necessary, and indicate how the binding problem can be solved in these cases using a special network architecture and attention to reaccess and make explicit the required low-level information.

7.1.3 Spotlight models

Spotlight models (SM) of attentional processing constitute a special class of models that operate on a saliency map specifying *where* things and locations of interest are, but not exactly *what* they are. A focus of attention (FOA, see section 7.1.1) is computed preattentively, and attention then operates on the features inside the FOA for preferential processing. Features inside the FOA are routed to higher processing centers. Binding is accomplished by defining that all features inside the FOA belong together. This is an easy task for such models, because the stimuli outside of the FOA are neglected or suppressed.

Problems appear at several points in these models. They imply an architectural division into a preattentively working area (the saliency map area) and the object identification area, which relies on attentive processing. They also imply a strict order of operations, in the sense that spatial segregation (the computation of the FOA) has to precede binding and both have to precede object identification. This means that the attentional processing areas are obliged to wait for the outcome of the saliency map. Experimentally, the order of the operations as imposed by spotlight models is not always confirmed. It is known that identification cues can strongly influence the segregation of a stimulus (take e.g., the well-known picture of a dalmatian dog against a spotty black and white background). In addition, experimental data shows that targets can pop up in a visual search display before they are captivated by the focus of attention, even if they are defined by a conjunction of elementary attributes [8, 9, 29, 56]. This reverts the order of operations, since in this case binding occurs before attentional processing takes place. A similar question is that of detection and localization. Since the focus of attention has to have locked onto an object for attentional processing to occur, localization also precedes identification. But there is evidence that there can be preattentive detection that precedes localization [46]. Thus, the division between segmentation, attentional processing, and binding is not as clear cut as spotlight models suggest.

---

[1]Meaning here that it can be processed or made explicit in such a way that other network areas can make use of it.

The basic question is how the attentional spotlight itself knows where to be directed to, i.e., how the saliency map computes the most sensible FOA. The computations in the saliency map are e.g. modeled using a series of feature detectors working along different feature dimensions and at several spatial scales, and then computing a "winner" (i.e., most salient) location (see also the chapter by Niebur et al. for an implementation of the FOA computation using a saliency map). Since the attentional processing network itself is also assumed to work based on a series of maps containing similar types of feature detectors, spotlight models require a duplicate network architecture.

Another evidence found in psychophysical experiments that shows the complex nature of attentional processing indicates that more than one stimulus participates in attentional processing even if all of the stimuli can be bounded by nonoverlapping concave regions (see e.g. [11]). This suggests a simultaneous processing of a limited number of objects with attention distributed over a wider area that overlaps several stimuli, contrary to the object-per-object manner of strictly serial spotlight models. Similarly, an attentional interaction between several stimuli that are positioned in or near the focus of attention can be observed (see e.g. [32]). And considerable processing even continues to occur for nonattended objects when attention is directed elsewhere.

Finally, there is evidence for attentional mechanisms that operate both with a high spatial resolution (which is incompatible with the FOA notion) and on non-spatial, feature-based attentional mechanisms. The saliency maps of spotlight models have to segment objects based purely on low level feature analysis (i.e., without object identification). This makes it impossible to isolate a single interesting spot in the case of mutually overlapping objects. Further support for high spatial resolution in combination with attentional processing comes from experiments with spatially coincident patterns, such as transparent random dots [2, 4, 27, 35, 52]. Nonspatial, feature-based attentional mechanisms have also been observed in the case of color [31] and motion [36, 48, 49, 50]. In this case, the response of a cell to a stimulus is enhanced when the animal directs its attention towards features that the cell prefers.

7.1.4   The discussion forum metaphor

Attentional processing as it is understood by spotlight models has been compared with a theater stage on which a play is taking place and which includes a single, bright spotlight that is directed towards the stage from outside and which serves to highlight a selected area. In this case, there is need for a central "spotlight manager" that directs the light always to one person or place on the stage at one time. This is the role taken by the saliency map.

The model we propose is quite contrary. It is based on a handful of central assumptions. First of all, we think that the selective function common to the different types of attention (see section 7.1.1) expresses in a common functional architecture, meaning that all types of attention can be explained within a basic framework. Second, coarse locationally guided attentional processing like that of spotlight models is but one (special) type of attention that fits into this framework. Third, there is no single executive center that indicates where (in space, or to what features) attention is directed to. Instead, attention can originate in a series of different centers, explaining the different types of attention. Fourth, attention is not exclusively dedicated to a single object, totally suppressing all non-attended ones, but instead causes a concentration of resources on some aspects of the scene that are then processed preferentially.

To illustrate the operation of an attentional network under such assumptions, we develop in the following a new metaphor, opposed to the (in our view, misleading) theater stage idea. We need no central "spotlight manager" because the decision, which parts of the scene are important, is taken locally by the actors of the play themselves. We say that attentional processing is like the organization of a public discussion in a democratic or artistic forum, with a mixed hierarchy of speakers and a number of available microphones to speak out loud. The microphones constitute a limited resource. Without having to rely upon a central moderator, groups of people with a fluctuating number of participants will concentrate around the microphones, compete with each other for statements and reorganize into new constellations. Sometimes a single person (perhaps, and very likely, of a high hierarchical ranking) will assert himself and dominate temporarily the discussion; at another time, a large group of persons will gather and coordinate themselves to a speaking choir, or different persons will speak one after another. This allows the system to dynamically organize according to the momentary situation, and to adjust optimally to the actual needs.

A consequence of this view is that there are two grouping processes going on at the same time at different levels of organization, one between single persons at the submicrophone level and the other between the different groups of people around each microphone. These two processes correspond to preattentive and attentive processing. In addition, attention is never allocated exclusively, but always distributed among several contenders. The concentration of attentional resources in the brain that can be observed in experiments is only the manifestation of this dynamical organization process (in the discussion forum metaphor, the words that you can hear aloud thanks to the microphones), while the organization itself occurs largely unnoticed. Such a view also implicates that there is no sharp boundary between preattentive and attentive processing in terms of segregated and consecutive processing stages. Instead, preattentive processing can be seen as the processing that takes place in the head of each of the participants and between single persons of the discussion forum, while attentional processing is everything related to the organization of speakers around the microphones and their speaking out loud using them. Therefore, preattentive and attentive processing takes place *on the same substrate*, in this case, the participants of the discussion forum. The difference between preattentive and attentive resides in the modes of operation. In a model implementation, this means that the same architectural structures are used both for preattentive and attentive processing (with all the advantages for the network generation, learning, etc., and without requiring a duplicate network architecture, as in the FOA case, see section 7.1.3). Preattentive and attentive processes could be implemented using different neuronal codes, e.g., synchronized activity for attentional processing that is imposed on top of the non-synchronized, preattentive signal. The discussion forum metaphor also implies that attention is distributed along the entire visual processing system, and that there are multiple distributed areas where the attentional signal can have its origin. These multiple areas coordinate with each other so that the system keeps running smoothly and efficiently. In the following, we present a network model for the implementation of attentional processing in the visual system according to the "discussion forum" metaphor.

7.2   Activity–Gating Networks

7.2.1   Working hypotheses about the coding of information

We propose that groups of neurons in the visual cortex can be joined into neuronal *ensembles* (also called *assemblies*, *pools* or *neuronal cliques*). This is motivated by the cortical columnar organization, which presents numerous dendrites running orthogonally to the layers and thus potentially enables neurons to collect input from all the layers they cross on their way and so to gather information available in all layers of their column. A further motivation to group neurons is the experimental observation that cortical neurons of the same type that are located near to each other tend to receive similar inputs. In experiments one often finds that neurons of the same type that are located close to each other are activated simultaneously, or in a correlated fashion. In cortical networks, this may be due to reciprocal connections and common convergent input.

In modeling studies it therefore seems sensible to consider all neurons of the same type in a small cortical volume as a building block of a neuronal network. All pool neurons have to be equivalent in the sense that they have the same input/output connection characteristics and, additionally, the same dynamics parameters. This is explained in Fig. 7.2. All neurons that constitute a pool feel a common synaptic input field, but still, each neuron evolves according to its own internal dynamics.

In the visual cortex, we can for example assume that all neurons of the same type located in the same layer and in the same cortical column, and which additionally have a similar stimulus selectiveness, form an assembly. Assemblies constitute one of the basic building blocks of our attentional network. In the following, we explain some of the dynamical properties of neuronal assemblies.

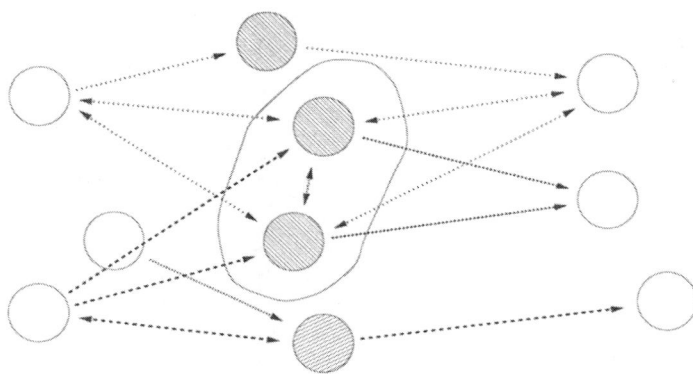

FIGURE 7.2. A "pool" or "assembly" of neurons. Neurons belonging to the same pool or assembly are characterized by having the same input and output connectivity pattern. Furthermore, all neurons of the same pool have the same parameters. In the figure, different types of neurons and connections are characterized by different textures (white neurons are of any type). According to the assembly definition, only the two neurons inside of the closed contour belong to the same pool.

The response of neuronal assemblies is quantified as a macroscopic activity $A(t)$, which has the dimensions of spikes per time. The quantity $A(t)\Delta t$ is the assembly-averaged number of spikes elicited by the pool in the time interval $(t, t + \Delta t]$. Typically, isolated assemblies settle at a constant *stationary activity* if a constant external drive is applied. The activity vs. drive curve then resembles a sigmoidal gain function, commonly encountered in assembly-averaged pool models.

In addition to stationary solutions, self-coupled assemblies can develop stable, self-sustained *oscillatory activity*, e.g., in the case of assemblies composed of neurons of the spike–response or integrate–and–fire type (for a review, see [16]) if the neuronal parameters fulfill the requirements of the locking-theorem [17]. Oscillatory activity has been found in visual cortex in a variety of experiments (see e.g. [42]) and appears particularly in relation with states of pronounced cortical activation and attention (see e.g. [20]).

An *oscillation threshold* often prevents single neurons or groups of neurons to develop oscillatory activity below certain activations, such as it is the case for the "chattering cells", which exhibit normal spiking behavior for low stimulation but elicit bursts if the stimulation is strong enough [18]. For groups of neurons, the existence of an oscillation threshold can be verified with simulations of assemblies of spike–response or integrate–and–fire neurons, both for excitatory and inhibitory reciprocal couplings (see e.g. [17]).

In states of oscillatory activity, an *amplitude coding* scheme enables assemblies to remain sensitive to stimulus changes even if they are working in an oscillatory regime with strong coupling loops and high activity peaks. Amplitude coding means that increasing stimulation strengths are expressed in gradually increasing oscillation amplitudes. Amplitude coding can be observed again both at the single cell level for "chattering cells" which elicit bursts of increasing length in response to increasing stimulation [18], and in simulations of assemblies of spike–response neurons. The amplitude-coding scheme has been verified experimentally comparing optimally and suboptimally driven neurons [26].

The oscillation threshold above is defined for stationary stimulation. Interestingly, although assemblies of spiking neurons with oscillation threshold will not develop oscillatory activity if the stimulation is too weak, they will readily *follow an oscillatory stimulation* (see e.g. [15] for a theoretical account). This means that in this regime, assemblies develop oscillatory activity, e.g. if they receive a modulated input from another assembly with a pronounced oscillation amplitude.

Finally, connections between assemblies can be broadly classified into *driving* and *modulating* types. Driving connections have a direct effect on a cell's response, enhancing or suppressing its activity. Modulating connections have an indirect effect, modifying a cell's activity only if it has already been activated previously by some driving sources. Although the evidence is by no means conclusive, the two types of connections and the corresponding source neurons of driving and modulating signals seem to have been confirmed experimentally to some extent (see e.g. [6, 41]). They may serve to avert an activity explosion in a large network composed of neuronal assemblies (in particular the modulating connections, in a network with no strong driving loops, [6]).

7.2.2 Implementation of neuronal ensembles

There exists a variety of models for describing the activity of neuronal assemblies that have the properties enumerated in section 7.2.1. Since at the assembly level (and because of the computational costs), we cannot cope with all the details of neuronal arborization and dendritic structure, we neglect them and start at the descriptional level of threshold models of spiking neurons. Motivated by simulations and analytical calculations with assemblies of the spike–response type we introduce an abstract model for describing assembly dynamics. The model can be mapped quantitatively with assemblies of spiking neurons in the regime of oscillatory activity and it can be derived from phase oscillator neurons, but we will not go into the analytical details here.

The model is defined as follows. The dynamics of each assembly m is determined by three variables. These variables are the *mean activity* $S_m(t) \in [0, 1]$, the oscillation *amplitude* $r_m(t)S_m(t) \in [0, 1]$ and the *phase*

$\Psi_m(t) \in [0, 2\pi)$ [2]. The oscillation amplitude is composed by the oscillation *coherence* (or *centroid*) $r_m(t) \in [0, 1]$ and the mean activity $S_m(t) \in [0, 1]$. The mean activity, oscillation amplitude, and oscillation phase correspond to the minimum of three variables needed to describe the joint activity of a neuronal assembly with periodic or quasiperiodic firing, i.e., that exhibits coherent oscillatory activity. In a stable oscillatory solution with period T_m, they correspond to the 0th Fourier-component and the coefficient and phase of the first Fourier component of the activity $A_m(t)$.

The mean activity $S_m(t)$ indicates how many neurons participate in spiking during a time period $(t - T/2, t + T/2]$. A value of $S_m(t) = 0$ indicates that the entire assembly remains silent and no spikes are elicited during the period. A value of $S_m(t) = 1$ indicates that all neurons of the assembly release spikes. Neurons that do not spike may nevertheless participate in the assembly dynamics. They participate in the phase dynamics since they still exhibit a pronounced *subthreshold oscillation* in their synaptic input, but they will not participate in driving other neurons, since they do not fire. Therefore, they do not contribute to the macroscopic variables such as the activity or the amplitude and phase. The magnitude of $S_m(t)$ thus indicates the proportion of suprathreshold neurons in an assembly m, and also the mean spike-rate or mean activity of a pool during $(t - T/2, t + T/2]$. For experimental evidence of in vivo subthreshold oscillations in cortical tissue see [44].

The coherence $r_m(t)$ indicates the proportion of suprathreshold neurons that participate in the oscillatory activity. A value of $r_m(t) = 0$ indicates that all suprathreshold neurons fire incoherently, so that no oscillatory activity is present. A value of $r_m(t) = 1$ indicates that all suprathreshold neurons fire coherently, so that the oscillation amplitude $r_m(t)S_m(t)$ is at its maximum for fixed $S_m(t)$. Therefore, $r_m(t)$ is the *degree of locking of the spiking neurons*. The phase of the oscillatory peak is expressed by $\Psi_m(t)$. Around $\Psi_m(t) = 0$ we find the maximum of the peak of the oscillatory activity.

The full assembly dynamics for a network with L assemblies connected by coupling and feeding links with strengths J_{mn}^{coup} and J_{mn}^{feed} then reads

$$
\frac{d}{dt} r_m(t) = \begin{cases} \frac{r_m(t)}{\tau_m(t)}\left[1 - \frac{r_m(t)}{r_m^*(t)}\right] & \text{if } J_m^{\text{eff}}(t) > J_m^{\text{crit}} \\ -\frac{r_m(t)}{\tau_m(t)} & \text{otherwise} \end{cases}
$$

$$
\frac{d}{dt} \Psi_m(t) = \Omega_m - \frac{h_m^{\Psi}(t)}{r_m(t)}
$$

$$
\frac{d}{dt} S_m(t) = -\frac{1}{\tau_m^S}\{S_m(t) - G_m[h_m^S(t)]\} \tag{7.1}
$$

---

[2] All phases in this work are used modulo 2π.

with the effective self-coupling $J_m^{\text{eff}}(t) := h_m^r(t)/r_m(t)$, the fields

$$h_m^r(t) = \sum_{n=1}^{L} J_{mn}^{\text{coup}} r_n(t) S_n(t) \cos[\Psi_m(t) - \Psi_n(t)]$$

$$h_m^\Psi(t) = \sum_{n=1}^{L} J_{mn}^{\text{coup}} r_n(t) S_n(t) \sin[\Psi_m(t) - \Psi_n(t)]$$

$$h_m^S(t) = \sum_{n=1}^{L} J_{mn}^{\text{feed}} F_m[S_m(t), r_m(t), \Psi_m(t), S_n(t), r_n(t), \Psi_n(t)] \,.(7.2)$$

the "stationary" centroid strength (defined for $J_m^{\text{eff}}(t) > J_m^{\text{crit}}$)

$$r_m^*(t) = \sqrt{1 - \frac{J_m^{\text{crit}}}{J_m^{\text{eff}}(t)}} \tag{7.3}$$

and the time parameter for centroid growth

$$\tau_m(t) = \frac{1}{2} |J_m^{\text{eff}}(t) - J_m^{\text{crit}}| \,. \tag{7.4}$$

The function $G_m(h_m^S)$ in Eqn. (7.1) is a sigmoidal gain function. The function $F_m(\ldots)$ in Eqn. (7.2) depends on the *macroscopic* states of the pool m and the other pools n of the network. This completes the equations for the *Graded Oscillator Model (GOM)*.

The 3 fields $h_m^r(t)$, $h_m^\Psi(t)$ and $h_m^S(t)$ drive the centroid strength $r_m(t)$, the centroid phase $\Psi_m(t)$ and the mean activity $S_m(t)$, respectively. The centroid strength field $h_m^r(t)$ enters the dynamics equation for $r_m(t)$ in form of the effective self-coupling $J_m^{\text{eff}}(t)$. The oscillatory threshold is given by J_m^{crit}. Only if $J_m^{\text{eff}}(t) > J_m^{\text{crit}}$, the centroid strength or coherence $r_m(t)$ (and thus, the oscillation amplitude) grows. The frequency $\Omega_m = 2\pi/T_m$ is determined by the oscillation period T_m. Two different types of connections, quantified by the strengths of the links between assemblies, J_{mn}^{coup} and J_{mn}^{feed}, influence separately the oscillation dynamics (J_{mn}^{coup}, *coupling* connections) and the dynamics of the mean activity (J_{mn}^{feed}, *feeding* connections). In addition to the dynamics as defined by 7.1 and 7.2, noise plays an important role. We will not go into further details here but indicate that the effect of noise on $S_m(t)$, $r_m(t)$ and $\Psi_m(t)$ can be derived from microscopical considerations and mainly influences $r_r(t)$ and $\Psi_m(t)$ at small oscillation amplitudes $r_m(t) S_m(t) \to 0$. In this case, both the coherence $r_m(t)$ and the phase $\Psi_m(t)$ exhibit large fluctuations.

Pools modeled by (7.1) have a series of properties that are found in pools modeled explicitly using spiking neurons, and behave in a qualitatively similar manner. They develop coherent oscillations of the activity in response to large inputs of the feeding field $h_m^S(t)$. They exhibit a pronounced oscillatory threshold, but nevertheless follow a modulating external stimulation (i.e., they develop a nonvanishing oscillation amplitude if another assembly with a pronounced oscillation provides input conveyed by means of the coupling connections J_{mn}^{coup}). In addition, the strength of the feeding signal is coded in the oscillation amplitude.

The dynamics 7.1 behave in a controllable way, since they can be demonstrated to have a Lyapunov function for symmetrical connections $J_{mn}^{\text{coup}} = J_{nm}^{\text{coup}}$ and $J_{mn}^{\text{feed}} = J_{nm}^{\text{feed}}$. They code explicitly an assembly's coherence and amplitude signals, thus allowing an easy decoding of such signals. And finally, they introduce in a natural way two different types of connections J_{mn}^{coup} and J_{mn}^{feed}, which access different types of macroscopic variables of the assembly.

7.2.3   Computational units

The dynamics of a network composed of interconnected assemblies modeled according to (7.1) behave in a much more complex way than a single isolated pool. For a single pool m, the oscillatory behavior is determined by two parameters: its critical coupling strength J_m^{crit}, and its self-coupling J_{mm}. This is easy to understand since in the single pool case the effective self-coupling reduces to $J_m^{\text{eff}}(t) = J_{mm}S_m(t)$, and the critical condition for the development of coherence $[r_m(t) > 0]$ is $J_m^{\text{eff}}(t) > J_m^{\text{crit}}$. The centroid $r_m(t)$ thus remains zero until the feeding field $h_m^S(t)$ raises a sufficient proportion of assembly neurons $S_m(t)$ to a suprathreshold level so that $J_{mm}S_m(t) > J_m^{\text{crit}}$, i.e., the oscillatory threshold is overcome. For more than one pool, the effective coupling strength $J_m^{\text{eff}}(t)$ of each pool is determined by the network connections and the centroids of the rest of the pools. In the case of one pool having a non-zero centroid strength, the effective coupling strengths of all other pools it connects to are modified. This has the consequence that a pool will practically always have a non-zero centroid if any other pool that is connected to it (regardless of the sign of the connections) also has a non-zero centroid. (Note that for $r_m(t) > 0$, we need $J_m^{\text{eff}}(t) = h_m^r(t)/r_m(t) > J_m^{\text{crit}}$, and this can always be fulfilled for a positive field $h_m^r(t) > 0$ if $r_m(t)$ is small enough.) A pool with non-zero centroid will influence all other pools in such a way, that although they would not have been able to develop a locking state by virtue of their internal couplings alone, they now have a non-vanishing centroid.

This effect is easy to understand intuitively since any non-constant external driving force pulls and pushes the neurons of a pool into some preferred phase, resulting in a coherence $r_m(t) > 0$. The critical coupling at which the entire network begins to have locking components is an implicit function of the network topology, i.e., it may be regarded as an emergent property of the network. Below that critical coupling, stationary activities dominate. Above it, oscillations can be used information processing.

After oscillations emerge, the fixed connections J_{mn}^{coup} and J_{mn}^{feed} are modified by amplitudes and phases of the presynaptic pools (7.2), generating temporarily new effective connections by strengthening or weakening the existing ones. This has two important consequences. First, it causes the appearance of *"dynamical assemblies"*, which are *groups of assemblies* that can be labeled as belonging together and separated from others because of their strong reciprocal effective connections[3] (compare the organization into dynamical assemblies with the self-organization of persons around the microphones in section 7.1.4). The idea of dynamical assemblies that work in such a way goes back as far as Hebb [19]. The second consequence is that, since the new effective connection strengths modify the network's hardware, they can remodel it in a specific, task- and goal-oriented way. Instead of an all-purpose machine, we then get a refined hardware that adapts dynamically to the momentary processing needs.

In terms of the variables $r_m(t)$ and $\Psi_m(t)$, the GOM dynamics (7.1) try to maximize the oscillation amplitudes and to adjust the relative phases of the assemblies to each other according to the couplings J_{mn}^{coup}. Assemblies connected by excitatory connections $J_{mn}^{\text{coup}} > 0$ "feel" each other stronger when their phases are near together, in the sense that they stabilize and enhance each other's oscillation amplitudes. The same happens with inhibitory couplings $J_{mn}^{\text{coup}} < 0$ for phases that are far away from each other. The phase is therefore a sort of grouping label, indicating which assemblies should belong together in view of the internal network knowledge and the sensory input that is actually applied.

We differentiate four types of information that have to be processed in the network. First of all, *sensory input* about the external world arrives at each assembly by means of feeding connections J_{mn}^{feed} and influences mainly its mean activity $S_m(t)$. As a second type of very different information, *hypotheses, expectations* and *attentional signals*[4] are conveyed to the assemblies, also mediated by feeding connections J_{mn}^{feed}. This is the information determined to a great extent by previously learned knowledge, stored in the network's connections. Using the sensory and the attentional data, the assemblies try to reconcile the prior knowledge with the sensory data of the current task.

As a third type of information, *grouping hypotheses* (or grouping labels) are processed which are coded in the phase labels $\Psi_m(t)$ and communicated between assemblies using the coupling connections J_{mn}^{coup}. As a fourth and last type, *grouping certainty* (or grouping strength) is coded in the oscillation amplitude $r_m(t)S_m(t)$ and communicated between assemblies using the same coupling connections J_{mn}^{coup} as the grouping hypotheses. Grouping hypotheses as well as grouping certainty depend on both the sensory input and the expectational signals.

---

[3] Keep in mind the following difference. An assembly is a *group of neurons* that process similar information. A dynamical assembly is a *group of assemblies* that temporarily (for a specific processing task or sensory stimulation) organize themselves as belonging together.

[4] We will use this nomenclature interchangeably.

7.2.4   Working hypotheses about the network function

We propose that the function of a network that processes and makes use all these different types of information is that of using the prior information to constrain the processing space, thus allowing the system to generalize, to eliminate conflicts and to resolve ambiguities. The network accomplishes this task by continually generating hypotheses which are tested against the sensory data and used to gate the information flow through the network.

In a network that uses both sensory and attentional data in the same architecture, it has to be made sure that the data is kept separately on a local scale and does not corrupt each other. Otherwise, the attentional data would be able to influence the sensory input to such a point that it cannot be discerned any more as such, meaning e.g., that the network would simply detect the objects it is hypothesizing of, corresponding to a situation of uncontrolled visual imagery. Therefore, a single computational unit of our network is basically composed of two functionally distinct assemblies. One is responsible for sensory data and the other one for attentional information. In addition, both assemblies receive information from other processing units about grouping processes. Fig. 7.3 A shows the basic computational unit and the four types of information that can influence its behavior. *SP* is the *sensory processing assembly* and *OG* the oscillation generating, and *attentional processing assembly*. Two assemblies SP and OG always work together, processing data that corresponds to the same feature selectivity. (Meaning that, if an SP assembly shows a strong selectivity for a specific stimulus, its corresponding OG assembly manages the hypotheses about that particular stimulus.)

The expectational data is used to locally gate the information flow through the network. This is explained in Fig. 7.3 B, C, and D, which correspond to the situations of a unit receiving sensory data alone, of receiving both sensory and expectational data, and of receiving expectational data alone (thick and thin arrows indicate connections carrying weak and strong signals, respectively). In the first case (B), the SP assembly responds in a standard way, according to its input specificity. In the second case (C), the output of the processing unit is enhanced in relation to the first case, because *the expectation (or local hypothesis) has been confirmed by the sensory data*. The enhancement is coded as a modulation of the resulting SP activity, and indicated in the figure as an outgoing coupling connection carrying a strong signal. The confirmation of the SP activity occurs by modulation through coupling OG → SP connections, i.e., an enhancing occurs only if there is sensory data present. This is shown in the third case (D), where nothing occurs at the SP assembly, although the OG assembly indicates that the network would expect input on the corresponding SP counterpart (strong coupling OG → SP signal).

In case that the sensory and the expectational data confirm each other, the activity of the corresponding SP pool is modulated in such a way that postsynaptically connected units detect a stronger signal, indicating that the unit's output is to be processed preferentially. In our implementation,

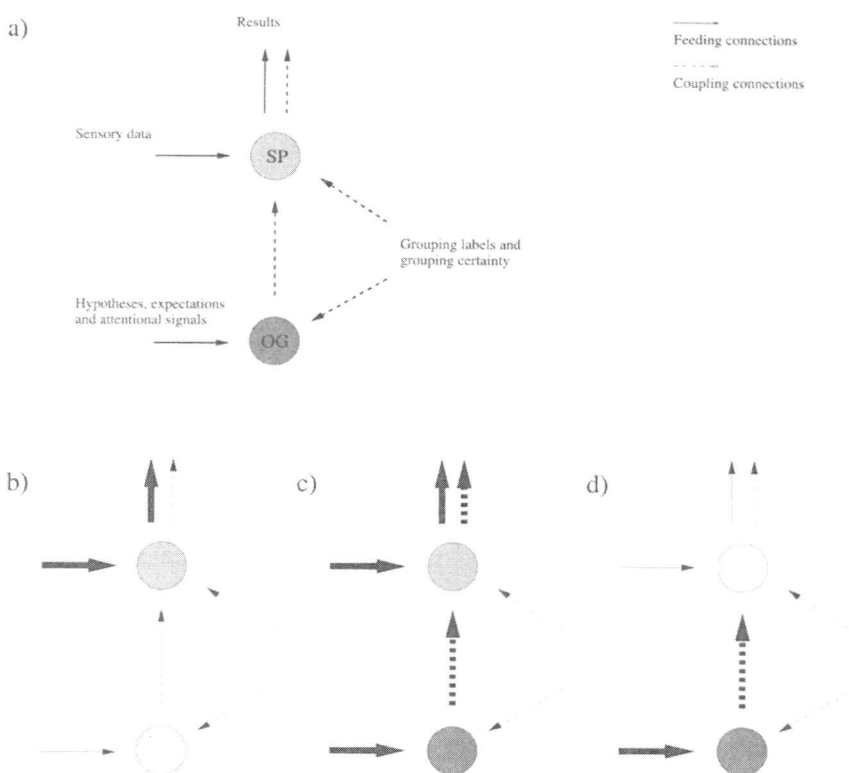

FIGURE 7.3. Four types of information are processed in a single computational unit. First of all, sensory data is processed independently from hypotheses, expectations and attentional data. This is accomplished by two functionally specialized assemblies, called SP (sensory processing) and OG (oscillation generating and attentional processing). In A), the two assemblies with the corresponding feeding connections are shown. In addition, both assemblies process grouping hypotheses and grouping certainties, provided by coupling connections, and coded in the oscillation phase and the oscillation amplitude, respectively. The attentional data gates the information flow through the network. The gating occurs locally at those computational units at which sensory data is confirmed by attentional data, as shown in C). In B), the unit receives sensory input only, in this case the data is propagated in a standard way through the SP assemblies exclusively. To the contrary, in D), the unit receives attentional input only. In this case, no further propagation of data occurs, since the expectations are not confirmed by sensory data.

the modulation occurs using a *coherent oscillation of the assembly activity that is imposed on the otherwise nonmodulated activity evoked by sensory input* on SP. This means that the information flow is gated through the network using an activity modulation, therefore we call this type of networks *activity–gating* networks. The strength of the activity modulation (and thus, the "gating saliency" of a unit) is coded in the oscillation amplitude $r_m(t)S_m(t)$. In short, from two units with the same sensory input but different oscillation amplitudes, the one with the largest amplitude will have a better chance to pass on the results of its computations to further processing levels.

Postsynaptically connected units are affected by the gating signal either directly using the coupling connections, or they can work explicitly as *coherence detectors*[5], converting the modulated activity back to a field $h_m^S(t)$ that influences the mean activity $S_m(t)$ (for theoretical accounts on coherence detecting capabilities of spiking neurons see e.g. [24]). In this case, the function $F_m(\ldots)$ from (7.2) expresses as:

$$h_m^S(t) = \sum_{n=1}^{L} J_{mn}^{\text{simple}} S_n(t)$$

$$+ \sum_{n=1}^{L} J_{mn}^{\text{intra}} S_n(t)r_n(t)$$

$$+ \sum_{n=1}^{L} J_{mn}^{\text{inter}} S_n(t)r_n(t) \cos[\Psi_m(t) - \Psi_n(t)] \qquad (7.5)$$

The first term on the right detects the mean activity, while the second and the third terms are sensitive to the intrapool coherence and the joint intra/interpool coherence, respectively.

A single computational unit that works according to the activity–gating principle is basically composed of 2 GOM assemblies, corresponding to the SP and OG pools from Fig. 7.3. It is shown in Fig. 7.4 with its internal connections, and feedforward, lateral and feedback connections arriving from and going to other units of the system. This completes the microarchitecture of the network.

From the dynamics (7.1) and Fig. 7.4, we see that the effect of OG pool activity on its SP counterpart is to influence SP's oscillation amplitude $r_m^{\text{SP}}(t)S_m^{\text{SP}}(t)$ [6]. More precisely, in the microcircuit architecture of Fig. 7.4 it *influences only the coherence* $r_m^{\text{SP}}(t)$ of the SP pool, so that its effect on the oscillation amplitude is *purely modulating*. Without sensory input (coded in the mean activity $S_m^{\text{SP}}(t)$ of the SP pool), the influence of the OG pool vanishes. With sensory input, it can be calculated from the dynamics

---

[5]We speak of *coherence detection* instead of *coincidence detection* to indicate the population character.

[6]From now on, we label the computational units with m, n and indicate additionally the assembly with SP and OG.

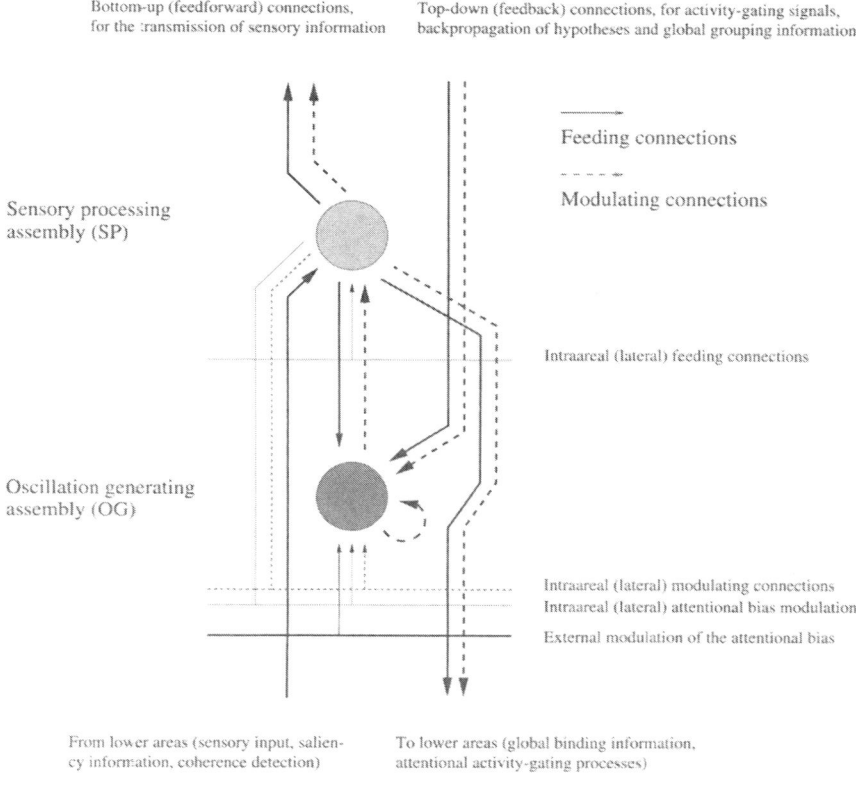

To higher areas   From higher areas

Bottom-up (feedforward) connections,   Top-down (feedback) connections, for activity-gating signals,
for the transmission of sensory information   backpropagation of hypotheses and global grouping information

Feeding connections

Modulating connections

Sensory processing
assembly (SP)

Intraareal (lateral) feeding connections

Oscillation generating
assembly (OG)

Intraareal (lateral) modulating connections
Intraareal (lateral) attentional bias modulation

External modulation of the attentional bias

From lower areas (sensory input, salien-   To lower areas (global binding information,
cy information, coherence detection)   attentional activity-gating processes)

From lower areas   To lower areas

FIGURE 7.4. The computational unit of the activity-gating network as it is implemented in the model. The proposed microarchitecture is able to process separately sensory input on one hand and hypotheses, expectations and attentional signals on the other. It is composed of two GOM pools. The sensory processing (SP) pool receives feeding sensory input from lower areas and relays feeding and coupling information on to both higher and lower areas. The oscillation generating (OG) pool receives feeding and coupling attentional input, mainly from higher areas. Only the OG pool is able to intrinsically generate and sustain oscillations, if it is activated strong enough. The oscillatory modulation is then imprinted on the SP pool and used for activity–gating and grouping processes in higher and lower areas. The result is a modulated activity of the SP pool, with an oscillation amplitude $r_m^{OG}(t)S_m^{SP}(t)$, i.e., with a coherence determined by the hypotheses, expectations and attentional signals, multiplied by the mean activity determined by sensory input.

that, for equal J_m^{crit} for the SP and the OG pools and equal strength of the coupling connections OG $\rightarrow$ OG and OG $\rightarrow$ SP, it is $r_m^{\text{SP}}(t) = r_m^{\text{OG}}(t)$ and therefore

$$r_m^{\text{SP}}(t)S_m^{\text{SP}}(t) = r_m^{\text{OG}}(t)S_m^{\text{SP}}(t) , \qquad (7.6)$$

so that the coherence of the SP pool is determined by that of its OG counterpart, and the amplitude of the SP pool is a *multiplication* of the sensory certainty, coded in $S_m^{\text{SP}}(t)$, and the confirmation of the sensory data by internal expectations, coded in $r_m^{\text{OG}}(t)$. This multiplicatory character arises in our model as a consequence of the assembly dynamics and the microarchitecture, and not by the introduction of special multiplicative connections at the neuronal level (see e.g. [10, 25]).

7.2.5   Network architecture: Complementary processing streams

The large-scale network architecture uses the function of the computational units as hypothesis generating and testing devices (additionally to the processing of sensory information), and provides the framework so that the different types of data can be accessed.

In the network, each unit receives sensory input and passes it on to other units. This means that sensory information is directly propagated through the network using only the SP assemblies as relay stations, and in a *fast, feedforward sweep of sensory activity*. To the contrary, the influence of the OG assemblies on processing is indirect, in the sense that they can only influence other units by their SP intermediary, so that the information flow goes OG $\rightarrow$ SP $\rightarrow$ OG $\rightarrow$ SP $\rightarrow$ OG $\ldots$. In addition, they use oscillatory activity for their modulatory signals, which take time to build up. It must also be considered that the attentional signals are mainly sent to hierarchically lower subsystems, because they provide global hypotheses. Therefore, opposed to the fast, feedforward sweep of sensory activity, we have *a slower building up of expectational and attentional signals using lateral and feedback connections*. In summary, *two complementary processing streams* process differently sensory vs. attentional information, and influence each other mainly at those units at which the information flows meet. Fig. 7.5 shows the implementation of the complementary processing streams using the computational units from Fig. 7.4.

The relay on the intermediary SP units for attentional processing prevents, in combination with the modulatory (i.e., multiplicatory) character of the OG $\rightarrow$ SP connections, a proliferation of hypotheses that are not confirmed by sensory input. This is a sensitive side effect of the architecture, since it averts a *hypothesis explosion* on the attentional processing stream. Without such an architecture, a hypothesis explosion occurs because there are many different and ambiguous ways to compose a high-level feature from low-level features.

The overall picture then is the following. Instead of introducing a division between preattentive and attentive processing on an architectural

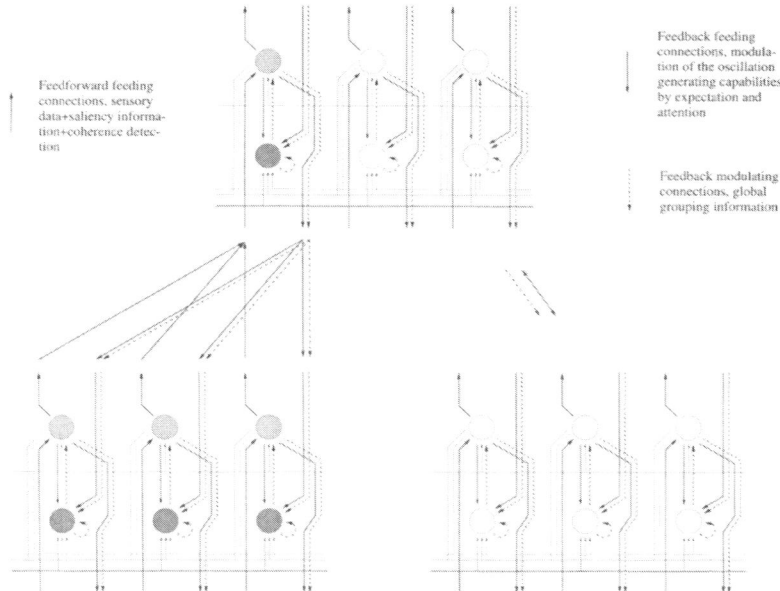

Feedforward feeding connections, sensory data+saliency information+coherence detection

Feedback feeding connections, modulation of the oscillation generating capabilities by expectation and attention

Feedback modulating connections, global grouping information

FIGURE 7.5. Two complementary counterstreams serve to process sensory vs. attentional information. The sensory information is processed in a mainly feedforward stream that comprises the SP pools (light grey circles). The attentional information is processed in a stream that uses lateral and feedback connections and comprises the OG pools (dark grey circles). While the feedforward stream directly communicates different hierarchical areas, in the feedback stream, the attentional information (represented in the activity of the OG pools) has to be confirmed by the activity of the corresponding SP pools. There is *no direct* link of feeding connections in the feedback stream, instead, the feedback information always crosses at least one modulatory OG → SP connection, meaning that there are no purely feeding loops in the system, and preventing a *hypothesis explosion*.

scale (see section 7.1.3 on FOA models), the difference between preattentive and attentional processing is more subtle. It resides in the different coding schemes and the different time scales at which the processing of information occurs. On one hand, preattentive processing is inherently fast, occurs in parallel and serves to trigger hypotheses and categorize sensory input in the entire network. It is responsible for detecting fast onsets, novel appearing stimuli and imposed synchronization (e.g., by external sources), and for coding the presence of sensory features. On the other hand, attentive processing modulates existing sensory data, testing hypotheses and organizing the network into dynamical assemblies that are specialized for task-dependent processing. Since attentional processing uses grouping labels it is intrinsically capacity-limited, imposing a serial processing at the highest hierarchical stages of the network. If there is not sufficient time for the generation of modulatory signals, attentional processes cannot develop, (re)access to low-level information and grouping processes are impeded and ambiguity errors (such as false conjunctions) appear (see e.g. [47]).

In summary, in the activity–gating framework, attention controls, both locally and globally, the information flow through the network.

7.2.6   Overall network organization

The overall network is organized into several different processing pathways (see also section 7.1.2). Each pathway concentrates on specific stimulus attributes, such as location, form, color or texture, and is composed of a hierarchy of areas which show an increasing specialization for higher hierarchical levels. Between pathways, crossconnections exist. Most connections are bidirectional, i.e., there are both feedforward and feedback links between areas as well as between pathways.

This means that the network has neither a strictly serial hierarchical organization, nor that it is strictly feedforward. It is rather a hierarchical organization of areas in a broad sense, with higher and lower areas, concurrent processing streams, parallel processing among the streams and extensive feedback.

The organization of the network crudely mimics the so-called "what", or ventral, and "where", or dorsal, pathways of the visual cortex of primates with some of its identified processing areas, such as V1 and V2. The information arrives the network through a common area (corresponding to the LGN) and is then processed in parallel by the different pathways. Feature detectors of increasing complexity are implemented in all areas of the dorsal and the ventral pathways. In a pathway, each area innervates its successor with feedforward connections, and sends back information to its predecessor using feedback connections. In the upper left diagram of Fig. 7.6 a scheme of the main pathways, areas and connections is shown.

In our model, the areas denominated as V1 and V2 serve as the detecting areas for such simple features as points, simple lines, end-stopped lines and edges, while areas C1, C2, and C3 serve to detect color/texture. In all these areas, the visual field is sampled into discrete subregions, and for each subregion there exists a number of units that detect simple features. Units with the same feature selectivities repeat over the entire visual field. The result is a sort of grid arrangement of feature detectors. All feature detectors responsible for the same retinotopical position form a *hypercolumn*. All processing units that compose a single feature detector are grouped into a *column*. The organization of areas V1 and V2 into columns and hypercolumns is shown on the left half of Fig. 7.6. A similar organization is implemented for the areas of the dorsal pathway.

The specialization of the processing pathways increases with increasing hierarchical level. At the same time, each area gets increasingly invariant against information that it is not specialized to process. For example, in the form processing pathway, the features get more form-specific but less

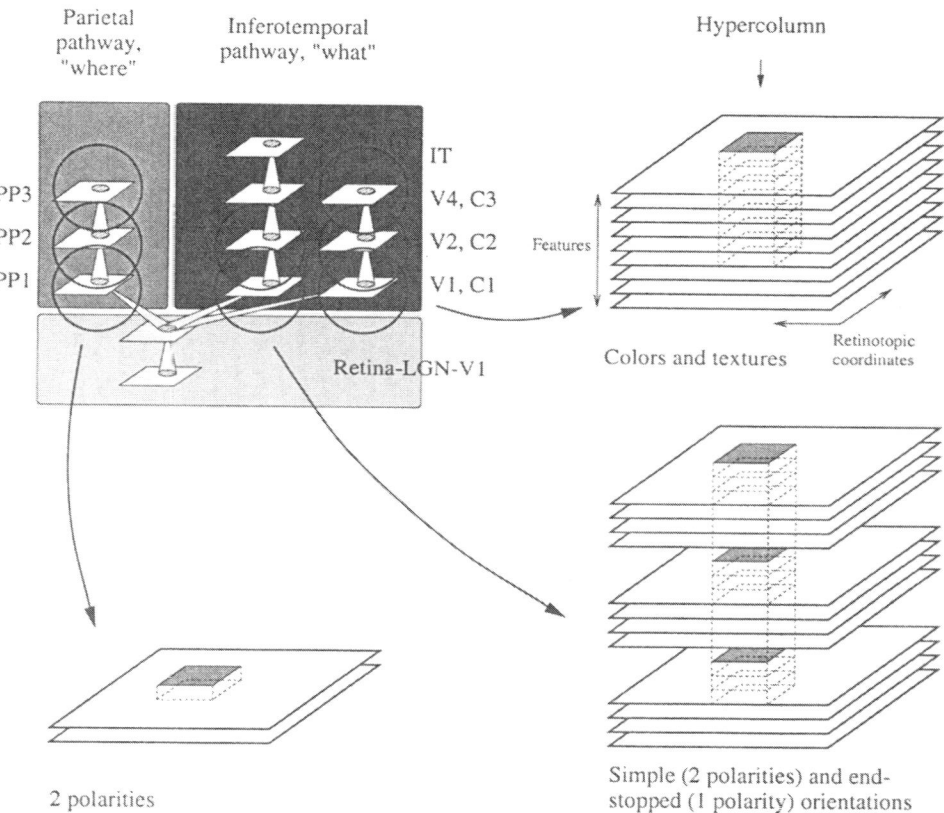

FIGURE 7.6. Scheme of the attentional network. There are different process-ing pathways (what and where), specialized to analyze location (where-pathway, areas PP1, PP2, and PP3), form (what-pathway, areas V1, V2, V4, and IT) , and color/texture (areas C1, C2, and C3) of the stimulus. Each pathway has an ascending hierarchy of areas with retinotopical organization. In each area, "hy-percolumns" of feature detectors analyze in parallel the stimulus at the same retinal position. Also shown: The convergence resp. divergence pattern of feed-forward and feedback connections. Moving towards higher processing areas, the convergence causes a gradual increase of generalization capabilities and a loss of low-level information that is irrelevant for each pathway.

specific to the exact location of the stimulus. This occurs by an increase of the complexity and size of receptive fields, a pooling of the inputs from lower levels and a coarser retinotopic sampling of the visual field. To the contrary, in the where pathway, only the receptive field sizes increase, whereas the form and the retinotopic sampling of the receptive fields keep unchanged. The result is an increasingly translation-invariant form detection along the what pathway, and an increasingly form-invariant location detection along the where pathway.

In the entire network, an area is composed of the units of Fig. 7.4. The corresponding feedforward and feedback connectivity is as depicted in Fig. 7.5. In addition, lateral connections connect units of a single area. We will not explain it in detail here, but in short, lateral connections have an intermediate connectivity pattern between feedforward and feedback links. They serve the multiple purposes of transmitting hypotheses laterally inside an area, redundancy reduction, regulation of biased competition and enhancement of rare feature conjunctions.

7.3   Results

7.3.1   Biased competition

The typical suppression of responses seen in experiments related with attention is caused in our network by biased competition effects. Attentionally gated signals not only are processed preferentially using activity modulation and coherence detectors, but they also favor the affected units in the sense that attention enables them to elicit stronger suppressing signals on other units they compete with. This attentionally biased competition causes processing resources to be focused on less items, which is necessary because of the limited resources of the system (see section 7.1.2).

In the experiment of Moran and Desimone ([30], see Fig. 7.7), single cells of monkey visual cortical areas V4 and IT were recorded. A cell responding effectively to one type of visual stimulus and ineffectively to another was selected from one of the mentioned visual areas. In addition, the region of the visual field that influenced its response, its receptive field (RF), was determined. The monkey was trained to attend to stimuli at one location in the visual field and to ignore stimuli at another, while the fixation point in the visual field remained cued to the same location. The monkey was presented simultaneously an effective (one that the cell would selectively respond to under single stimulus presentations) and an ineffective sensory stimulus. Because the sensory conditions were identical for the two types of trials, during which the monkey directed its attention to any one of the two stimuli locations, the difference in the response characteristic of the selected cell has to be attributed to the effects of selective attention. When the locations of both stimuli fell inside the receptive field of the selected cell, and the animal attended to the location of the effective stimulus, the cell gave a strong response. But when the animal attended to the ineffective

Both stimuli inside RF

Attention to effective stimulus:
Strong response

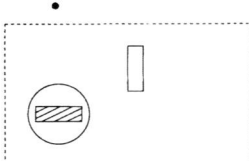

Attention to ineffective stimulus:
Attenuated response

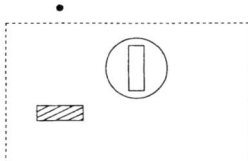

One stimulus inside RF and one stimulus outside

Attention to effective stimulus:
Strong response

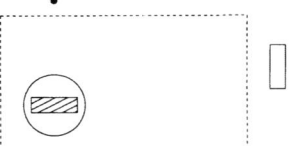

Attention to ineffective stimulus:
Strong response

FIGURE 7.7. The experiment of Moran and Desimone [30]. When both the effective and the ineffective stimulus are inside the RF of a V4 cell, and the animal attends to the location of the ineffective stimulus, the cell gives almost no response. This effect does not appear if the ineffective stimulus is outside the RF. (Small black spot: fixation point.)

stimulus, the cell gave almost no response, although the effective stimulus was still within its receptive field. In this case, the response of the cell was greatly reduced (by more than half) by the attentional effects. On the other hand, if the ineffective stimulus was placed outside the receptive field of the selected cell, the neuron showed similar responses to the effective stimulus for both attentional conditions.

Fig. 7.8 shows the simulation results of the Moran and Desimone experiment within our network. The attentional effect is based purely on the spatial location of the effective stimulus, so that attention biases all OG assemblies of those units that are close to its position. Fig. 7.8 A) shows a simplified network of the three units (corresponding to the effective unit, an ineffective unit inside of the effective unit's RF and an ineffective unit outside of the effective unit's RF) and the connections. In all cases, we concentrate on the left (i.e., the effective) unit. In Fig. 7.8 B), spatial attention is directed to the left unit's position ("attention to effective stimulus" in Fig. 7.7). Although the middle unit also receives a sensory input of the same strength, it loses the competition and is suppressed by the attended stimulus. The suppressive effect of the middle unit on the left unit is therefore diminished. In Fig. 7.8 C), the situation is reverted and spatial attention is

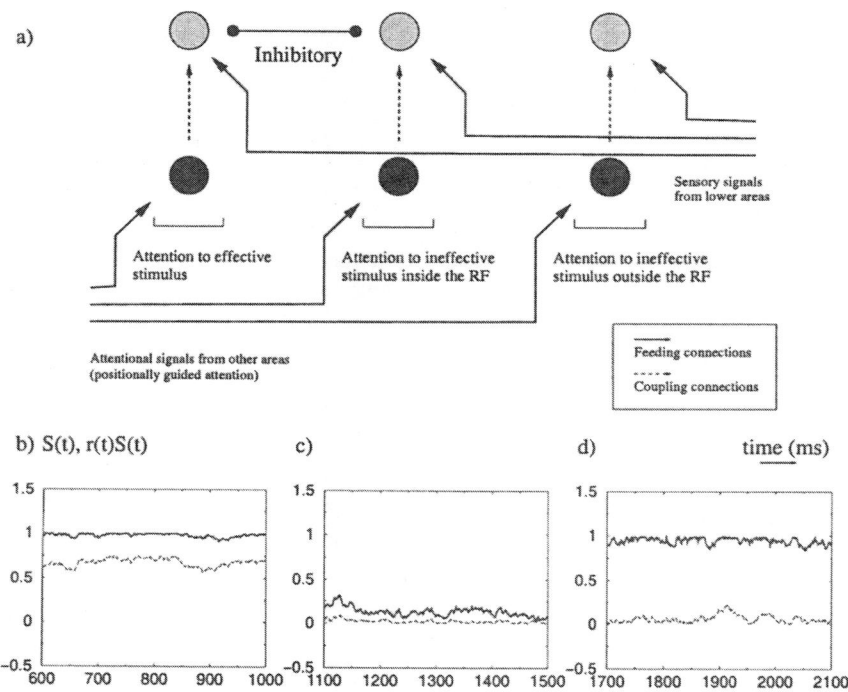

FIGURE 7.8. Suppression of activity that appears in relation with attentional processing is a result of biased competition. Attentional processing evokes a modulation of the activity at attended units, which then get a competitive advantage. This causes the network to concentrate the previously broadly distributed processing resources on fewer units. At the top (A), a simplified diagram of the involved units and connections is shown. The bottom row of three graphs (B, C, and D) shows the leftmost unit's SP assembly activity (dark solid line) and oscillation amplitude (dashed line), in different attentional conditions. The three conditions are: attention to the effective stimulus (B), to an ineffective stimulus located inside of the effective stimulus' RF (C), and to an ineffective stimulus located outside of the effective stimulus' RF (D). It can be seen that the unit's activity is supressed in the second case (C). In this case, attention provides the middle unit of A) with a competitive advantage, so that it has greater suppressive effects on the left unit than viceversa.

directed to the middle unit's position ("attention to ineffective stimulus" in Fig. 7.7). Now the middle unit is favored in the competition, so that the suppressive effect on the left unit is strong, therefore its response is attenuated. In Fig. 7.8 D), spatial attention is directed to the right unit's position, outside of the left unit's receptive field. In this case, the units do not compete with each other, and there is no influence of attention on the left unit's response. This demonstrates the effects of biased competition using lateral inhibition that is triggered by activities with large amplitudes.

Of course, spatial attention may not only suppress but also enhance activity, because hierarchically lower areas are also influenced by attentional gating, rising the activity of units that are selective for features at the attended location. This in turn strengthens the activity of the attended units in higher areas.

7.3.2   Contour integration by laterally transmitted expectation

As explained in section 7.2.6, in addition to enhancing biased competition, lateral connections serve to propagate hypotheses and expectations laterally inside an area. An example is contour integration in V1/V2. This is implemented by using lateral SP → OG connections. If a feature detector receives a large sensory activation, indicating e.g. the presence of an oriented line segment, it sends a signal to all other feature detectors that it expects to be activated as well, according to the prior information about stimulus properties stored in the network. In the example shown in Fig. 7.9, the lateral connections reflect the psychophysical Gestalt laws of good continuation, similarity and proximity, so that continued smooth lines are preferred. At those units in V1/V2 at which the laterally transmitted expectations are met by sensory data, activity is enhanced, oscillation amplitudes grow and grouping processes take place. This can be seen in the time development of of the network activity in Fig. 7.9. The lateral connections cause a grouping of those feature detectors that correspond to a "good" line according to the Gestalt laws. This in turn gates feedforward connections, so that grouped feature detectors now have a larger impact on further processing.

Of course, lateral propagation of hypotheses can only contribute proposals of limited range about grouping and segmentation, and may include conflicting signals. These cannot be resolved but by further analysis in higher processing stages and using the backpropagation of global hypotheses. This is accomplished by the connections of units as shown in Fig. 7.5.

7.3.3   Different types of attention

Attentional gating signals can originate at any point of the network at which large oscillation amplitudes of the SP pools are evoked by a convenient match of sensory and expectational input. According to Fig. 7.4, we see that feeding input arrives at the OG pool by means of a direct connection from its SP counterpart, and by indirect connections providing expect=20 ms

t=100 ms

t=300 ms

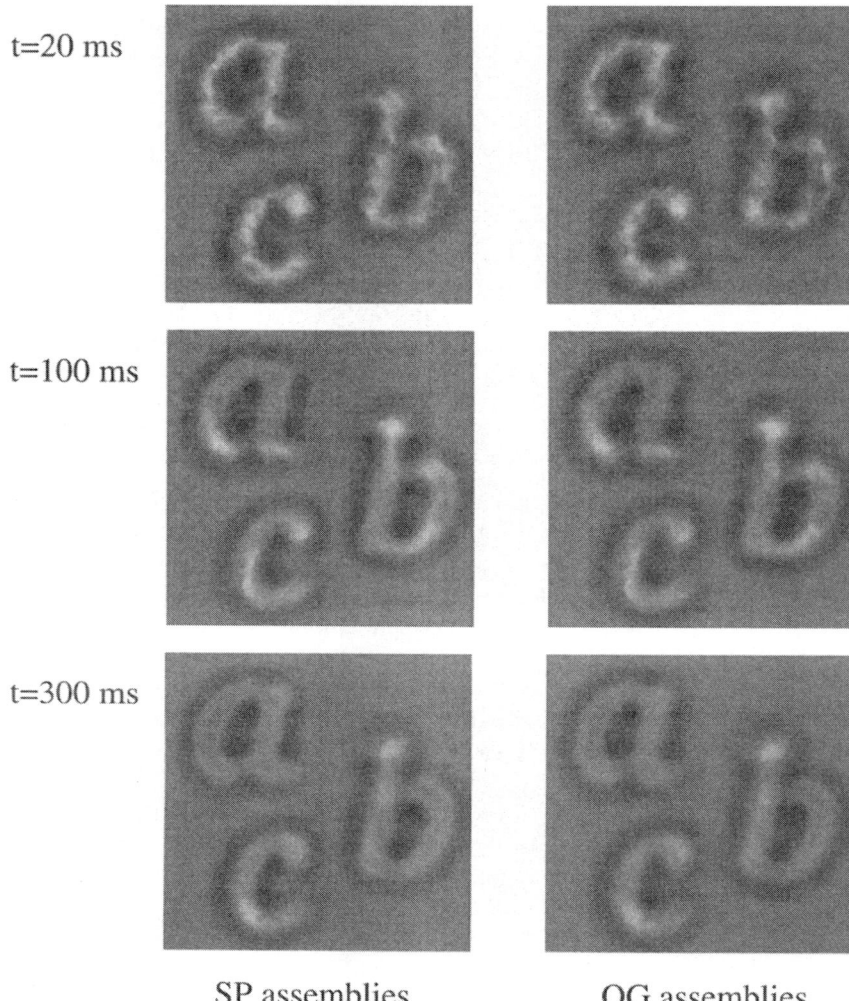

SP assemblies   OG assemblies

FIGURE 7.9. Contour integration by laterally transmitted expectation. The network continuously creates hypotheses that evoke expectations and attentional signals at other units, where they have to be confirmed by sensory data. Particularly in the lower areas V1/V2, lateral SP → OG connections cause an increase of the amplitudes at attended units and provide grouping signals. The figure shows the amplitude and phase of the V1 units coded as color saturation and color hue. It can be seen that in the time course from top to bottom, the phases adjust so that a grouping of smooth contours occurs (in greyscale pictures, a smoothing of the roughness can be distinguished). Preferred contours are those in accordance with the Gestalt principles of good continuation, similarity and proximity, reflecting the stimulus statistics expected by the network.

tations and hypotheses using mainly lateral and feedback connections from other units. Therefore, there are two sources that facilitate the generation of large amplitudes. First, they can be evoked directly by sensory input, in this case the attentional signals are *saliency, or stimulus driven*. This is e.g. the case for highly detectable stimuli, such as a red spot against a background of another color [7]. Second, facilitating signals can be evoked by extraneous sources, in this case the attentional processing is *task driven*. This is the case if the task demands the system to detect a particular feature. In this case, attentional resources are directed to the units that represent that feature, and thus gating signals in form of large oscillation amplitudes are more likely to originate at those units. Usually, both types of signals will be present, so that a strong task-determined attentional signal can overwhelm the stimulus saliency or viceversa.

According to the saliency characteristics of the stimulus and the task that the system has to solve, different areas will be the origin of the strongest attentional gating signals. Furthermore, according to the area of origin, we may distinguish the network behavior as corresponding to *different types of attention*. In our network of Fig. 7.6, the attentional signals can originate in any of the pathways, so that we can differentiate *locationally-guided attention* (areas PP1, PP2, and PP3), *feature-guided attention* (e.g., areas V1/V2, color and texture processing areas) and *form or memory-guided attention* (higher areas of the form-processing pathway). This list is by no means exhaustive, since further processing pathways, such as a pathway responsible for motion processing which is not implemented in the current version of our network, are known to exist and should provide further sources of attentional signals.

In Fig. 7.10, we show the results of network simulations for feature-guided attentional processing. Attention was drawn to a tilted segment (/), meaning that the corresponding OG pools in V1 were biased accordingly by top-down connections. As a consequence, attentional gating signals in form of large oscillation amplitudes are generated preferably in units with strong sensory input and the right orientation. This in turn allows the further propagation of attentional signals to other areas and pathways. In the figure, this causes an enhancement of activity in area PP1, which is sensitive to the stimulus location. Area PP1 thus indicates the positions where tilted segments were found.

---

[7]In section 7.1.4 we indicated that instead of separating architecturally the saliency map from attentive processing areas, it may be an advantage to integrate saliency and attentional computations into the very same network. In this case, the stimulus-driven gating signals are gained by internal saliency computations.

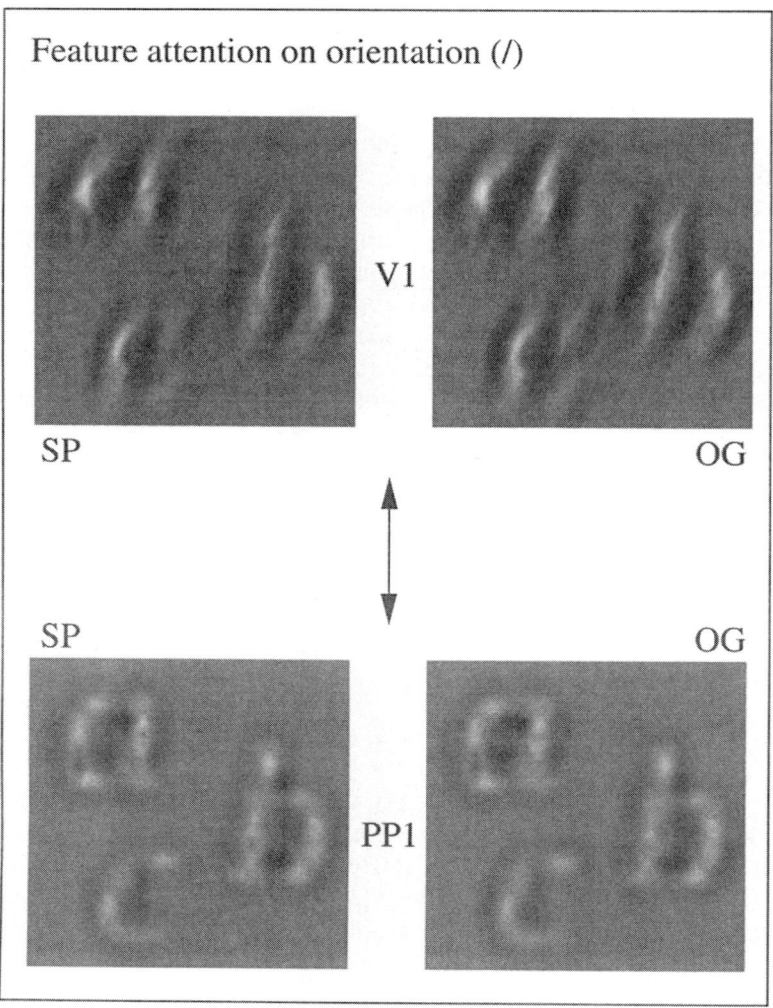

FIGURE 7.10. Feature-guided attention and top-down propagation of signals. In this figure, attention is drawn to a tilted segment (/) of a fixed orientation by using top-down signals that bias specific orientation-sensitive feature-detecting units. Shown are areas V1 and PP1 from the network of Fig. 7.6. Attentional gating signals are generated preferably in the feature-selective units of V1. This allows a propagation of attentional signals to other areas and pathways. In the figure, this is shown for area PP1, from where the location of the tilted segments can be extracted. Locationally-guided attention works in the opposite direction, with attentional signals that originate in the parietal areas and are propagated from there to the inferotemporal areas, say PP1 → V1.

7.4 Discussion

7.4.1 Microarchitecture and concurrent processing streams

The architecture of the system is based on physiological and anatomical data and on the dynamics of assemblies of spiking neurons observed in detailed simulations. Of course, the mapping on real biological structures is to a great extent a matter of debate and requires further experimental evidence, so that we intend this network as a first approximation to the very complex visual system. However, we have introduced a network which fully incorporates feedback processing in a realistic framework, which may have major implications for a further understanding of information flow in the cortex.

For simplicity, we will not discuss the experimental evidence in detail, but mention only a few important points. The organization of the visual system is fundamentally hierarchical [13, 39, 53, 55]. But instead of a strict pyramidal hierarchy, parallel processing in concurrent pathways seems to be the norm (see e.g. [54] for an account). The dorsal pathway along with the posterior parietal cortex has been implicated in the analysis of spatial relationships, movement and locationally-guided attention, while the ventral pathway seems to be involved in the analysis of form, color, texture and disparity. There are massive feedback projections to virtually any region of the visual cortex, they seem to be involved in gating, signal-to-noise enhancement and the improvement of figure-ground segmentation (see e.g. [21]). There is a marked asymmetry between feedforward and feedback connections regarding the layers of origin and destination of the connections [13], suggesting a functionally different role for the two types of connections. In addition, connections and neurons have been found that can be crudely categorized as having a driving and a modulating function [41].

A strategy of bidirectionally processing counterstreams with some characteristics similar to our model has been proposed previously by Ullman [51]. It has to be regarded, however, that no concrete model implementation is proposed, and the precise roles and functions of feedback, processing pathways and attention are not specified in his model.

A coherent modulation of the activity to enhance signals and increase feedforward inhibition at an attended location of the visual field is used in models of Niebur et al. [34], and of Niebur and Koch [33]. Their models are based on the architectural division between preattentive saliency computation and attentive postprocessing as explained in section 7.1.3. Their work shows that very simple neurons are able to enhance coherent activity if driven at the right parameters. In comparison, the proposed framework of our model does not rely on a segregated saliency map and the usage of activity modulation plays a fundamental, more complex role in gating signals in the entire network. Moreover, our model is able to account for a variety of effects related to attention (see section 7.3.3) and feedback.

7.4.2   Biased competition

The model unifies two central characteristics of attentional processing that have been proposed previously. One is that attention acts by imposing a biased competition among attended units [7, 38]. The resulting effect of attention thus is to favor certain units against others.

The second characteristic is that the origin of the attentional signal need not be a single area, such as the saliency map of spotlight models, but that the entire network continuously produces saliency signals for the control of attention. R. Desimone already conjectured from his experimental results that "we believe that there is no single saliency map in the attentional control system, but rather a series of maps (or possibly a distributed map) ..." [7].

In our network, competition is triggered and biased by attention because attention means increased oscillation amplitudes and it is assumed that coherent activities evoke stronger signals on the postsynaptic side than non-coherent ones. In the case of biased competition, these signals are inhibitory and the signals are indeed stronger because of the intra and interpool coherence detecting contributions to (7.5), which are sensitive to the amplitude $r_m(t)S_m(t)$.

This is a realistic assumption, since synchronized and coherent activity is more effective in driving postsynaptic neurons over their firing thresholds because of a temporal summation of postsynatic potentials. Therefore, we consider biased competition in conjunction with coherent activity to be a biologically plausible contribution to attentional mechanisms.

In our model, attention expresses both as a suppression, caused by increased lateral competition among attended units, and enhancement, caused by activity-modulated signals arriving from lower areas that are gated using the coherence detectors. Both suppression and enhancement have been found in connection with attentional processing (see e.g. [49]), so that attentional effects cannot be restricted to either enhancement or suppression alone.

7.4.3   Origin of the attentional signal

The view of a distributed attentional control (or, in the sensory-driven case, distributed attentional saliency computation) implies that there may be different points of origin of an attentional signal. This different origin correlates in our model with the different types of attention. For example, attentional signals can originate in the parietal area, in hierarchically high areas of the form processing pathway, or in an orientation, texture or color processing area, corresponding to locationally-guided, high-level form (or form-memory-guided), or feature-based attention, respectively.

The question remains how to determine the origin of the attentional signal for an applied stimulus. In our model, two criteria are used for this purpose. One is how well the functionality of an area matches the current task that has to be resolved. This implies e.g. that tasks that require localization of a stimulus slightly bias the parietal areas (in their entirety, non-specifically) so that attentional effects are enhanced and originated in these areas. The second is that, in a single area, attentional signals should originate in those units whose response selectivity is matched best by the stimulus, i.e., those that are strongest activated by sensory input. This means that, e.g., locationally-guided attention will originate in that special area of the parietal pathway with a spatial resolution that best matches the input.

7.4.4 Saliency and the focus of attention

As explained, the computation of salient parts of the scene is accomplished by the very same network units that also process attentional data. The focus of attention of the spotlight models relies mainly on the computation of a bounded region of interest in retinotopic space. It can therefore be compared with locationally-guided attention that originates in our network in the areas PP1, PP2, or PP3. These can be thought of constituting a spatial "saliency map" that indicates the position of interesting parts of the scene. Undoubtedly, locationally-guided attentional processing is crucial in scanning objects that are easy to segregate by their positions. Nevertheless, in our model this is but one source of attentional signals, and it may as well be the case that, instead of sending attentional signals to the other pathways it may receive attentional signals from them. This is e.g. the case if an object is identified by some prominent features, and afterwards its exact location has to be computed by the parietal areas.

7.4.5 Predictions of the model and conclusions

A series of predictions arise from the model. One of the most important predictions is that there is spatially specific attentional modulation right down to the very low areas with high spatial resolution, where the different pathways originate. Furthermore, the model predicts that the attentional effects in these areas are not limited to locationally-guided attention, but can as well be feature or object based. In fact, experimental data seemed to support that attention modulates higher (extrastriate) object recognition areas, but not the primary visual cortex. Only recently, detailed studies have shown that there are robust attentional effects in the primary visual cortex [14, 40, 43], confirming the predictions.

A second important prediction is that visual processing occurs on two timescales, corresponding to a first, mainly feedforward activity propagation which lasts for some tens of ms and a second processing step that deploys on a slower timescale, relying on feedback processing and involving attentional processing. This division into early and late responses, with the late responses acquiring an increasing specificity with time, should be observed experimentally. In a recent paper [45], this has been confirmed for neurons of the IT cortex. An information analysis showed that IT neurons encode different stimulus attributes in their early and late responses to the same stimulus.

Finally, in our model it makes no difference whether the source of synchronicity is internal to the network, or if it is imposed by externally induced time cues. All that the units feel is an input with a temporal modulation. Therefore, the external temporal structure can be as strong an indicator for visual processing as internally generated synchronicity. That external time cues can help the brain in object processing has been confirmed experimentally in [1, 28].

We do not claim that binding occurs exclusively using synchronicity cues, nor that coherent oscillations must be present in every task that involves binding. Our claim is more moderate, stating that the possibility to label units is a very powerful tool in the framework of the activity–gating architecture, and coherent oscillations are a simple method to implement such labels. Furthermore, binding by synchronicity is often understood as 'every parts of an object must be in synchrony for the object to be identified'. This is not the case in our model. Instead, a great deal of the recognition processes can occur without synchronicity. Nevertheless, synchronicity is able to provide additional information required for specific tasks, specially those involving several processing pathways.

7.5 References

[1] D. Alais, R. Blake, and S.-H. Lee. Visual features that vary together over time group together over space. *Nature Neuroscience*, 1(2):160–164, 1998.

[2] H.-U. Bauer. Is there parallel binding of distributed objects? Preprint, 1994.

[3] D. E. Broadbent. Task combination and selective intake of information. *Acta Psychologia*, 50:253–290, 1982.

[4] A. Chauduri. Modulation of the motion aftereffect by selective attention. *Nature*, 344:60–62, 1990.

[5] F. Crick. Function of the thalamic reticular complex: the searchlight hypothesis. *PNAS*, 81:4586–4590, 1984.

[6] F. Crick and C. Koch. Constraints on cortical and thalamic projections: the no–strong–loops hypothesis. *Nature*, 391:245–250, 1998.

[7] R. Desimone. Neural circuits for visual attention in the primate brain. In G. A. Carpenter and S. Grossberg, editors, *Neural Networks for*

Vision and Image Processing, Bradford Book, pages 344–364. MIT Press, Cambridge, Massachusetts, 1992.

[8] R. Desimone and J. Duncan. Neural mechanisms of selective visual attention. *Annu. Rev. Neurosci.*, 18:193–222, 1995.

[9] J. Duncan and G. W. Humphreys. Visual search and stimulus similarity. *Psychol. Rev.*, 96:433–458, 1989.

[10] R. Eckhorn, H. J. Reitboeck, M. Arndt, and P. Dicke. Feature linking via synchronization among distributed assemblies: Simulations of results from cat visual cortex. *Neural Comput.*, 2:293–307, 1990.

[11] C. W. Eriksen. Attentional search of the visual field. In B. David, editor, *International Conference on Visual Search*, pages 3–19, 4 John St., London, WC1N 2ET, 1988. Taylor and Francis Ltd.

[12] C. W. Eriksen and J. D. St. James. Visual attention within and around the field of focal attention: A zoom lens model. *Percept. Psychophys.*, 40:225–240, 1986.

[13] D. J. Felleman and D. C. van Essen. Distributed hierarchical processing in the primate cerebral cortex. *Cerebral Cortex*, 1:1–47, 1991.

[14] S. P Gandhi, D. J. Heeger, and G. M. Boynton. Spatial attention affects brain activity in human primary visual cortex. *PNAS*, pages 3314–3319, 1999.

[15] W. Gerstner. *Populations of Spiking Neurons*, chapter 10, pages 261–296. MIT Press, Cambridge, MA, 1998.

[16] W. Gerstner and J. L. van Hemmen. Coding and information processing in neural networks. In E. Domany, J. L. van Hemmen, and K. Schulten, editors, *Models of Neural Networks II*, pages 1–93, Berlin, 1994. Springer.

[17] W. Gerstner, J. L. van Hemmen, and J. D. Cowan. What matters in neuronal locking? *Neural Comput.*, 8(8):1653–1676, 1996.

[18] C. M. Gray and D. A. McCormick. Chattering cells: Superficial pyramidal neurons contributing to the generation of synchronous oscillations in the visual cortex. *Science*, 274:109–113, 1996.

[19] D. O. Hebb. *The Organization of Behavior*. Wiley, New York, 1949.

[20] S. Herculano-Houzel, M. H. J. Munk, S. Neuenschwander, and W. Singer. Precisely synchronized oscillatory firing patterns require electroencephalographic activation. *The Journal of Neuroscience*, 19(10):3992–4010, 1999.

[21] J. M. Hupé, A. C. James, B. R. Payne, S. G. Lomber, P. Girard, and J. Bullier. Cortical feedback improves discrimination between figure and background by V1, V2 and V3 neurons. *Nature*, 394:784–787, 1998.

[22] W. James. *The Principles of Psychology*, volume 1. Henry Holt, New York, 1890.

[23] B. Julesz. Toward an axiomatic theory of preattentive vision. In G. M. Edelman, W. E. Gall, and W. M. Cowan, editors, *Dynamic Aspects of Neocortical Function*, pages 585–612. Neurosciences Research Foundation, 1984.

[24] R. Kempter, W. Gerstner, J. L. van Hemmen, and H. Wagner. Extracting oscillations: Neuronal coincidence detection with noisy periodic spike input. *Neural Comput.*, 10:1987–2017, 1998.

[25] C. Koch and T. Poggio. Multiplying with synapses and neurons. In T. McKenna, J. Davis, and S. F. Zornetzer, editors, *Single Neuron Computation*, pages 315–345. Academic Press, London, 1992.

[26] P. König, A. K. Engel, P. R. Roelfsema, and W. Singer. How precise is neuronal synchronisation? *Neural Comput.*, 7:469–485, 1995.

[27] M. J. M. Lankheet and F. A. J. Verstraten. Attentional modulation of adaptation to two-component transparent motion. *Vision Res.*, 35:1491-1412, 1995.

[28] S.-H. Lee and R. Blake. Visual form created solely from temporal structure. *Science*, 284:1165–1168, 1999.

[29] P. McLeod, J. Drive, and J. Crisp. Visual search for a conjunction of movement and form is parallel. *Nature*, 332:154–155, 1988.

[30] J. Moran and R. Desimone. Selective attention gates visual processing in the extrastriate cortex. *Science*, 229:782–784, 1985.

[31] B. C. Motter. Neural correlates of feature selective memory and pop-out in extrastriate area V4. *Journal of Neuroscience*, 14:2190–2199, 1994.

[32] T. D. Murphy and C. W. Eriksen. Temporal changes in the distribution of attention in the visual field in response to precues. *Perception and Psychophysics*, 42:576–586, 1987.

[33] E. Niebur and C. Koch. A model for the neuronal implementation of selective visual attention based on temporal correlation among neurons. *Journal of Computational Neuroscience*, 1(1):141–158, 1994.

[34] E. Niebur, C. Koch, and C. Rosin. An oscillation-based model for the neural basis of attention. *Vision Research*, 33:2789–2802, 1993.

[35] K. M. O'Craven, B. R. Rosen, K. K. Kwong, A. Treisman, and R. L. Savoy. Voluntary attention modulates fMRIactivity in human MT-MST. *Neuron*, 18:591–598, 1997.

[36] D. R. Patzwahl, U. J. Ilg, and S. Treue. Switching attention between transparent motion components modulates responses of MT and MST neurons. *Soc. Neurosci. Abstr.*, 24:649, 1998.

[37] M. I. Posner. Orienting of attention. *Q. J. Exp. Psychol.*, 32:3–25, 1980.

[38] J. H. Reynolds, L. Chelazzi, and R. Desimone. Competitive mechanisms subserve attention in macaque areas V2 and V4. *The Journal of Neuroscience*, 19(5):1736–1753, 1999.

[39] K. S. Rockland and D. N. Pandya. Laminar origins and termination of cortical connections of the occipital lobe in the rhesus monkey. *Brain Research*, 179:3–20, 1979.

[40] P. R. Roelfsema, V. A. F. Lamme, and H. Spekreijse. Object-based attention in the primary visual cortex of the macaque monkey. *Nature*, 395:376–381, 1998.

[41] S. M Sherman and R. W. Guillery. On the actions that one nerve cell can have on another: Distinguishing "drivers" from "modulators". *PNAS*, 95:7121–7126, 1998.

[42] W. Singer and C. M. Gray. Visual feature integration and the temporal correlation hypothesis. *Annu Rev Neurosci*, 18:555–586, 1995.

[43] D. C. Somers, A. M. Dale, A. E. Seiffert, and R. B. H. Tootell. Functional MRI reveals spatially specific attentional modulation in human primary visual cortex. *PNAS*, 96:1663–1668, 1999.

[44] E. A. Stern, D. Jaeger, and C. J. Wilson. Membrane potential synchrony of simoultaneously recorded striatal spiniy neurons in vivo. *Nature*, 394:475–478, 1998.

[45] Y. Sugase, S. Yamane, S. Ueno, and K. Kawano. Global and fine information coded by single neurons in the temporal visual cortex. *Nature*, 400:869–873, 1999.

[46] A. Treisman and G. Gelade. A feature–integration theory of attention. *Cognitive Psychology*, 12:97–136, 1980.

[47] A. Treisman and H. Schmidt. Illusory conjunctions in the perception of objects. *Cognitive Psychology*, 14:107–141, 1982.

[48] S. Treue and J. H. R. Maunsell. Attentional modulation of visual motion processing in cortical areas MT and MST. *Nature*, 382:539–541, 1996.

[49] S. Treue and J. H. R. Maunsell. Effects of attention on the processing of motion in macaque middle temporal and medial superior temporal visual cortical areas. *The Journal of Neuroscience*, 19(17):7591–7602, 1999.

[50] S. Treue and M. Trujillo. Feature-based attention influences motion processing gain in macaque visual cortex. *Nature*, 399:575–579, 1999.

[51] Sh. Ullman. Sequence seeking and counterstreams: A model of bidirectional information flow in the cortex. In C. Koch and J. L. Davis, editors, *Large–Scale Neuronal Theories of the Brain*, chapter 12, pages 257–270. MIT Press, Cambridge, Massachusetts, 1994.

[52] M. Valdez-Sosa, M. A. Bobes, V. Rodriguez, and T. Pinilla. Switching attention without shifting the spotlight: Object-based attentional modulation of brain potentials. *J. Cog. Neurosci.*, 10:137–151, 1998.

[53] D. C. van Essen, C. H. Anderson, and D. J. Felleman. Information processing in the primate visual cortex: An integrated systems perspective. *Science*, 255:419–423, 1992.

[54] D. C. van Essen and E. A. DeYoe. Concurrent processing in the primate visual cortex. In M. S. Gazzaniga, editor, *The Cognitive Neurosciences*, chapter 24, pages 383–400. MIT Press, Cambridge, MA, 1995.

[55] D. C. van Essen and J. H. R. Maunsell. Hierarchical organization and functional streams in the visual cortex. *Vision Research*, 24:429–448, 1983.

[56] J. M. Wolfe, K. R. Cave, and S. L. Franzel. Guided search: An alternative to the feature integration model for visual search. *J. Exp. Psychol.*, 15:419–433, 1989.

8

Timing and Counting Precision in the Blowfly Visual System

Rob de Ruyter van Steveninck and William Bialek[1]

ABSTRACT We measure the reliability of signals at three levels within the blowfly visual system, and present a theoretical framework for analyzing the experimental results, starting from the Poisson process. We find that blowfly photoreceptors, up to frequencies of 50-100 Hz and photon capture rates of up to about $3 \cdot 10^5/\text{s}$, operate well within an order of magnitude from ideal photon counters. Photoreceptors signals are transmitted to LMCs through an array of chemical synapses. We quantify a lower bound on LMC reliability, which in turn provides a lower bound on synaptic vesicle release rate, assuming Poisson statistics. This bound is much higher than what is found in published direct measurements of vesicle release rates in goldfish bipolar cells, suggesting that release statistics may be significantly sub-Poisson. Finally we study H1, a motion sensitive tangential cell in the fly's lobula plate, which transmits information about a continuous signal by sequences of action potentials. In an experiment with naturalistic motion stimuli performed on a sunny day outside in the field, H1 transmits information at about 50% coding efficiency down to millisecond spike timing precision. Comparing the measured reliability of H1's response to motion steps with the bounds on the accuracy of motion computation set by photoreceptor noise, we find that the fly's brain makes efficient use of the information available in the photoreceptor array.

8.1 Introduction

Sensory information processing plays a crucial role in the life of animals, including man, and perhaps because it is so important it seems to happen without much effort. In contrast to this, our subjective experience suggests that activities of much lower urgency, such as proving mathematical theorems or playing chess, require substantial conscious mental energy, and this seems to make them inherently difficult. This may deceive us into

---

[1]NEC Research Institute, 4 Independence Way, Princeton NJ 08540

thinking that processing sensory information must be trivially easy. However, abstract tasks such as those mentioned are now routinely performed by computers, whereas the problems involved in making real-life perceptual judgments are still far from understood. Playing chess may seem much more difficult than discerning a tree in a landscape, but that may just mean that we are very efficient at identifying trees. It does not tell us anything about the "intrinsic" difficulty of either of the two tasks.

Thus, to find interesting examples of information processing by the brain we do not need to study animals capable of abstract thinking. It is sufficient that they are just good at processing sensory data. Partly for this reason, sensory information processing by insects has been an active field of study for many years. Undeniably, insects in general have simpler brains than vertebrates, but equally undeniably, they do a very good job with what they do have. Noting that insect brains are very small, Roeder (1998) remarks:

> Yet insects must compete diversely for their survival against larger animals more copiously equipped. They must see, smell, taste, hear, and feel. They must fly, jump, swim, crawl, run, and walk. They must sense as acutely and act as rapidly as their predators and competitors, yet this must be done with only a fraction of their nerve cells.

Spectacular examples of insect behavior can be found in Tinbergen (1984) . and Berenbaum (1995). Brodsky (1994) describes the acrobatics of fly flight: "The house fly can stop instantly in mid flight, hover, turn itself around its longitudinal body axis, fly with its legs up, loop the loop, turn a somersault, and sit down on the ceiling, all in a fraction of a second." An example of some of this performance is shown in Fig. 1, and clearly the acrobatic behavior displayed there must be mediated by impressive feats of sensory information processing.

In this chapter we will look at the tip of this iceberg, and study some aspects of visual information processing in the blowfly. Insects lend themselves well for electrophysiological and behavioral studies, especially for quantitative analysis, and the emphasis will be on quantifying the *accuracy* of neural signals and neural computations, as advocated explicitly by Bullock (1970). This is a fundamentally probabilistic outlook, requiring a suitable statistical description of signal and noise. It also is the principal description from the point of view of representation and processing of information. For example, if you (or the brain) are asked to estimate the light intensity in the outside world based on the reading of a photoreceptor voltage, then of course you need to know the gain, i.e., the conversion factor from light intensity to photoreceptor voltage. To find the light intensity you just divide the voltage by this gain. The accuracy of your estimate depends on the magnitude of the cell's voltage noise, and this translates into an uncertainty in light intensity through the same gain factor. But then the uncertainty in the estimate depends on the ratio of the signal voltage and

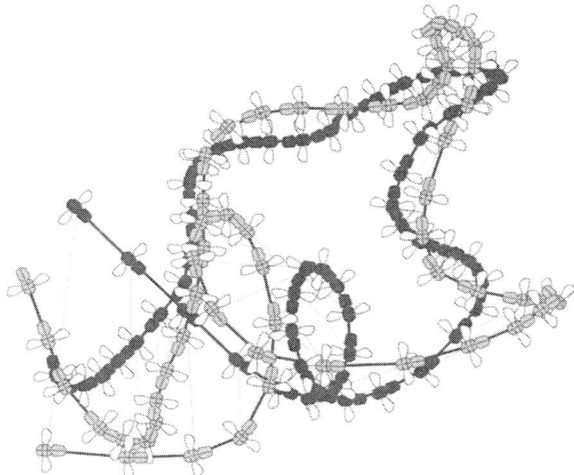

FIGURE 8.1 Two flies in a chase. The black fly is chased by the gray one for a duration of about a second. Flies (*Fannia canicularis*) were filmed from below, and their positions estimated every 20 ms. The chasing fly estimates the position of the leading one purely by means of visual input. Modified from Fig. 4 in Land and Collett (1974).

the noise voltage, and not on the specific value of the gain. Note that the gain is a "biological" parameter—why does the cell produce millivolts and not hundreds of millivolts?—while the accuracy of estimates is measured in the same units as the physical stimulus in the outside world. Further, we will see that there are often absolute limits to accuracy that are set by physical principles. Thus by studying the accuracy of estimates we move from a characterization of neurons that is specific to the their biological context toward a characterization that is more absolute and universal.

The example of estimating light intensity by looking at the voltage in a photoreceptor admittedly is very simple. Nonetheless, it illustrates an important point, namely that we can quantify on an absolute scale the performance of neurons as encoders and processors of sensory information even without a complete understanding of the underlying physiological mechanisms. The challenge that we address in later sections of this chapter is to give a similarly quantitative description for neurons at higher stages of processing, where the interpretation of the neural signals is much more difficult.

We begin at the beginning of vision, namely light. Because light is absorbed as an irregular stream of photons, it is an inherently noisy physical signal. This is modeled mathematically by the Poisson process, of which we give some mathematical details which will be used later on. Then we will focus on the performance of the photoreceptor cells of the fly's retina which capture light and convert it into an electrical signal. These cells encode signals as graded, or analog, variations of their membrane potential. This representation is noisy in part due to the fluctuations in the photon flux, and in part due to limitations imposed by the cell itself, and we can tease apart these contributions. Then we will look at the accuracy of chemical synaptic transmission between the photoreceptor cell and the LMC (Large Monopolar Cell) by comparing signal and noise in the presynaptic and postsynaptic cells. A chemical synapse releases discrete vesicles, and therefore the Poisson process is a natural starting point for a mathematical description of signal transfer across the synapse.

Of course, having a representation of light intensities in the outside world does not in itself constitute vision. It is the task of the brain to *make sense* of the ever fluctuating signals in the photoreceptor array. As an example we consider the estimation of wide-field motion from signals in the photoreceptor array, which is a relatively simple neural computation. We will analyze the limits to the reliability of this computation, given the reliability of the photoreceptor signals, and we will compare this to the reliability of performance of H1, a spiking, wide field motion sensitive cell in the fly's brain.

By comparing the measured reliability of cells with the physical limits to reliability set by the noise in the input signal we get an idea of the statistical efficiency of nerve cells and of neural computation. This also makes it possible to quantify rates of information transfer, which, although by no means the whole story, nevertheless captures the performance of nerve cells in a useful single number.

8.2   Signal, Noise and Information Transmission in a Modulated Poisson Process

A Poisson process is a sequence of point events with the property that each event occurs independently of all the others. We will treat some of the mathematics associated with Poisson processes and shot noise, but our emphasis is on an intuitive, rather than a rigorous presentation. Many of the issues discussed here can be found in a treatment by Rice (1944, 1945), which also appears in an excellent collection of classic papers on noise and stochastic processes by Wax (1954). A more comprehensive mathematical treatment is given by Snyder and Miller (1991).

The Poisson process is a useful model first of all because it describes the statistics of photon capture very well. There are exotic light sources which deviate from this (Saleh and Teich 1991), but biological organisms deal with normal thermal light sources. The photoreceptor response can then be modeled to first approximation as shot noise, which is a linearly filtered Poisson process. The filtering process itself may be noisy, as is the case in phototransduction. Consequently, we also treat the more general case in which the filter itself is a random waveform. The Poisson process is often used to model other point processes occurring in the nervous system, such as vesicle release at a chemical synapse, or trains of action potentials. We will present examples of both, and show that for vesicle release the Poisson process is probably not a good approximation, and for describing spikes in H1 it is certainly not good. But even then it is useful to understand the Poisson process, because it provides a simple example with which to ground our intuition.

8.2.1   Description of the Poisson process

Many of the notions treated in this section are illustrated in Fig. 2. Fig. 2B for example, shows a single realization of a point process on the time axis. This can be written as a series of events occurring at times $t_1, t_2, \cdots$. A useful mathematical abstraction is to represent each event as a delta

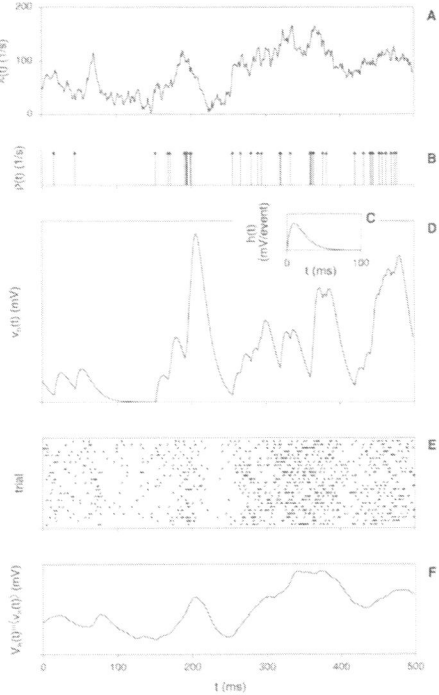

FIGURE 8.2. Illustration of some of the basic concepts used in the analysis. A: Time dependent rate, $\lambda(t)$, of a Poisson process. B: A single realization of statistically independent events on the time axis, generated from the rate function shown in A. The arrows represent delta functions. C: A temporal filter, $h(t)$, used to smooth the train of events in B to produce the example of a single shot noise trace, $v_h(t)$, shown in panel D. E: A raster representation of the outcome of 25 trials in which events are generated by the rate function $\lambda(t)$ shown in A. F: The average trace, $V_h(t) = \langle v_h(t) \rangle$, of a large number of trials filtered by $h(t)$.

function, $\delta(t - t_k)$, so that the full series is a sum of these functions:

$$\rho(t) = \sum_k \delta(t - t_k). \tag{8.1}$$

Here the delta function has the dimension of inverse time, and an area equal to 1, so that the integral of $\rho(t)$ over a time window will be equal to the number of events in that window. What distinguishes a Poisson process from other sequences of events is that the occurrence of each event is random and completely independent of the occurrence of all the others. Eqn. 8.1 describes one single outcome, which is analogous to observing the number on a die in one single throw. Although the particular result may be important in a game of dice, it is not the particular realization that we are interested in here. Instead we wish to derive what we can expect under "typical" conditions. A construction that helps in such derivations is to think not of one single outcome of the process, but of a great number of independent realizations. This is similar to the concept in statistical mechanics of an ensemble of independent systems, each in its own state, and all obeying the same macroscopic conditions (see for example Schrödinger 1952). The advantage of this mental picture is that it provides a convenient starting point for reasoning about typical or average behavior.

For the die we usually take it for granted that the chance of getting any particular number from 1 to 6 is just 1/6. But we could also imagine an immense number of dice thrown at random, and if we were to count the ones that landed on 6, we would find their proportion to the total to be very close to 1/6. Obviously, using the ensemble concept for reasoning about dice is a bit overdone. Later on we will look at experiments where the same time dependent stimulus sequence is repeated a large number of times, while neural responses are recorded. In that case it will be useful to think about a large set of outcomes—our independent trials—and to distinguish ensemble averages over that set of trials from time averages . Ensemble averages will generally be indicated by $\langle \cdots \rangle$, and time averages by $\overline{\cdots}$. Here we use the term "ensemble" in a loose sense, and primarily to make the distinction with time averages. A synthesized example is shown in Fig. 2E, while Fig. 17B shows a set of spike timing data from a real experiment.

Suppose that Eqn. 8.1 represents a realization of a Poisson process that has a constant, time independent rate of λ, that is on average we expect to see λ pulses in a one second time window. How does this relate to the sequence described by Eqn. 8.1? In the ensemble picture we construct a large number, M, of independent realizations of $\rho(t)$, denoted $\rho_1(t), \rho_2(t), \cdots, \rho_M(t)$, with total counts $N_1, N_2, \cdots, N_M$. The ensemble average, $r(t)$ of $\rho(t)$ is then:

$$r(t) = \langle \rho(t) \rangle = \lim_{M \to \infty} \frac{1}{M} \sum_{m=1}^{M} \sum_{k_m=1}^{N_m} \delta(t - t_{k_m}) = \lambda. \tag{8.2}$$

Obviously λ has the same dimension, inverse time, as the delta function. But how do we derive this result in the first place? Let us introduce a rectangular pulse, $[\Pi(t/\Delta t)]/\Delta t$. Here $\Pi(t/\Delta t)$ is by definition equal to 1 for $|t| < \Delta t/2$ and zero otherwise (see Bracewell, 1978, and Rieke et al. 1997, A.1), so that $[\Pi(t/\Delta t)]/\Delta t$ is Δt wide, and $1/\Delta t$ high. This obviously has unit area, just like the "real" delta pulse, and it becomes more like a delta pulse if we let Δt shrink to zero. If $\Delta t \ll 1/\lambda$, then such pulses will (almost) never overlap. If we now also imagine time to be discretized in bins Δt, then it becomes a matter of bookkeeping to count what proportion of the realizations in the ensemble has a pulse in a given time bin. The answer is $P_1 = \Delta t \cdot \lambda$. To normalize to units of time we must divide this by the binwidth, Δt, so the rate will be λ.

Thus we expect to see about $\mu = \lambda \cdot T$ events occurring in a window of T seconds in a single realization, independent of where we put the window. Of course, μ is an average, and we do not expect to observe exactly μ events each time, if only because μ can be any nonnegative number, while each event count value must be an integer. It is interesting to know how the values in each trial are distributed. This is described by the Poisson distribution:

$$P(N) = \frac{(\lambda T)^N}{N!} \cdot e^{-\lambda T}. \tag{8.3}$$

It is easy to show that this distribution has mean value $\mu = \langle N \rangle = \lambda \cdot T$, and variance $\sigma^2 = \langle N^2 - \langle N \rangle^2 \rangle$. Further, $\langle N^2 - \langle N \rangle^2 \rangle = \langle N \rangle$, and this means that $\sigma^2 = \mu$, which is an important property of the Poisson distribution. A distribution which is more narrow than a Poisson distribution has $\sigma^2 < \mu$, and is often called sub-Poisson. The broader distribution, with $\sigma^2 > \mu$, is called super-Poisson.

8.2.2 The modulated Poisson process

If we want to study visual information transfer we should not confine ourselves to looking at steady light levels, as constant visual stimuli have a zero rate of information transmission. One reading of a signal value can of course carry information, but if it does not change over time the information rate, in bits/s, goes to zero. Perhaps for that reason the visual system does not seem to care very much for constant input signal levels. Later we will see the example of Large Monopolar Cells which filter out practically all of the standing signal in the fly's photoreceptors. In real life, signals are changing all the time, and in real experiments we usually modulate signals and probe the response of nerve cells to these changes. To study the response of photoreceptors one may deliver flashes of light, sine waves of various frequencies, pseudorandom signals, and so on. All of these are really modulations of an underlying Poisson process, namely the rate at which photons are captured. We therefore must extend our description of the homogeneous Poisson process to an inhomogeneous, or modulated, version in which the rate is a function of time. So suppose that the rate is time dependent $\lambda = \lambda(t)$, as depicted in Fig. 2A. If we have a large number of outcomes of a Poisson process modulated with $\lambda(t)$, then in analogy to Eqn. 8.2 the ensemble average, $r(t)$, must equal $\lambda(t)$:

$$r(t) = \langle \rho(t) \rangle = \lim_{M \to \infty} \frac{1}{M} \sum_{m=1}^{M} \sum_{k_m=1}^{N_m} \delta(t - t_{k_m}) = \lambda(t). \qquad (8.4)$$

8.2.3   Correlation functions and spectra

Because the pulses in a Poisson process occur independently of one another it is easy to compute correlation functions and power spectral densities. These correlation functions are natural objects from a physical point of view, and are used widely in the analysis of experimental data. There is a difficulty, however, in that correlation functions are averages, and we have two different notions of averaging: averaging over time, and averaging over an ensemble of responses to identical inputs. In many ways the situation is similar to that in the statistical mechanics of disordered systems, where we can average over thermal fluctuations, over different realizations of the quenched disorder, or over both. Our discussion here and in the next section is brief, summarizing results that are largely well known. What we aim at ultimately is the derivation of Eqn. 8.23, connecting observable quantities to the underlying rates of Poisson events.

We recall that if we average the signal $\rho(t)$ over many trials in which the inputs are the same, then we obtain the Poisson rate $\lambda(t)$, $\langle \rho(t) \rangle = \lambda(t)$. Since different pulses occur independently of one another, once we know the rate $\lambda(t)$ there is no additional mechanism to carry correlations across time. On the other hand, since $\rho(t)$ is built out of delta functions at the moments of the point events, there will be singularities if we compute correlations at short times. It is straightforward to show that

$$\langle \rho(t)\rho(t') \rangle = \lambda(t)\lambda(t') + \lambda(t)\delta(t - t'). \qquad (8.5)$$

Thus if we compute the "connected" correlation function—subtracting off the means of the terms taken separately—all that remains is the delta function term:

$$\langle \rho(t)\rho(t') \rangle_c \equiv \langle \rho(t)\rho(t') \rangle - \langle \rho(t) \rangle \langle \rho(t') \rangle$$
$$= \lambda(t)\delta(t - t'). \qquad (8.6)$$

It is interesting that if we now average over time, the answer is insensitive to the time dependence of the rate. More precisely, the time average of the connected correlation is

$$\overline{\langle \rho(t)\rho(t + \tau) \rangle}_c \equiv \lim_{T \to \infty} \frac{1}{T} \int_0^{\infty} dt \, \langle \rho(t)\rho(t + \tau) \rangle = \bar{\lambda}\delta(\tau), \qquad (8.7)$$

and this is true whether or not the Poisson process is modulated. In this sense the fluctuations in the Poisson stream are independent of the modulations, and this is an important way of testing for the validity of a Possion description.

Instead of averaging over an ensemble of trials we can also average directly over time. Now we have no way of distinguishing between true correlations among successive events (which would indicate a departure from Poisson statistics) and correlations that are "carried" by the time dependence of the rate itself. The result is that

$$\overline{[\rho(t)\rho(t+\tau)]} = \overline{[\lambda(t)\lambda(t+\tau)]} + \bar{\lambda}\delta(\tau), \qquad (8.8)$$
$$\overline{[\rho(t)\rho(t+\tau)]}_c = \overline{[\Delta\lambda(t)\Delta\lambda(t+\tau)]} + \bar{\lambda}\delta(\tau), \qquad (8.9)$$

where the fluctuations in the rate are defined as $\Delta\lambda(t) = \lambda(t) - \bar{\lambda}$. Thus the (connected) autocorrelation of the pulse train, computed by averaging over time, has a contribution from the (connected) autocorrelation function of the rate and a delta function contribution.

We recall that the integral of $\rho(t)$ over a time window counts the number of events in that window. The variance of the count found in a window thus is equal to a double integral of the ρ—ρ correlation function (see, for example, Appendix 2 in Rieke et al. 1997). The delta function in the autocorrelation leads to a term in the count variance that grows linearly with the size of the window over which we count, much as a delta function in the velocity autocorrelation for a Brownian particle leads to diffusion; here we have a diffusion of the count. Since the mean count grows linearly with the size of the counting window, the variance and mean are growing in proportion to one another, and in fact the proportionality constant is one. This equality of variance and mean is one of the defining features of the Poisson process.

If we try to analyze the signal $\rho(t)$ in the frequency domain, then the first step is to subtract the time average and define $\delta\rho(t) = \rho(t) - \overline{\rho(t)}$. Note that $\overline{\rho(t)} = \bar{\lambda}$ if we average for a sufficiently long time. Then if we Fourier transform and compute the obvious correlation functions, we obtain:

$$\delta\hat{\rho}(f) = \int dt \exp(+2\pi i f t)\delta\rho(t), \qquad (8.10)$$
$$\langle\delta\hat{\rho}(f)\delta\hat{\rho}(f')\rangle = \delta(f+f')S_{\delta\rho}(f), \qquad (8.11)$$
$$S_{\delta\rho}(f) = S_{\Delta\lambda}(f) + \bar{\lambda}, \qquad (8.12)$$

where we use $\hat{}$ to denote Fourier transforms. Thus the spectral density of fluctuations in ρ, $S_{\delta\rho}$, is related to the spectral density of rate fluctuations, $S_{\Delta\lambda}$, plus a "white noise" term with an amplitude equal to the mean rate. This structure parallels that for the correlation functions in Eqn. (8.9) because the spectra and correlation functions form a Fourier transform pair,

$$\overline{[\rho(t)\rho(t')]}_c = \int df \exp[-2\pi i f(t-t')]S_{\delta\rho}(\omega); \qquad (8.13)$$

this is the Wiener–Khinchine theorem.

Notice that if we have an ensemble of many trials with the same input, then we can define a different "fluctuation" in the signal ρ by subtracting off the time dependent mean $\langle \rho(t) \rangle = \lambda(t)$. Thus if we write $\Delta\rho(t) = \rho(t) - \lambda(t)$, then the equations analogous to (8.11) and (8.12) above are as follows:

$$\langle \Delta\hat{\rho}(f)\Delta\hat{\rho}(f') \rangle = \delta(f + f')S_{\Delta\rho}(f) \tag{8.14}$$

$$S_{\Delta\rho}(f) = \bar{\lambda}. \tag{8.15}$$

Again we see that the connected fluctuations are independent of the modulation.

8.2.4   Shot noise

Idealized Poisson events have zero duration, or equivalently, infinite bandwidth. This is clearly unrealistic; real signals have finite rise and decay times. We can give the Poisson process some more meat by replacing every zero-duration event with a fixed waveform and having all the waveforms in the train be additive. Another way of saying this is that the train of events in Eqn. 8.1 is filtered by a linear filter $h(t)$ (see Fig. 2C), and this is often a fair first order model for the physics underlying shot noise, such as pulses of electrons from a photomultiplier tube. The output is found by convolving this filter with the sequence of delta functions:

$$v_h(t) = \sum_k \delta(t - t_k) \otimes h(t) = \sum_k h(t - t_k), \tag{8.16}$$

where we use v_h to denote the result, because in our applications the filtered process will always be a voltage. An example of a filtered process is given in Fig. 2D. If the width of $h(t)$ is very small compared to $1/\lambda$, then the shot noise still looks a lot like the original Poisson process in the sense that there are clearly separated events. If, however, $h(t)$ is much larger than the average separation between events then $v_h(t)$ will be a rather smooth signal, and its value will be distributed according to a Gaussian. This is easy to see: When at any one point in time, v_h is the sum of a large number of filtered events, spread randomly in time, the central limit theorem (Feller 1966) tells us that the probability distribution of that sum approaches a Gaussian, at least for reasonable forms of $h(t)$.

Convolution in the time domain is equivalent to multiplication of the Fourier transforms in the frequency domain, and so, using Eqn. 8.12, we can write down the power spectral density of the shot noise process as:

$$S_v(f) = \left[S_{\Delta\lambda}(f) + \bar{\lambda} \right] \cdot |\hat{h}(f)|^2, \tag{8.17}$$

where we use the subscript v in S_v to remind us that the noise power density will represent the power density spectrum of voltage fluctuations measured from a cell. Notice that the spectral density includes contributions from the "signal"—the time variation of λ—as well as from noise due to randomness in the event times. Below we will isolate the "true" noise component of the spectrum, which we will call $N_v(f)$.

Again these results are what we obtain by making a Fourier analysis, which is related to the correlation functions that are defined by averaging over time. On the other hand, if we can observe many responses to the same inputs, then we can compute an average over this ensemble of trials, to define the ensemble averaged output, $V_h(t)$:

$$
\begin{aligned}
V_h(t) \equiv \langle v_h(t) \rangle &= \lim_{M \to \infty} \frac{1}{M} \sum_{m=1}^{M} \sum_{k_m=1}^{N_m} [\delta(t - t_{k_m}) \otimes h(t)] \\
&= \left[\lim_{M \to \infty} \frac{1}{M} \sum_{m=1}^{M} \sum_{k_m=1}^{N_m} \delta(t - t_{k_m}) \right] \otimes h(t) \\
&= \lambda(t) \otimes h(t),
\end{aligned}
\tag{8.18}
$$

where the first step follows because convolution is a linear operation, so that the order of summation and convolution can be changed, and the second step is from Eqn. 8.4. The end result is illustrated in Fig. 2F. Equation 8.18 thus simply states that the ensemble averaged output of a filtered modulated Poisson process is equal to the filtered rate. This will be used in analyzing the transduction of contrast by photoreceptors and second order cells, which for moderate contrast fluctuations are reasonably close to linear. Thus from an experiment where we generate $\lambda(t)$ and measure $V_h(t)$ we can in principle derive $h(t)$.

It will be convenient to write $\lambda(t)$ as the product of a constant, λ_0, and a contrast modulation function $c(t)$: $\lambda(t) = \lambda_0 \cdot [1 + c(t)]$, where $c(t)$ represents contrast as a function of time. This also conforms closely to the experiment, where we use $c(t)$ as a signal to modulate a light source with an average photon flux equal to λ_0. In the experiment we will present the same stimulus waveform a large number of times to generate an ensemble of responses. Now, using the frequency representation of Eqn. 8.18 we get:

$$
\begin{aligned}
\hat{V}_h(f) &= \hat{\lambda}(f) \cdot \hat{h}(f) \\
&= \lambda_0 \cdot \hat{c}(f) \cdot \hat{h}(f) \qquad (f \neq 0).
\end{aligned}
\tag{8.19}
$$

Thus the Fourier transform of the ensemble averaged response equals the product of the Fourier transforms of the rate and the filter. In an experiment we set the stimulus, $\lambda(t)$ in the above equation, and we measure $V_h(t)$. Using these data and Eqn. 8.19 we can directly compute the transfer function $\hat{h}(f)$ that translates the stimulus into the average response. But we can also look at the fluctuations around this average,

$$
\Delta v_h(t) = v_h(t) - V_h(t),
\tag{8.20}
$$

and now we want to compute the power spectral density of these fluctuations. The key is to realize that

$$\Delta v_h(t) = h(t) \otimes \Delta\rho(t), \qquad (8.21)$$

which means that

$$\begin{aligned} N_v(f) &= |\hat{h}(f)|^2 S_{\Delta r}(f) \\ &= \lambda_0 |\hat{h}(f)|^2, \qquad (8.22) \end{aligned}$$

where we use the notation N_v because this truly is a spectral density of noise in the response v_h. Again the crucial result is that the noise spectral density is determined by the mean counting rate, independent of the modulation. There is even a generalization to the case where the filter $h(t)$ is itself random, as discussed in the next section.

These results give us a way of calculating the underlying average Poisson rate from the observed quantities in the experiment:

$$\lambda_0 = \frac{|\hat{V}_h(f)|^2}{N_v(f) \cdot |\hat{c}(f)|^2}. \qquad (8.23)$$

This of course is valid only for an ideal, noiseless, system, the only noise being that introduced by Poisson fluctuations. We could choose to apply Eqn. 8.23 to the quantities we measure, and then consider the result as a measurement of an *effective* Poisson rate (see also Dodge et al. 1968). This can then be compared to the rate we expect from an independent measurement of photon flux (which we can make—see the experimental sections) to assess how close the real system comes to being ideal. But we can make the decription a bit more realistic by introducing a random filter in the chain of events leading to the cell's response.

8.2.5   A Poisson process filtered by a random filter

The analysis presented above relies on the filter $h(t)$ having a prescribed, fixed shape. That assumption is not entirely realistic. More to the point, we are interested in characterizing the limitations of the system as a transmitter of information, and so it is precisely this deviation from strict determinacy that we wish to analyze. Phototransduction is a biochemical process, rooted in random collisions of molecules. Not surprisingly therefore, quantum bumps in photoreceptors are known to fluctuate (Fuortes and Yeandle 1964, Baylor et al. 1979, Wong et al. 1980, Laughlin and Lillywhite 1982), varying both in shape and in latency time. We would like to incorporate the effect of such fluctuations in our formulation, and we do that here for the simplest case. Suppose a Poisson event at time t_k is filtered by a filter $h_k(t)$, that the shape of this filter is drawn from a probability distribution of filter shapes, $\mathcal{P}[h(t)]$, and that these draws are independent for different times t_k. As before, the contributions of the filtered events are assumed to add linearly. We can picture this distribution of filter shapes again as an ensemble, and of course this ensemble has some average filter shape, $\langle h(t) \rangle$. Because everything is still linear, we can exchange orders of averaging, and so it is easy to see that we can replace the fixed shape $h(t)$ in Eqn. 8.19 by its ensemble average. In other words in the case of independently fluctuating filters we obtain:

$$\hat{V}_h(f) = \langle \hat{v}_h(f) \rangle = \hat{\lambda}(f) \cdot \langle \hat{h}(f) \rangle, \tag{8.24}$$

and instead of being able to measure the fixed shape of a filter we have to settle for characterizing its average.

Finally we should compute the effect of variable filter shapes on the power density spectrum, that is we want the analogue of Eqn. 8.22. Here it is useful to remember that the power density spectrum is the Fourier transform of the autocorrelation function. If bumps of variable shapes are generated at random times, then each bump correlates with itself. The correlations with the others, because their shapes and arrival times are assumed independent, will lead to a constant. The autocorrelation of the bump train is then the ensemble average of the autocorrelation of all individual bumps. Likewise the power density spectrum is the ensemble average of the power spectra of the individual bumps. The end result is then that:

$$N_v(f) = \lambda_0 \cdot \langle |\hat{h}(f)|^2 \rangle \tag{8.25}$$

$$|\hat{V}_h(f)|^2 = \lambda_0^2 \cdot |\langle \hat{h}(f) \rangle|^2 \cdot |\hat{c}(f)|^2. \tag{8.26}$$

We can now define an effective Poisson rate that we compute from the experimental data:

$$\begin{aligned}
\hat{\lambda}_{\text{eff}}(f) &= \frac{|\hat{V}_h(f)|^2}{N_v(f) \cdot |\hat{c}(f)|^2} \\
&= \lambda_0 \cdot \frac{|\langle \hat{h}(f) \rangle|^2}{\langle |\hat{h}(f)|^2 \rangle},
\end{aligned} \tag{8.27}$$

and this is in general a function of frequency, because when $h(t)$ fluctuates in shape, the behavior of $\langle |\hat{h}(f)|^2 \rangle$ is different from that of $|\langle \hat{h}(f) \rangle|^2$.

It is worthwhile fleshing out what effect fluctuations in different bump characteristics have on $\hat{\lambda}_{\text{eff}}(f)$. Here we consider variations in amplitude and in latency. There is good evidence that such variations in bump parameters occur almost independent of one another (Howard 1983, Keiper and Schnakenberg 1984). Variations in the amplitude of the bump can be modeled by assuming that individual bump shapes are described by $\beta \cdot h_0(t)$, where β is a random variable with mean $\langle \beta \rangle = 1$, and variance σ_β^2, and where $h_0(t)$ has a fixed shape. It is easy to derive that in that case we have $\hat{\lambda}_{\text{eff}}(f) = \lambda_0/(1 + \sigma_\beta^2)$. In other words, random variations in bump amplitude lead to a frequency independent decrease in the effective Poisson rate. Any form of noise that leads to a spectrally flat effective decrease in photon flux is therefore sometimes referred to as multiplicative (Lillywhite and Laughlin 1979). An important special case is that of a decrease in the photon flux. This can be described by taking β to be a random variable with value either 0 or $1/p_1$, with p_1 the probability of a photon being transduced (so that $\langle \beta \rangle = 1$, as required, and $\sigma_\beta^2 = 1/p_1 - 1$). This leads to $\hat{\lambda}_{\text{eff}}(f) = p_1 \cdot \lambda_0$, as expected.

Another important source of randomness is a fluctuating latency time from photon absorption to bump production. Suppose that we have a fixed shape, $h_0(t)$ as before, but that the time delay is distributed, independent for different bumps, according to $p(t_{\text{lat}})$. Displacing random events in a random way preserves the independence and the mean rate, so that there is no effect on the noise power density. The timing with respect to external modulations is compromised, however, and this mostly affects the reliability at high frequencies. Again, starting from Eqn. 8.27 it is easy to derive: $\hat{\lambda}_{\text{eff}}(f) = \lambda_0 \cdot |\hat{p}(f)|^2$—that is, the frequency dependence of the effective Poisson rate is given by the Fourier power transform of the latency distribution (de Ruyter van Steveninck and Laughlin 1996b). Because $p(t_{\text{lat}})$ is a probability distribution, its Fourier transform must go to 1 for $f \to 0$, and will go to 0 for $f \to \infty$ if the distribution is smooth. Thus, if bumps have random latencies the effective Poisson rate will be frequency dependent. Low frequencies are not affected whereas the effective rate goes to zero in the limit of very high frequencies.

Fluctuations in the duration of bumps as well as external additive noise in general have frequency dependent effects on $\hat{\lambda}_{\text{eff}}(f)$. If the aim is to describe the phototransduction cascade and the other processes occurring in the cell, then it is interesting to try and tease all the contributions apart, and this may not be easy. Here, however, our goal is more modest, in that we want to quantify the reliability of photoreceptors and the LMCs onto which they project; compare the results to the limits imposed by the stochastic nature of light; and explore some of the consequences of photoreceptor signal quality for visual information processing.

8.2.6   Contrast transfer function and equivalent contrast noise

It is useful to define two other quantities. The first is the contrast transfer function, defined by:

$$\hat{H}(f) = \frac{\hat{V}_h(f)}{\hat{c}(f)} = \lambda_0 \cdot \langle \hat{h}(f) \rangle, \qquad (8.28)$$

which expresses the cell's gain not as the translation from single photon to voltage, but from contrast to voltage. This is practical because photoreceptors and LMCs work mostly in a regime of light intensities where bumps are fused, and then it is natural to think of these cells as converting contrast into voltage In addition to this transduced contrast there is voltage noise. It is conceptually useful to express these noise fluctuations as an effective noise η_c added to the contrast itself,

$$\hat{v}(f) = \hat{H}(f)[\hat{c}(f) + \hat{\eta}_c(f)]. \tag{8.29}$$

The spectral density of this effective noise source is then the equivalent contrast noise power density $N_c(f)$, and has units of (contrast)^2/Hz. Since contrast itself is dimensionless this means that N_c has units of 1/Hz, and hence $1/N_c$ has the units of a rate; we will see below that for an ideal photon counter this rate is exactly the mean rate at which photons are being counted. To find $N_c(f)$ we inverse filter the measured voltage noise power density by the contrast power transfer function:

$$N_c(f) = \frac{N_v(f)}{|\hat{H}(f)|^2}, \tag{8.30}$$

and we can easily derive that $N_c(f) = 1/\lambda_{\text{eff}}(f)$. If we now have a cascade of elements, such as the photoreceptor and the LMC, and we measure the equivalent contrast noise power density at each stage, we would like to define the accuracy of the interposing element, in this case the array of synapses between photoreceptors and LMC. Using the equivalent contrast noise power density it is easy to do this: If we measure $N_{c1}(f)$ and $N_{c2}(f)$ for two elements in a cascade, then for all f we must have $N_{c2}(f) \geq N_{c1}(f)$, and the difference is the contribution of the element in between. In the particular case of photoreceptors and LMCs we have to be careful to include the effect of having six photoreceptors in parallel. This, assuming we may treat them as statistically independent but otherwise equal, is easy to do: When elements are combined in parallel we divide $N_c(f)$ for the individual one by the number of elements to get the equivalent contrast noise power density of the combination. The charm of working with the equivalent contrast noise power density is that it allows us to compute the result of combining elements in series and in parallel, in the same way that we calculate the net resistance of series and parallel combinations of resistors.

8.3   The Early Stages of Fly Vision

8.3.1   Anatomy of the blowfly visual system

A good proportion of the surface of a fly's head is taken up by its compound eyes, which is a direct indication that the eyes are very important to the fly. This is also clear from other considerations: The energy consumed by all the photoreceptors in the blowfly's retina in bright light is about 8% of the total energy consumption of a fly at rest (Laughlin et al. 1998). In the blowfly, each eye has about 5000 lenses, corresponding to 5000 pixels of visual input. These pixels are distributed over almost a hemisphere, so the two eyes provide the fly with almost complete surround vision. The male's eyes are somewhat larger than the female's, because the visual fields of the male's two eyes overlap in the dorsofrontal region of visual space. This region is used in detecting and chasing females.

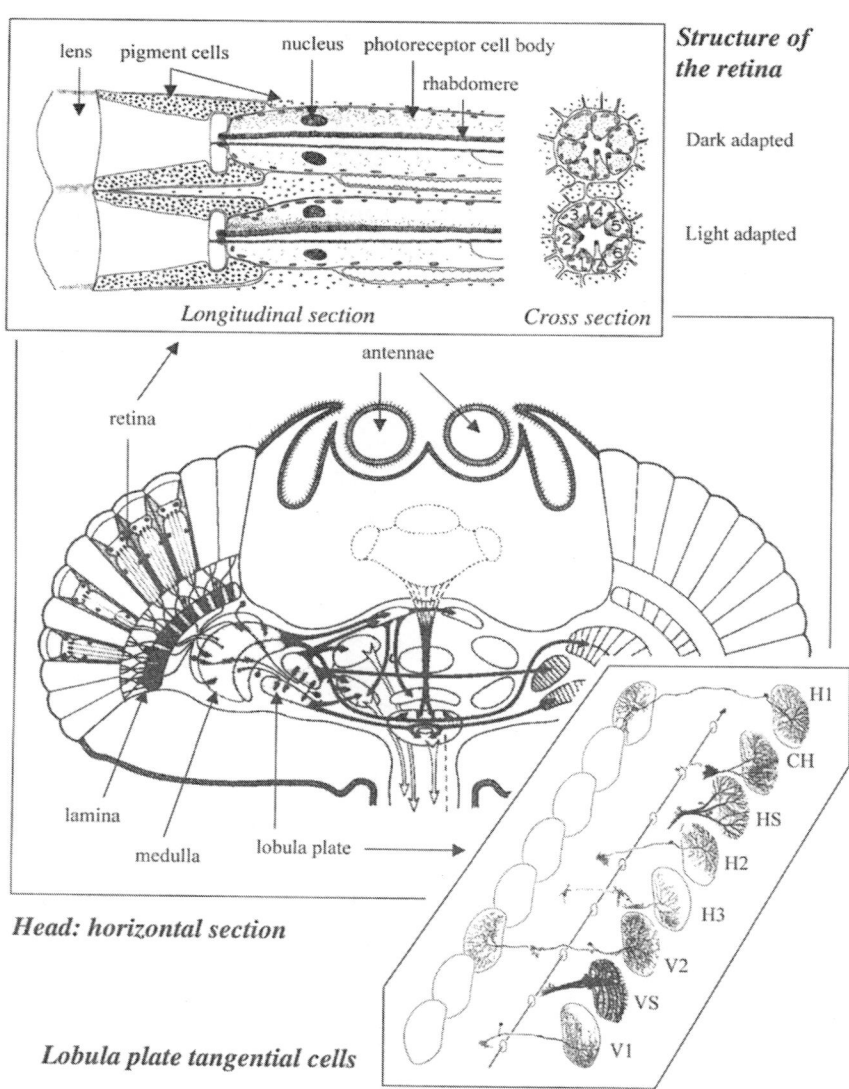

FIGURE 8.3.

FIGURE 8.3. Top: Part of the retina, modified from Stavenga (1974), showing two ommatidia, each containing one lens, several pigment cells, and 8 photoreceptor cells. Light enters the lens and is focused on the tips of the rhabdomeres. It then travels in a bound mode along the rhabdomere, in which it can be absorbed by a membrane bound rhodopsin molecule (Stavenga 1995). Through a series of biochemical steps this then leads to a measureable electrical response across the cell membrane. Center: Schematic horizontal cross section through the head of a typical fly, modified from Kirschfeld (1979). The areas mainly relevant for visual information processing are the retina with its photoreceptors, the lamina, where photoreceptor signals are combined and filtered, and the medulla and lobula complex, where more complex information processing takes place. Bottom: Exploded view of giant tangential cells of the lobula plate, modified from Hausen (1981). The outlines of the two lobula plates and the esophagus are drawn in thin lines. The lobula plate is an output station of the visual system, where tangential cells collect motion information over large areas of the visual field. From here signals are sent to the thoracic ganglion where they are used in the control of flight. Cells drawn in this figure collect their information in the right lobula plate; H1 and V2 project to the contralateral plate. The tangential cells of the lobula plate encode information about wide field motion. All are direction selective, that is their firing rate increases for motion in their preferred direction, while spike activity is suppressed for motion in the opposite direction. The H cells code horizontal, and the V cells vertical motion. CH has a more complicated directional selectivity. The labels HS and VS refer to groups of cells. The tangential cells are unique and identifiable, so that they can be numbered. H1 in particular is a good cell to record from both because it is very easy to find on the contralateral side, and because it is inward sensitive, responding preferentially to motion from the side toward the midline of the animal. The combination of this directional selectivity with the contralateral projection is unique, so that H1 can be identified unambiguously from the electrophysiological recording.

8.3.2   Optics

The compound eye of insects forms an image onto an array of photoreceptors through a large number of tiny lenses (with diameters ranging typically
from 10 to 30 μm; see Fig. 3). Each lens belongs to an ommatidium, which
typically contains eight photoreceptors, and a number of optical screening
cells. Part of the photoreceptor cell membrane is densely folded, and forms
a long ($\approx 100-200\mu$m), thin ($\approx 2\mu$m) cylinder, called the rhabdomere. The
membrane consists mainly of phospholipids and proteins so that its refrac-
tive index is higher than that of the surrounding watery medium. Therefore, the rhabdomere acts as a waveguide, and light can travel along its long
axis in a bound mode. The combination of a lens and the tip of an optical
waveguide in its focal plane forms a spatial filter (Snyder 1979): Only light
coming from a small angular region can enter the waveguide. The physical
limit to the resolution of this system is set by diffraction: $\Delta\phi \approx \lambda/D$, and
with $\lambda = 500$ nm and $D = 25$ μm, we have $\Delta\phi \approx 1/50$ rad $\approx 1°$. See Exner
(1891), Barlow (1952), and Feynman et al. (1963) for an analysis of the
optics of compound eyes. The physics of the lens-photoreceptor system is
well understood, and theory is in very good agreement with physiological
findings (van Hateren 1984).

Flies do not have an iris pupil or eyelids, yet they need to protect their
photoreceptors from excessively intense light. This is accomplished by an
elegant mechanism depicted in the top box in Fig. 3. The top ommatidium shows a dark adapted photoreceptor, which has tiny pigment granules
dispersed through its cell body. In the light adapted state shown in the
bottom ommatidium these granules have migrated close to the photoreceptor rhabdomere. The granules absorb light, and because they are close
to the light guiding rhabdomere, they act as a light absorbing cladding
that captures up to 99% of the incoming photon flux. This then prevents
the photoreceptor from saturating in bright daylight (Howard et al. 1987).
The effectiveness of the pupil as a light absorber is regulated by feedback
(Kirschfeld and Franceschini 1969, Franceschini and Kirschfeld 1976).

The photoreceptors in each ommatidium are numbered R1-R8, and arranged such that R7 lies on the optical axis of the lens. R8 lies behind
R7, while the rhabdomeres of R1-R6 lie in a trapezoidal pattern around
the center, as shown by the cross section in the top frame in Fig. 3. The
optical axes of neighboring ommatidia point in slightly different directions
and they differ by an amount that matches the angular difference among

the photoreceptors in a single ommatidium. Therefore, eight photoreceptor cells in seven neighboring ommatidia share a common direction of view. Receptors R1-R6 in all ommatidia have the same spectral sensitivity (Hardie 1985), and those R1-R6 receptors in neighboring ommatidia that share the same direction of view combine their signals in the next layer, the lamina. This is known as neural superposition (Braitenberg 1967, Kirschfeld 1967, 1979). Receptors R7/R8 are special, having a spectral sensitivity different from R1-R6, and bypassing the lamina, projecting directly to the medulla.

8.3.3   Reliability and adaptation

Because single photon responses are more or less standardized it seems a good idea to model the photoreceptor voltage as a shot noise process, consisting of a stream of photon induced events often referred to as "quantum bumps" (Dodge et al. 1968, Wong et al. 1980, de Ruyter van Steveninck and Laughlin 1996a,b). However, we are dealing with a highly adaptive system. As discussed in the previous section, the variance of shot noise should be proportional to the rate of the underlying Poisson process. Panels B and C of Fig. 4 show that the variance at a mean bump rate of $300/s$ is much higher than at a rate of $3 \cdot 10^5/s$. So shot noise does not appear to be a good model. The solution to this dilemma is that the bumps change shape so that both their amplitude and their width decrease when the photoreceptor is exposed to higher light intensity (see also Fig. 9). This is the gist of the "adapting bump model" formulated by Wong et al. (1980). Clearly, when bumps adapt, the shot noise model loses some of its generality in terms of predicting the expected response amplitude to a certain stimulus or of predicting the noise power spectral density at different light intensities. However, one much more crucial aspect remains—the signal to noise ratio. Even if bump amplitudes adapt to the ambient light intensity, the frequency dependent signal to noise ratio depends only on the rate of the underlying Poisson process. One last caveat is in place here: In the standard shot noise model, the "bump" shape is taken to be fixed. Surely, in any realistic system there is noise, and here one can distinguish two of its net effects: A variation in the amplitude of the bumps, which limits the reliability of the system at all frequencies, and a loss in timing precision, which affects the higher fequencies more than the lower frequencies, as treated in section 8.2.5. See Stieve and Bruns (1983) for more details on bump shape and de Ruyter van Steveninck and Laughlin (1996b) for its effects on the overall reliability of the photoreceptor.

The results we present below were obtained in experiments using a green light emitting diode (LED) as a light source, which was always modulated around the same mean light level. The mean photon flux into the fly's photoreceptors was set by using filters of known density in the optical path between LED and fly. Light intensities were calibrated by counting the rate of quantum bumps at a low light intensity, as depicted in Fig. 4a. This was

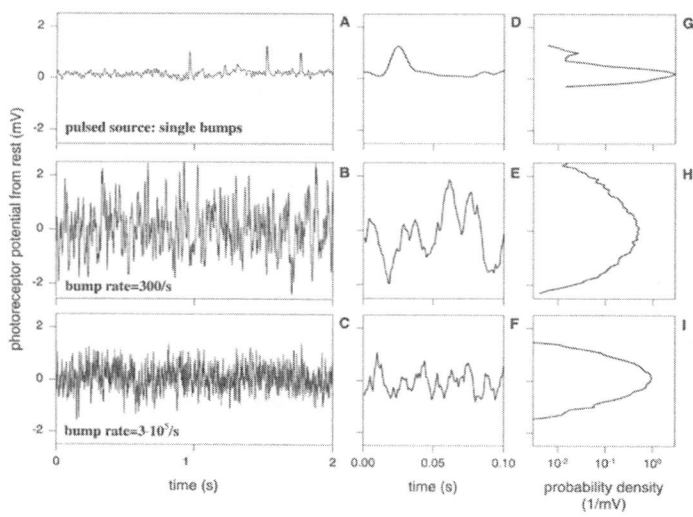

FIGURE 8.4. Intracellular photoreceptor recordings. A, B, C: samples of photoreceptor voltage, each 2 seconds long. A: Recording in the dark adapted state, with a light source that emitted brief flashes leading to about 0.5 photons captured per flash on average. B: The same cell, with a continuous light source, and a photon capture rate of 300 per second. C: As B, but now at a 100 times higher light intensity. Panels D, E, F show part of the traces in repectively A, B, C but at higher time resolution. G, H, I: Amplitude distributions of the signals in A, B, C. Note that these are drawn sideways, so that the vertical scales in all panels are the same. The probability densities are drawn on a logarithmic scale, which means that the curves should approximate parabolas for Gaussian amplitude distributions.

generally done with the LED delivering short (1 ms) flashes, using filters of optical density between 3 and 4, that transmit a fraction of 10^{-3} to 10^{-4} of the incident photons. From this low level, light intensities were increased in calibrated steps, and we can so define an extrapolated bump rate for each setting of light intensity in the experiments. Note that this procedure does not specify the absolute quantum efficiency of photon detection. At low light levels this efficiency, defined as the probability for an on axis photon of optimal wavelength (490 nm) to lead to a quantum bump, is about 50% (Dubs et al. 1981, de Ruyter van Steveninck 1986).

In most experiments the LED was modulated with a pseudorandom waveform of duration 2s, sampled at 1024 Hz with independent values at all sample times. This waveform was constructed in the frequency domain, by computing a series of 1024 random numbers on $[0, 2\pi)$, representing the phases of Fourier components. That list, concatenated with an appropriate list of negative frequency phase components, was inverse Fourier transformed, resulting in a sequence of real numbers representing a series of

2048 time samples that was used as the contrast signal $c(t)$. This procedure ensured that the amplitudes of all frequency components were equal. This is not necessary in principle, but it is very convenient in practice to have a signal that is free from spectral dropouts. The use of pseudorandom waveforms to study neural signal processing has a long history; for a review see Rieke et al. (1997). In the photoreceptors and LMCs of the fly, the first such experiments were done by French and Jährvilehto (1978).

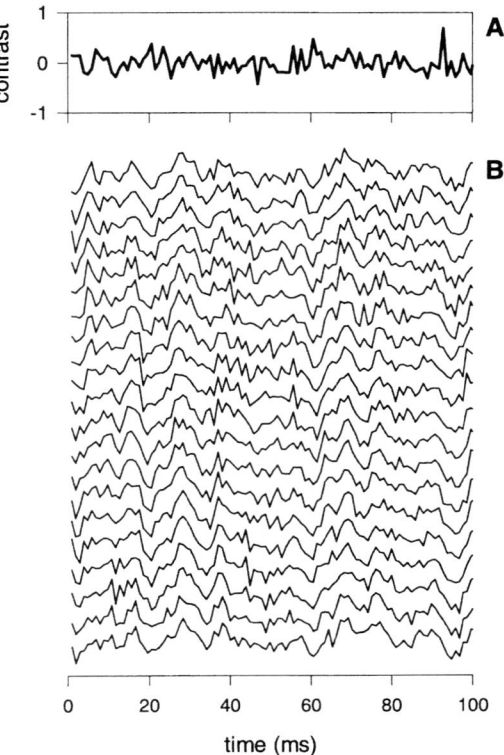

FIGURE 8.5. A: Modulation of the LED light source by a computer generated pseudorandom signal sampled at ≈1 ms intervals. This panel shows a small, 100 ms section of the total trace which was 2 s long. In the experiment the contrast waveform depicted here is presented repeatedly, and the response of the cell to all these repetitions is recorded. B: Twenty examples of individual traces recorded from a blowfly photoreceptor in response to the contrast waveform shown in A.

Fig. 5 summarizes the measurements made in a typical experiment on a photoreceptor. The same contrast waveform (top) was presented repeatedly and the response to each presentation recorded. From a large number of repetitions we obtain an ensemble of responses, from which we compute the ensemble average shown in Fig. 6A. Clearly the traces in Fig. 5B share the same overall shape, but are not identical. Each trace differs from the

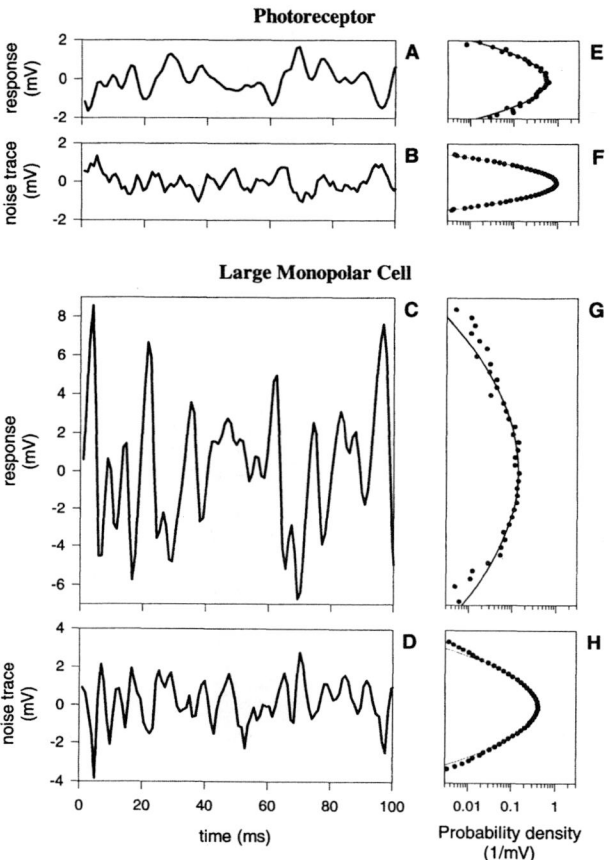

FIGURE 8.6. Sample traces of an experiment on a photoreceptor and an LMC, and amplitude distributions. A: A 100 ms segment of the ensemble averaged response of a photoreceptor to the modulation waveform shown in Fig. 5A. B: Example of a 100 ms noise trace, that is the difference between one trace in Fig. 5B and trace A in this figure. C, D As A, B above, but for an LMC. Note that the vertical scales in panels A-D are all the same: The LMC's signal is much larger than that of the photoreceptor. E: Amplitude distribution of the average voltage trace of the photoreceptor. Dots: measured values; line: Gaussian fit. F: As E, but now for the photoreceptor noise waveform. G, H: as E, F but for the LMC.

ensemble average, and these fluctuations represent the noise in the system. An ensemble of noise traces is obtained simply by subtracting the ensemble averaged response from each individual trace.

We first characterize the cell's linear response by computing the ratio of the Fourier transform of the ensemble averaged voltage waveform $V_h(t)$, and the Fourier transform of the contrast modulation waveform $c(t)$, as in Eqn. 8.28. The square of the absolute value of this ratio gives us the contrast power transfer function. The ensemble of noise traces is described by a power spectral density, which we find by averaging the Fourier power spectra of all the individual noise traces in the experimental ensemble, or equivalently by computing the variance of each Fourier coefficient across the ensemble of noise traces. Finally, the ratio of these two functions, as defined in Eqn. 8.27, is the effective Poisson rate. If the photoreceptor were an ideal photon counter, the ratio should not depend on frequency, and be numerically equal to the extrapolated bump rate. Fig. 7 shows that the ratio does depend on frequency, and that it goes down at the higher frequencies, notably so above 100 Hz. This is a consequence of the limited temporal resolution of the transduction cascade which after all consists of a sequence of chemical reactions, each of which depends on random collisions between molecules. However, at $3.8 \cdot 10^3$ incoming bumps per second the photoreceptor acts essentially as an ideal photon counter up to about 50 Hz. The deviation from this at low frequencies is due to excess low frequency noise in the noise power density (see Fig. 7B,E) which is almost certainly due to $1/f$ noise in the equipment. At the higher extrapolated bump rate the photoreceptor's efficiency is constant up to almost 100 Hz, and is about a factor of 2 from ideal. The behavior of the LMC is a bit different in that it acts as a high pass filter, and transmits low frequencies less reliably than would an ideal detector. However, for intermediate frequencies the LMC stimulated at a bump rate of $7.5 \cdot 10^4$ per second comes close to ideal. At the highest bump rate, $7.5 \cdot 10^6$ per second, the LMC deviates from ideal by a factor of 8 at its best frequency (≈ 50 Hz). At this frequency the absolute performance of the LMC is quite impressive, being equivalent to an effective photon flux of about 10^6 events per second.

Fig. 8 presents an overview of the cell's best performance, and compares it to the theoretical limit. Here we show the maximum of the effective Poisson rate for six photoreceptors and three LMCs, each at multiple light levels. These data are plotted as a function of the photoreceptor's extrapolated bump rate, which for LMCs is the extrapolated LMC bump rate divided by six, because it receives input from six photoreceptors. For the photoreceptors, at low light levels, the measured maximum and the extrapolated bump rate are very close, and they are at most a factor of two apart at the highest light intensities measured here. This indicates that the photoreceptor is designed to take advantage of each transduced photon, up to fluxes of about $3 \cdot 10^5$ extrapolated bumps/s, and up to about 100 Hz.

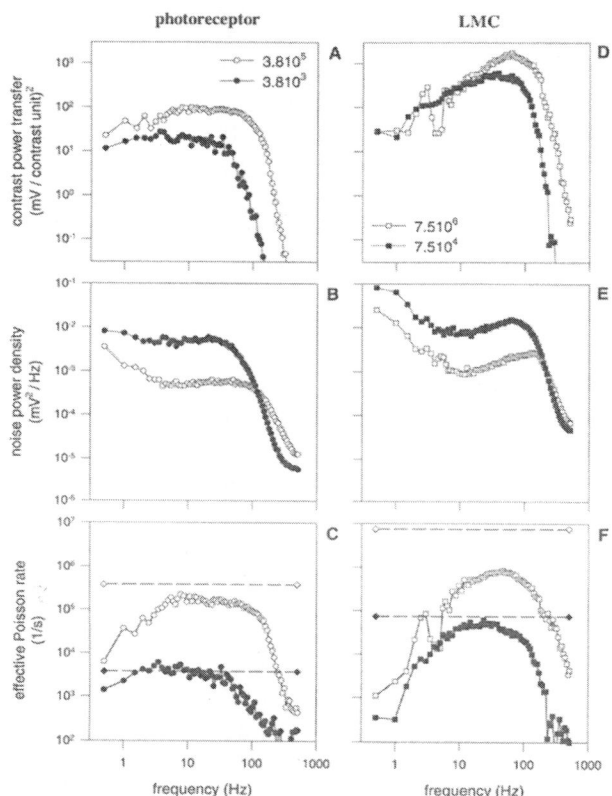

FIGURE 8.7. Characterization of a photoreceptor and an LMC in the frequency domain, each for two light intensities. The legends of A and B give these intensities, expressed as extrapolated bump count rates. A: Photoreceptor contrast power gain as a function of frequency. B: Power spectral densities of the ensemble of noise traces represented by Fig. 6B. C: The effective Poisson rate, calculated as the ratio of contrast power gain (panel A) and the power spectral density (panel B) at each frequency. If the photoreceptor were an ideal photon counter, not adding any noise itself, then the effective Poisson rate should be spectrally flat, at a level of the extrapolated bump rate given by the legend in panel A. This is depicted by the dashed lines in C. D, E, F: as A, B, C, but now for an LMC.

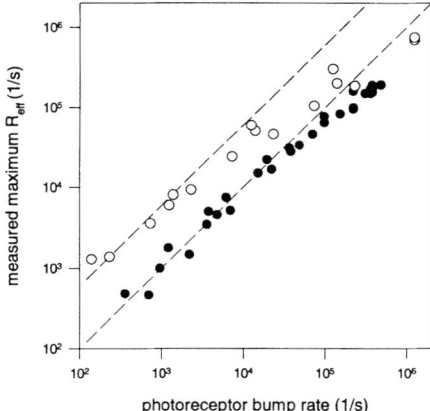

FIGURE 8.8. Comparison of the best measured statistical performance of photoreceptors and LMCs to the theoretical limit imposed by photon shot noise. The peak values of the effective Poisson rate curves, such as those in Fig. 7C,F are plotted as a function of the rate at which photons are absorbed by a photoreceptor. The rate of photon absorption is extrapolated from a dark adapted experiment, in which individual bumps are counted (see Fig. 4A). The dashed lines represent the behavior of an ideal photon counter for photorecptors (lower line) and LMCs (upper line). Along the abscissa the bump count calibration values of the LMC are divided by 6 to get the bump rate of each of the presynaptic photoreceptcrs. Statistically efficient neural superposition requires that the LMC uses all photons from all six photoreceptors from which it receives input, which means that the LMC data points should then follow the upper dashed line.

At bump rates around 10^5/s and higher the efficiency of the photoreceptor begins to decline. This is caused primarily by the action of the pupil (Howard et al. 1987; see also section 8.3.2), which attenuates the photon flux propagated through the rhabdomere and thus increases the effective contrast noise as described in section 8.2.5.

One reasonably expects that if the fly's brain is well designed, it would put the relatively high quality of the photoreceptor signals to good use. This would then imply that the accuracy of visual information processing is not too far from the photon shot noise limit at the light levels studied here. A first check is to see if neural superposition is efficient in this statistical sense. Comparing the measurements of maximal LMC effective photon flux with the ideal (open symbols and top dashed line in Fig. 8), we see that this is indeed the case up to photoreceptor bump rates around 10^4–10^5 per second. For higher light intensities the LMC, as the photoreceptor, becomes less efficient. There is a hint in the data that the LMC declines somewhat faster than the photoreceptor, perhaps due to limitations in the reliablity of synaptic transmission, as was noticed by Laughlin et al. (1987).

Given that neurons have a dynamic range much smaller than typical sensory signals, the question arises of how the nervous system copes with the input it receives in order to transmit and process information efficiently. Ultimately this is a matter of optimal statistical estimation, and the result will therefore depend on the statistical and dynamical characteristics of the signals that the animal encounters in its environment. It is well known, and we have seen examples already, that photoreceptors adapt their gain to the ambient light intensity, which may vary over many orders of magnitude. The usefulness of this adaptation seems obvious: At higher light intensity the absolute gain should be brought down to keep the transduced contrast fluctuations within the cell's voltage operating range. But implicit in this explanation is that the average light levels must change relatively slowly compared to the contrast fluctuations. If the sun flickered unpredictably on time scales of a second or so, and with an amplitude equal to that of the day night cycle, then it would be of no use at all to adapt the photoreceptor gain. Only because large changes tend to be slow does it make sense to design a system that tracks their mean, and changes its gain so as to encode the faster fluctuations more efficiently.

The effect of adaptation is easier to appreciate when the responses are plotted in the time domain. Fig. 9 shows impulse responses of both cell types, at four light levels spanning three orders of magnitude. All the curves are derived from inverse Fourier transforms of the contrast transfer function $\hat{H}(f)$ defined earlier. The top two panels show the absolute gain—that is the response normalized to a single photon capture. In the bottom panels the response is normalized to a 1 ms lightflash with an amplitude equal to the mean light level, so these are normalized to contrast. It is clear from the figures that there is a large range of gain control, as the shapes of the responses normalized to single photons vary dramatically. Expressed as contrast gain, the responses become higher in amplitude and sharper in time as the light intensity increases.

The photoreceptor primarily seems to scale down its photon conversion gain, both in amplitude and in time course. The LMC filters out the photoreceptor's DC level, except perhaps at the lowest light intensities, and it also scales the gain (Laughlin and Hardie 1978). The combined effect leads to a scaling of the LMC amplitude fluctuations. Remarkably, this scaling is such that when stimulated with naturalistic, and very non-Gaussian, sequences, the LMC produces a voltage output that follows a Gaussian quite closely (van Hateren 1997). There is evidence that adaptation of this system is set so as to optimize information transmission rates under different conditions (van Hateren 1992).

The data of Fig. 9 show the behavior of the cells while they are in their adapted states, as care was taken to to let the system adapt before the measurement was done. It is also interesting to study the time course of adaptation, and here we will look in particular at adaptation of synaptic transmission.

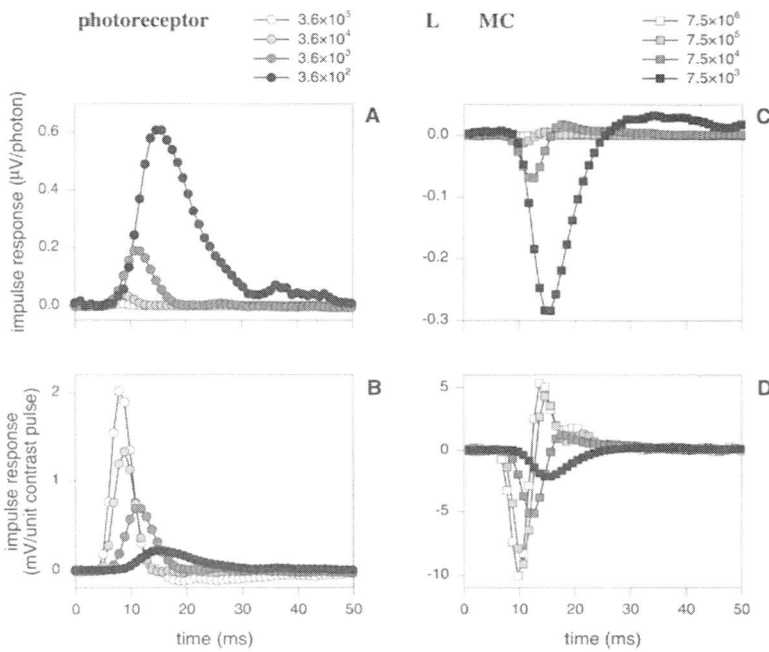

FIGURE 8.9 Impulse responses of blowfly photoreceptors and LMCs, measured in the adapted state at different light intensities. The average light intensity, expressed as the extrapolated rate of effective photon capture in bumps/s, is given in the legends above A and C. The responses are scaled in two different ways. A: Photoreceptor impulse response, scaled to represent the electrical response of a photoreceptor to the capture of a single photon. B: The same data as in A, but here expressed as the electrical response to a contrast pulse of 1 ms wide, with an amplitude equal to the mean light intensity. C, D: as A, B, but for an LMC.

8.3.4   Efficiency and adaptation of the photoreceptor-LMC synapse

The link between the photoreceptors and the LMC is a parallel array of chemical synapses. The detailed anatomy of this projection is well known (reviewed by Shaw 1981), and counts have been made of the number of active zones between photoreceptors and LMCs. The total number of synapses between one photoreceptor and an LMC is on average about 220 (Nicol and Meinertzhagen 1982, Meinertzhagen and Fröhlich 1983), so that the total number of active zones feeding into an LMC is close to 1320. Although this number was obtained from the housefly *Musca*, which is smaller than the blowfly, the total number is unlikely to be far off. The photoreceptor-LMC synapses are tonically active, just like the synapses of vertebrate retinal bipolar cells. This means that even in the absence of contrast fluctuations, they release a stream of vesicles (Shaw 1981, Uusitalo et al. 1995, Lagnado et al. 1999). Also they pass on graded potentials, and it thus seems a reasonable first approximation to model them as units that release vesicles with a rate depending on the presynaptic potential. One interesting question is then whether vesicle release can be modeled as a modulated Poisson process. At first sight this would seem wasteful, in the sense that if the synapse would have better control over its vesicle release, it could emit vesicles in a much more deterministic way. One might imagine that the synapse functions somewhat as a voltage controlled oscillator, releasing vesicles in a regular stream at a rate determined by the presynaptic voltage. One way to get at this issue is to measure the reliability of the synapse, and estimate from this a lower bound on the release rate, assuming that release is Poisson. If, through other independent methods, one can make a good estimate of the average total release rate, then one can compare the two rates. If the rate estimated from the Poisson assumption were found to be much higher than the total average rate, one would have a strong indication for tight control of vesicle release.

As argued in section 8.2.6, one can find the equivalent contrast noise power of a cascaded system by adding the equivalent contrast noise power of its separate elements $N_c(f) = N_{c1}(f) + N_{c2}(f)$. What we would like to do here is to infer the equivalent contrast noise power for the synapse from measurements of the equivalent contrast noise of the photoreceptor and the LMC. We have the data, because the equivalent contrast noise power is just the inverse of the effective Poisson rate plotted in Fig. 7C,F: $N_c(f) = 1/\lambda_{\text{eff}}(f)$ (see section 8.2.6). The measurements of effective Poisson rates in photoreceptors and LMCs, together with the given number of active zones, allow us to make an estimate of the effective Poisson rate of a single synaptic contact. This cannot be done directly because it has not been possible in practice to make a simultaneous recording from a photoreceptor and its postsynaptic LMC in vivo, so we have to interpolate. From a large number of experiments on different cells at different light levels we compute $N_c(f)$, and we do this separately for photoreceptors and LMCs. We interpolate each set of curves to find a smooth surface describing the overall behavior as a function of both frequency and bump rate (see de Ruyter van Steveninck and Laughlin 1996a). We estimate the synaptic contribution, $N_{c_{\text{Syn}}}(f)$ by subtracting the interpolated values obtained for photoreceptors, divided by 6 to account for the parallel projection of 6 photoreceptors, from those describing LMCs: $N_{c_{\text{Syn}}}(f) = N_{c_{\text{LMC}}}(f) - N_{c_{\text{PR}}}(f)/6$. The differences are small, and not so easy to estimate, which already indicates that the synaptic array itself cannot be much less reliable than the photoreceptor. Here we only present data from the highest light levels used in the experiment, mainly because there all signals are most reliable, and the effect of internal noise sources is most conspicuous.

Note that $N_{c_{\text{Syn}}}(f)$ describes the equivalent contrast noise of the full array of 1320 synapses in parallel. Each synapse is driven by the same photoreceptor voltage fluctuations, and each modulates its vesicle release rate accordingly. We assume that apart from this common driving force, all synapses release vesicles in a statistically independent way, and that all are equally effec:ive and reliable. Then we can simply divide $1/N_{c_{\text{Syn}}}$, as defined above, by 1320 to get the effective Poisson rate for the single synapse, as argued in Sect. 8.2.6. This then provides a lower bound on the reliability of a single synapse. If the assumptions mentioned here are invalid, then there must be at least one synapse in the array that does better than this average.

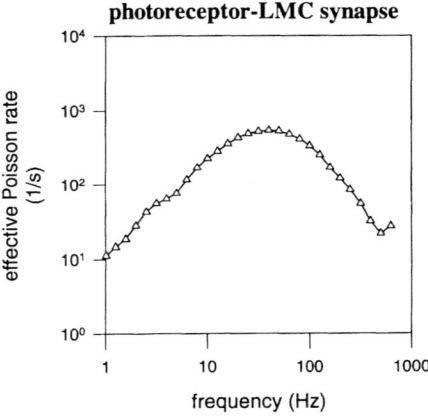

FIGURE 8.10. Effective Poisson rate for a single synaptic active zone as a function of frequency, computed as described in the text.

Fig. 10 shcws the result of the calculation, expressed as $1/N_c(f)$, the equivalent Poisson rate for one single active zone. The curve has a maximum of about 540 events per second per synaptic zone. Unfortunately we cannot directly identify this number with the supposed vesicle release rate. This can be done for photon flux modulations because in that case we know the proportion, $c(t)$, by which we modulate the flux: $\lambda(t) = [1 + c(t)] \cdot \lambda_0$. For the photoreceptor-LMC synapse, one may describe the modulation of the release rate by a gain factor g (which may also be frequency dependent) that converts photoreceptor voltage into vesicle flux (Laughlin et al. 1987). If g is high, then the synapse encodes relatively reliably at a low mean vesicle rate. The price is that the operating range will be small, as the rate cannot go below zero. If g is low, on the other hand, then the operating range is large, but the reliability of transmission is relatively poor. In other words, we measure an effective rate λ_{eff}, from which we wish to estimate a physical Poisson event rate. This means we must get an estimate of g, and we must understand how g affects our measurement. To begin with the latter, if we just apply Eqn. 8.23 to the case of a real rate λ_0 modulated by $g \cdot c$, then we would measure an effective rate λ_g depending on g:

$$\lambda_g = \lambda_0 \cdot |g|^2, \tag{8.31}$$

while release would shut down at contrast values below $c_{min} = -1/g$. To keep the argument simple we neglect the possible frequency dependence of g. That is justified here because we will not reach precise conclusions anyway, and the frequency dependence, being rather smooth, is not likely to affect the final result of the analysis too much. We need to find a way to estimate the operating range of the synapse, and we can try to get at that by estimating c_{min}. One hint is that LMCs are reasonably linear when the contrast fluctuations are not too large, perhaps of order 20% to 30%, but we would like to make this a bit more precise.

Fig. 11 presents data suggesting that we can see the synapse shutting down. The stimulus (Fig. 11A) is a 200 ms square wave of 95% contrast, repeated 360 times. Panel B shows the average photoreceptor response which follows that stimulus with a bit of sag. The LMC (panel C) responds phasically and with inverted sign. When the photoreceptor voltage makes its downward transition, the vesicle release rate decreases, and because the neurotransmitter (histamine, Hardie 1988) opens chloride channels the LMC depolarizes. In panel D we plot the fluctuations (8 samples) of the LMC potential around its average waveform. These show a rather dramatic effect just after the light to dark transition. It seems that synaptic transmission is completely shut down when the photoreceptor hyperpolarizes, and bounces back about 15 ms later, very similar to results reported by Uusitalo et al. (1995). This is confirmed by the standard deviation of the fluctuation waveforms shown in panel E. The fluctuations during constant light, say from 20-100 ms and from 130-200 ms, are due to a combination of photoreceptor noise amplified by the synapse and intrinsic noise of the vesicle release itself. We can also add a little probe signal to the large square wave stimulus, and we see a similar effect in the gain of the synapse: The photoreceptor fluctuations are not transmitted during the same 15 ms window, as can be seen in Fig. 12. The apparent shutdown coincides with the photoreceptor voltage being halfway between the light and dark adapted value, which in turn is induced by an intensity drop of almost 100%. Shutdown thus seems to correspond to about 50% modulation, in other words, $g \approx 2$. Combined with our earlier estimate of an effective rate of 540/s, from Eqn. 8.31 we get a rate of $540/2^2 = 135$ vesicles per second, based on the measured reliability. This is lower than the rate of 240 per second reported by Laughlin et al. (1987), but in view of the errors the discrepancy is not too surprising. We would like to compare the more conservative estimate of this number with a more direct measurement of average release rate. Unfortunately there are no conclusive data on synaptic release rates of the photoreceptor-LMC synapse in the fly. There are experimental estimates of tonic release rates for goldfish retinal bipolar cells which, like the fly photoreceptor, transmit graded potentials across a chemical synapse. In a recent paper, Lagnado et al. (1999) report 23 vesicles/s in this system, a factor of 6 lower than what we estimate here.

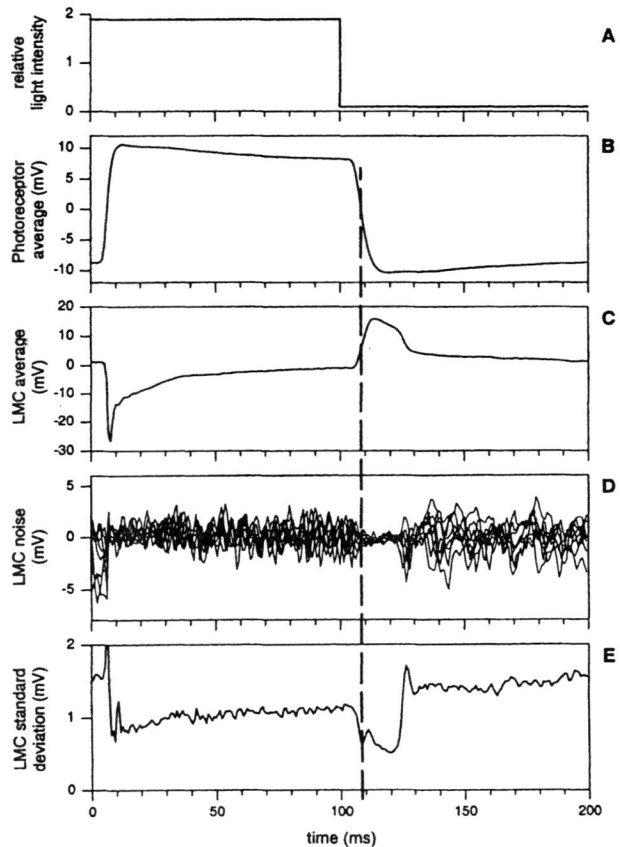

FIGURE 8.11. Averaged responses of a photoreceptor and an LMC, and LMC fluctuations in response to a large amplitude modulation. A: The stimulus contrast sequence is a square wave of amplitude 0.95, and duration 200 ms, sampled in 256 bins at 1280 Hz. This waveform is presented 380 times, while the response of a photoreceptor or an LMC is recorded. B: Ensemble averaged photoreceptor response. C: Ensemble averaged LMC response. D: Example of 8 traces showing fluctuations of the LMC response around its average waveform. E: Time dependent standard deviation of the LMC fluctuation traces. The standard deviation plotted here at each instant of time is the standard deviation across the ensemble of LMC voltage fluctuations, all taken at the same phase of the square wave stimulus.

The numbers we derive here are certainly not precise, and a comparison between very different species is always tenuous. Although no hard conclusions can be drawn, the comparison points to an interesting possibility. The discrepancy between the high Poisson release rate required to explain the reliability on the one hand, and the lower measured release rates in goldfish on the other, is large. The most interesting explanation for this, in our view, is that the Poisson assumption is not valid, and that the synapse would be capable of releasing vesicles with much higher precision than expected from that.

Of course Fig. 11 points out another interesting aspect of transmission by the synapse, namely that it is highly adaptive. The synapse seems to reset itself to follow the large swings in the DC component of the photoreceptor voltage, presumably to be able to encode fluctuations around this average more efficiently. It has been known for a long time that photoreceptor-LMC transfer is adaptive (Laughlin and Hardie 1978, Laughlin et al. 1987), and this has been interpreted as a resetting of parameters to optimize information transmission (van Hateren 1992, Laughlin 1989). Here we take a look at how fast this type of adaptation takes place.

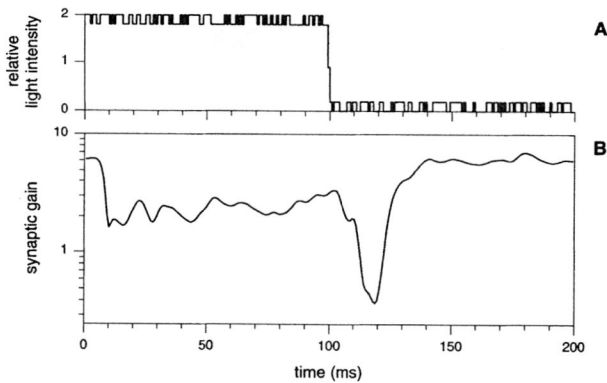

FIGURE 8.12. Characterization of the time dependent synaptic gain when the fly is stimulated with a 200 ms square wave pattern. A: Example of a combined waveform. To measure the gain of the photoreceptor and LMC along the 200 ms stimulus of Fig. 11, a small binary probe signal was added to the square wave. An example of the combined contrast waveform is shown here. B: Gain of synaptic transmission between photoreceptor and LMC, as the system cycles through the 200 ms period. See text for further explanation.

The experiment consists of the same square wave contrast modulation as shown in Fig. 11, but now a small amplitude random binary probe is added to the large waveform (see Fig. 12). At each presentation of the 200 ms square wave the probe is a series of 256 binary random values, played out at a sampling rate of 1280 Hz, and its random sequence is different at each presentation. The responses of the cell are recorded, also at 1280 Hz sampling rate. After the experiments we correlate the probe stimulus with the voltage variations $\Delta v_m(t) = v_m(t) - V(t)$ it induces in the response, at consecutive points along the 200 ms square wave. That is, we compute:

$$\Phi_{p\Delta v}(t, t') = \lim_{M \to \infty} \frac{1}{M} \sum_{m=1}^{M} p_m(t) \times \Delta v_m(t') \qquad (8.32)$$

where the m stands for the different presentations, and $\times$ represents an outer product. Thus $\Phi_{p\Delta v}(t, t')$ is the crosscorrelation between probe stimulus $p(t)$ taken at time t, and voltage response to the probe, $\Delta v(t')$, taken at t', which we find by computing the ensemble averaged outer product of the probe with the response. In practice we treat $p_m(t)$ and $\Delta v_m(t)$ as vectors of 256 elements each, spanning the 200 ms repeat period, and $\Phi_{p\Delta v}(t, t')$ is then a 256×256 crosscorrelation matrix. We make the same measurement in a photoreceptor and an LMC, and so get two crosscorrelation matrices. As before, we regard the photoreceptor and the LMC as a cascaded system, and, as long as things are linear, one should be able to describe the cascade as the following matrix multiplication:

$$\Phi_{p\Delta v_{\mathrm{LMC}}} = \Phi_{p\Delta v_{\mathrm{PR}}} \cdot \Phi_{p\Delta v_{\mathrm{Syn}}}. \qquad (8.33)$$

Note that we do not suggest that the system is linear in the response to the large amplitude square wave. It definitely is not. But what we try to characterize here are the small fluctuations due to the probe around the large amplitude average waveform induced by the square wave. That can be reasonably assumed linear, but nonstationary as a result of the square wave. This nonstationarity then naturally leads to two time indices, and they are both represented in the matrix formulation. Ideally, from Eqn. 8.33 we should be able to derive the synaptic cross correlation matrix by inverting the photoreceptor matrix, and multiplying it with the LMC matrix. Unfortunately, this procedure is rather unstable, both because of experimental noise, and because of the strong high frequency components in the LMC signal. Therefore we will settle for something more modest here, and calculate the probe induced variance. The stimulus induced time dependent variance is computed as the diagonal of the probe–response crosscorrelation matrix multiplied by its transpose:

$$\sigma^2_{p\Delta v}(t) = \left[\Phi_{p\Delta v}(t, t') \cdot \Phi_{p\Delta v}(t, t')^T \right] \cdot \delta(t - t'), \qquad (8.34)$$

and to compute the synaptic contribution we now take the ratio of the LMC and the photoreceptor diagonals, or if we want to express the linear gain we take the square root of this quantity:

$$\sigma_{p\Delta v_{\mathrm{Syn}}}(t) = \frac{\sigma_{p\Delta v_{\mathrm{LMC}}}(t)}{\sigma_{p\Delta v_{\mathrm{PR}}}(t)}, \qquad (8.35)$$

as shown in Fig. 12B. The figure again suggests a shutdown of synaptic transmission during the falling phase of the photoreceptor voltage. Further, the overall gain as defined here switches from about 2.5 during the bright phase to about 7 during the dim phase. The data indicate that this switch in gain is also accompanied by a change in shape of the synaptic impulse response, which seems to become sharper and more biphasic during the bright phase (see Juusola et al. 1995). To some extent these effects are also seen in the photoreceptor response, whose gain decreases and speeds up. In the photoreceptor that is presumably due in large part to the change in membrane conductance accompanying the change in membrane potential. In the case of the synapse it seems likely that the dynamics of vesicle release changes, and this interpretation is supported by the apparent shutdown in transmission.

8.4   Coding in a Blowfly Motion Sensitive Neuron

Thus far we have considered the reliability and precision of phototransduction and synaptic transmission, two of the first steps in vision. Now we want to look "deeper" into the brain, to a point where some nontrivial computations have been done. There are two very different questions. First, we are interested in the precision of the computation itself: Is the brain making use of all the information available in the array of photoreceptor signals? Second, we want to understand the way in which the brain represents the results of its computations: What is the structure of the "neural code"? For an accessible example of these issues we turn to the visual motion sensitive neurons in the fly's lobula plate.

The fly's lobula plate contains a number of motion sensitive cells, shown in Fig. 3, that are direction selective and have wide visual fields. They are thought to achieve wide field sensitivity by adding the contributions of a large number of small field motion sensitive cells from the fly's medulla (Single and Borst 1998, reviewed by Laughlin 1999). One important function of these lobula plate tangential cells is to provide input information for course control (Hausen and Wehrhahn 1983, Hausen and Egelhaaf 1989, Krapp et al. 1998). A distinct advantage of these cells in this preparation is that they allow long and stable recording. When care is taken to do minimal damage to the fly, and to feed it regularly, the same cell can be recorded from for several days on end. This is important in many of our studies of neural coding because there the general aim of the experiment is to characterize probability distributions of stimulus response relations, rather than only averages.

The data we present here are all obtained by extracellular recording with conventional techniques (see de Ruyter van Steveninck and Bialek 1995 for more experimental details). The nature of the signal is drastically different from what we saw before. We are now dealing with spikes, as depicted in

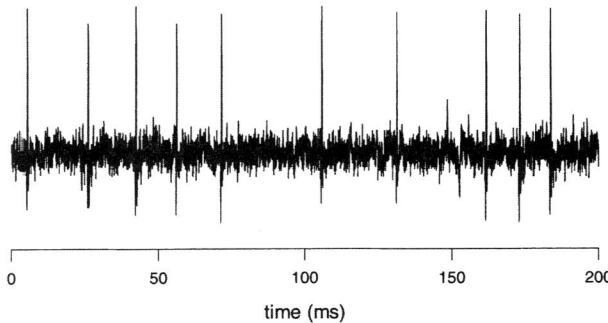

FIGURE 8.13. Example of a 200 ms segment of spike activity recorded extracellularly from H1. The spikes shown here are timed at $10\mu s$ precision and stored for off line analysis.

Fig. 13, instead of analog voltages, and one of the important issues is how we must interpret their temporal sequences. Spikes are transmitted by the nervous system at rates of usually not more than a few hundred per second, and certainly not at rates typical for photons entering photoreceptors. In contrast to photons, spikes are placed on the time axis, one by one, by physiological processes. It is therefore not unnatural to think that their position likewise could be read out and interpreted by other physiological processes. Indeed, one long standing issue in understanding neural coding is whether this is the case: Does the timing of individual spikes matter, or can we afford to coarse grain time and average spike counts over large windows? Here we will address that question for the case of H1, a motion sensitive neuron in the fly's brain. It is sensitive to horizontal inward motion (see also the legend to Fig. 3), and its visual field covers just about a full hemisphere.

In most of the experiments the fly watches an extended pattern generated on a display oscilloscope (Tektronix 608), written at a 500 Hz frame rate. One drawback of this setup is that the stimulated area of the visual field is only a fraction of the total field of the cell, and that the light levels are rather low, corresponding roughly to those at dusk. At the end of this chapter we will present data on a fly that was rotated around the vertical axis by a computer controlled stepper motor. Doing this experiment outside in a wooded environment the fly is stimulated with natural scenes of high light intensity, and by playing a rotational motion sequence derived from measured flight data, shown in Fig. 1, one expects to get closer to the ideal of presenting naturalistic stimuli.

8.4.1   Retinal limitations to motion estimation

As we have seen earlier, flies are quick and acrobatic, so it seems entirely reasonable to assume that the components of the flight control system are optimized to perform as accurately and quickly as possible. This should then obviously be true for the lobula plate tangential cells as well. Of course some general principles apply, and, like all sensory neurons they must work within the limits set by the reliability of their input signals.

It is instructive to make a rough estimate of what precision we can expect from a wide field cell that takes as input photoreceptor signals with realistic amounts of noise. Let us try to compute an estimate of the limits to timing precision of the response with respect to stimulus, as set by the photoreceptor signal quality. This is relatively easy to do, and it is relevant in a discussion of coding by spike timing. In many experiments we stimulate the fly with a large, high contrast bar pattern that moves randomly, jittering back and forth in the horizontal direction. The power density spectrum of the signal we, and presumably H1, are interested in is the velocity power density, given by $S_{vel}(f)$. This has dimensions $(°/s)^2/Hz$, because we are dealing with angular velocity, and this is customarily given in $°/s$.

The photoreceptors in the fly's retina have a profile of angular sensitivity that is determined by the optics of the lens and the waveguide behavior of the receptor cell itself; this profile is often approximated by a Gaussian. For the blowfly frontal visual field, its halfwidth is $\approx 1.4°$ (Smakman et al. 1984), corresponding to a "standard deviation" $\sigma_{PSF} \approx 0.5°$. Now suppose we have a contrast edge with intensity stepping from $I = I_0 \cdot (1 + c_0)$ to $I = I_0 \cdot (1 - c_0)$ aligned on the optical axis of this Gaussian point spread function. Then, if the edge moves by a small amount δx, the contrast step in the photoreceptor is:

$$\delta c = \frac{\delta I}{I_0} \approx \frac{1}{I_0} \cdot \frac{2 \cdot I_0 \cdot c_0 \cdot \delta x}{\sqrt{2\pi} \cdot \sigma} = \frac{c_0 \cdot \delta x}{\sqrt{\pi/2} \cdot \sigma}, \tag{8.36}$$

which converts a position change δx into a contrast change δc. This allows us to derive a contrast power spectrum, if we can convert the velocity power spectrum into the appropriate position power spectrum. Position is the integral of velocity, which means that we must divide the Fourier transform of velocity by frequency to get the Fourier transform of the position signal (see Bracewell 1978). Here the relevant quantities are power spectra, so we must use f^2 to make the correct conversion: $S_{pos}(f) = S_{vel}(f)/f^2$. The contrast power density spectrum is now:

$$S_c(f) = \left[\frac{\delta c}{\delta x}\right]^2 \cdot S_{pos}(f) = \frac{2 \cdot c_0^2 \cdot S_{vel}(f)}{\pi \cdot \sigma^2 \cdot f^2}. \tag{8.37}$$

In the experiment we stimulate a large number, M, of photoreceptors and when a wide field pattern moves rigidly then all photoreceptors are stimulated in a coherent way. That means that the total power of the signal available in the photoreceptor array scales as M^2. Finally, in the experiment we control the velocity stimulus, and thus $S_{vel}(f)$. All this combined leads to a contrast signal power spectrum:

$$S_{c_M}(f) = \frac{2M^2 \cdot c_0^2 \cdot S_{vel}(f)}{\pi \cdot \sigma^2 \cdot f^2}, \tag{8.38}$$

for the set of M photoreceptors stimulated by coherent, that is, rigid, motion.

To derive a limit of timing precision, or equivalently a limiting frequency, we must compare this available signal spectrum to the relevant noise power spectrum. In section 8.2.6 we defined the equivalent contrast noise power of a single photoreceptor cell. In a pool of M photoreceptor cells, their independent noise powers add, and we have:

$$N_{c_M}(f) = M \cdot N_c(f). \tag{8.39}$$

To define a limit to time resolution we determine the frequency at which $S_{c_M}(f)/N_{c_M}(f)$, the signal to noise ratio, crosses one:

$$\frac{S_{c_M}(f)}{N_{c_M}(f)} = \frac{2M^2 \cdot c_0^2 \cdot S_{vel}(f)}{\pi \cdot \sigma^2 \cdot f^2} \cdot \frac{1}{M \cdot N_c(f)} = \frac{M \cdot S_c(f)}{N_c(f)}. \tag{8.40}$$

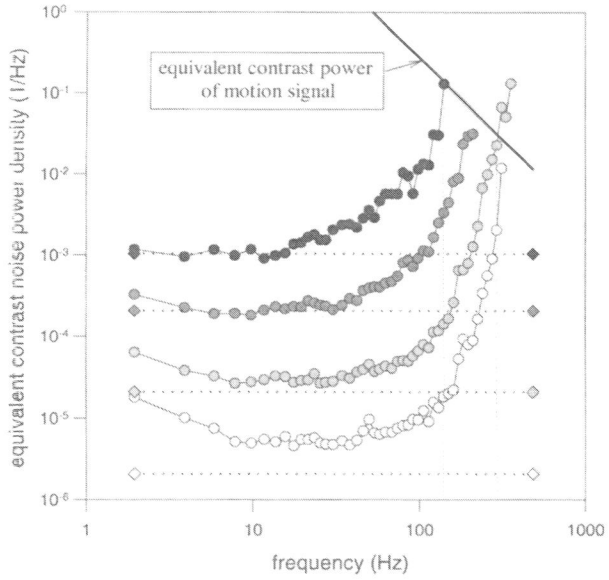

FIGURE 8.14. Circles: Equivalent contrast noise power spectral density of blowfly photoreceptors at different light intensities. The input light intensities, expressed in extrapolated bumps per second, are indicated by the diamonds connected by dotted lines. Solid thick line: Equivalent contrast noise power density of a motion signal calculated for a wide field pattern, under conditions typical for our experiments. See text for further explanation.

In Fig. 14 we therefore plot $N_c(f)$ at four light levels for a single photoreceptor, along with $M \cdot S_c(f)$, for conditions typical of our experiment:

$c_0 = 0.3$, $M \approx 3800$, $S_{\text{vel}}(f) \approx 10(°/s)^2/\text{Hz}$, and a small ($\approx 0.3$) correction for the fact that the edge can not be expected to be exactly at the center (the correction factor averages over edge position). The crossover frequency, f_{cross} of signal to noise ratio can be read from the figure directly, and lies between approximately 150 to 400 Hz, depending on light intensity. That means that when we stimulate H1 with the typical stimuli described here, we may expect to observe timing precision $\delta t \approx 1/(2\pi \cdot f_{\text{cross}})$ in the order of one to several milliseconds if the fly's brain effectively uses all the motion information present in the photoreceptor array. We should note that the approximation we make here is for small δx in Eqn. 8.36.

In Fig. 15 we show raster plots of a 500 ms segment of responses of H1 to a repeated dynamical motion trace. The same experiment was done at different light intensities, ranging over 4 orders of magnitude, from an estimated 5 to $5 \cdot 10^4$ transduced photons per photoreceptor per second. It is clear by visual inspection that spikes tend to line up better when the light intensity is higher, up to the highest light intensity used in the experiment, suggesting that external noise may be the limiting factor (see also Fermi and Reichardt 1963). That impression is confirmed by the histograms of spike arrival times to the right of the rasters, which have standard deviations ranging from 1.4 to 7.1 ms. These histograms describe the probability distributions of the first spike that follows $t = 410$ ms. The highest light intensity in this experiment corresponds roughly to the second-highest (light gray dots) intensity in the photoreceptor data of Fig. 14. At five per second, the lowest bump rate shown here, bumps in a single photoreceptor typically are nonoverlapping, and we see that there is still a modulation of the response of H1. Dubs et al. (1981) showed that flies respond behaviorally to moving patterns at light levels where single photon absorptions are nonoverlapping, and from the classical work of Hecht et al. (1942) we know that humans can perceive light flashes under these conditions.

One can think of other parameters likely to affect the quality of the input signal for a wide field motion sensitive cell, for example contrast, field of view, and stimulus velocity amplitude. In experiments where these parameters are varied we see effects on spike jitter qualitatively similar to what is shown here. It thus seems that the precision of spike timing in H1 is close to being determined by the information available to it in the photoreceptor array, i.e., bump latency jitter sets the threshold under the photon capture rate conditions of the experiments in Fig. 15. From Fig. 14 we can also read that the photoreceptor equivalent contrast noise has a very steep frequency dependence at high frequencies. This means that once the conditions of the experiment are such that the output accuracy is in the millisecond regime, only relatively large changes in stimulus parameters will lead to appreciable improvements in timing precision.

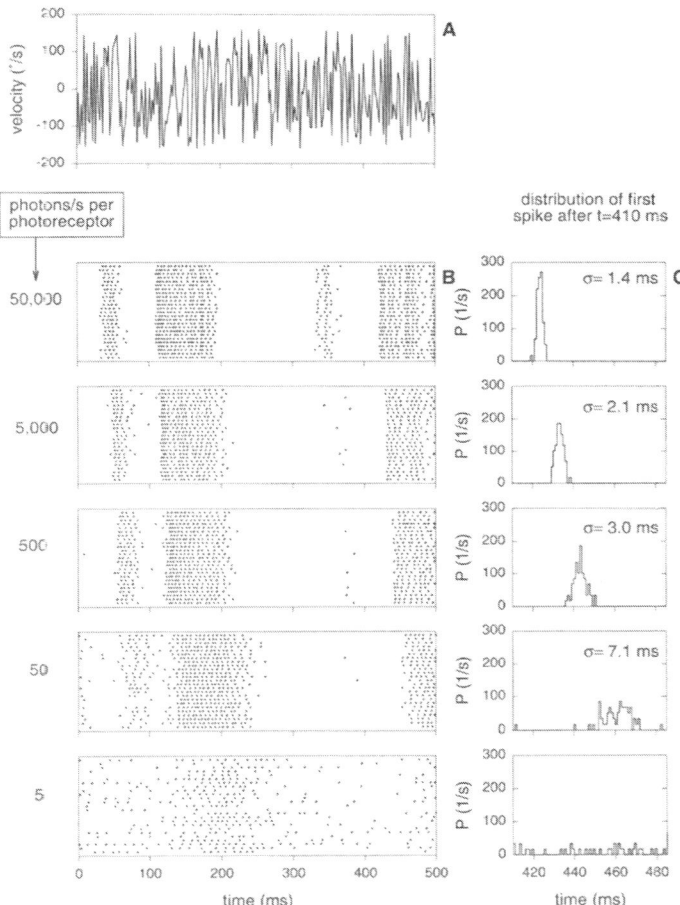

FIGURE 8.15. Responses of a blowfly H1 neuron to movement of a wide field pattern at different mean light levels. In this experiment the fly looked at a moving pattern through a round diaphragm with a diameter corresponding to 30 horizontal interommatidial angles. The velocity was random with a flat power spectral density. A: A 500 ms sample segment of the stimulus velocity. B: Raster plots of H1 spikes obtained at five different light intensities given by the estimated average photon flux for each single photoreceptor. C: Histograms of the timing of the first spike fired after $t = 410$ ms. With decreasing photon flux the response latency increases. Moreover, the peaks become wider (with σ the standard deviation of a fitted Gaussian), suggesting that timing precision may be limited by photoreceptor noise. At the lowest light level shown here there is still a visible modulation of H1's rate, but the timing of a single event is too spread out to produce a clear peak.

We can ask a related question, namely, how reliably can the arrival time of a spike tell us something about the strength of the stimulus that preceded it, and how close does that get to the photoreceptor limits. If we know what message the neuron encodes, we can use a computational model that retrieves that message from the sensory periphery. The model relevant for H1 is the Reichardt correlator model (Reichardt 1961, Reichardt and Poggio 1976), which describes a specific functional computation for extracting motion information from an array of photoreceptors. Its basic interaction is a multiplication of filtered signals originating from neighboring directions of view. The Reichardt model was formulated heuristically, but it has been shown to be the optimal solution to a general problem of motion estimation in the presence of noise, in the limit where the signal to noise ratio is low (Potters and Bialek 1994).

The experiment to measure reliability is very simple (de Ruyter van Steveninck and Bialek 1995). We present the fly with a pattern that makes sudden motion steps of several sizes, and record the responses of H1. From a large number of presentations we obtain a histogram of arrival times of the first and the second spike following the stimulus. If we compare two such histograms in response to two different step sizes we can compute the discriminability of those two stimuli as a function of time. This is done by framing the question as a decision problem (Green and Swets 1966). A stimulus is presented once, and the cell generates a response which is observed by some hypothetical observer. The question then is: Given the cell's response, what can the observer say about the identity of the stimulus, and how often is that assessment right? This is meaningful only if the distributions of responses to different stimuli are known to the observer, and are also different. In the analysis presented here, the hypothetical observer can judge the spike train in real time, starting at the moment of stimulation (or rather 15 ms after that, to minimize effects of spontaneous rate). The assumption that the time of stimulus presentation is known does of course not correspond to any natural situation, as there the timing of stimuli must be inferred from the sensory input as well. It is, however, still a valid characterization of the precision with which the fly's visual brain performs a computation.

Fig. 16a shows two conditional distributions, $p(\tau_0 | S = 0.24°)$ and $p(\tau_0 | S = 0.36°)$ for the first spike arrival after stimulus presentation, for two step stimuli of different size. For the large step, the first spike tends to come earlier than for the small step. This means that the observer should choose the large step when he or she sees a short interval, and the small step for a relatively large interval. The crossover for this case is at about 32 ms. How accurate will this judgement be on average? As can be read from the cumulative probability distributions in Fig. 16B, the probability for having a spike before 32 ms with the large step is 0.78. Thus if the large step was presented then the observer will make the right choice with a probability of 0.78. The cumulative distribution for the small step

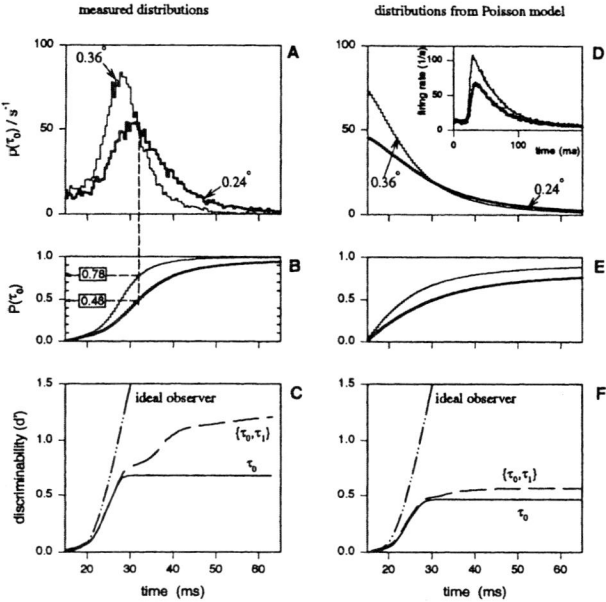

FIGURE 8.16. Statistics of H1's responses to small motion steps of a wide field pattern, and equivalent Poisson statistics. A: Histograms, normalized as probability densities of the first spike fired starting 15 ms after the stimulus step, for a 0.24° (thick line) and a 0.36° (thin line) step. B: Cumulative distributions for the same data as in panel A. C: Time dependent discriminability parameter, $d'(t)$, computed for the step size pair depicted in A, both for single spikes (τ_0), and for spike pairs ($\{\tau_0, \tau_1\}$). Also shown is the discriminability computed for an ideal observer using realistic photoreceptor signals as inputs. D: First spike histograms, as in A, but now for a modulated Poisson process. These data were computed using the measured PSTHs (inset in D) for the two step sizes as the rate of the underlying modulated Poisson process. E and F: As B and C, but now for the modulated Poisson case.

at 32 ms is 0.48, which means that if the small step was presented, the observer will choose correctly in a proportion of $1 - 0.48 = 0.52$. If each step has a prior chance of 0.5, then in an experiment where the two steps would be mixed at random, the proportion of correct decisions would be $P_C = 0.5 \cdot 0.78 + 0.5 \cdot 0.52 = 0.65$. It is convenient to translate P_C into a distance measure, and Green and Swets (1966) propose to use the distance between Gaussian distributions of unit standard deviation that would give rise to the same value of P_C. This measure is known as d' and is very widely used in the psychophysical literature. There is a simple 1:1 correspondence between d' and P_C, and for our case we find $d' = 0.68$. From the given spike timing distributions we can also construct a continuous function $d'(t_{\mathrm{obs}})$ that describes the equivalent distance as it evolves over the observation time interval since stimulus presentation. To see that, we should simply divide the distribution in a part that is described by the measured distribution up to t_{obs}, assign the probability that is as yet "unused" to one total remaining probability, and then treat this constructed distribution in the same way as the previously defined spike timing distributions. The time dependent $d'(t)$ based on the first spike only is shown as the solid line in Fig. 16C. This line plateaus at $t = 32$ ms, as from that moment on the choice will always be fixed. Instead of considering only the first spike we can also look at the combination of the first and the second spike arrival time, that is $\{\tau_0, \tau_1\}$. The distributions for these are not shown, but the reasoning is entirely similar. The end result, d' as a function of time after presentation, is shown in the same panel as the dashed line. It is clear in this case that the second spike carries substantial extra information about stimulus identity.

The comparison of the measured data to the ideal motion detector model can now be made. We measure representative photoreceptor power spectral densities, and the number of photoreceptors stimulated in the H1 experiment, and then apply the Reichardt model to compute its average step response as well as its output power spectral density (de Ruyter van Steveninck and Bialek 1995). From these we compute a time dependent $d'_{model}(t)$, which is plotted as the dash-dot line in Fig. 16C. The crucial comparison is between the slope of the measured and the computed $d'(t)$. In the range of 23-28 ms, $d'_{model}(t)$ rises about twice as fast as the measured $d'(t)$. This shows that H1 approaches, within a factor of two, the performance of an optimal motion detector limited only by noise in the photoreceptor array. Given that the signal passes through at least four synapses to be computed, this precision is quite remarkable. In this case we measure how accurately H1 represents the amplitude of motion steps. The estimation takes place over a somewhat extended time interval, and is therefore not limited by the bump jitter that sets the timing resolution of the photoreceptor array, but rather by low frequency (roughly below 100 Hz) accuracy of the photoreceptor. As this latter is close to the photon shot noise limit (Figs. 7,14) we are reminded that the precision of neurons in a functioning brain is not just given by the physiology, but is determined in part, or in this case maybe even dominated, by the statistical properties of the stimulus (see also Bialek et al. 1991, de Ruyter van Steveninck and Bialek 1995, Rieke et al. 1997).

Motion discrimination using the spike train output of H1 thus provides us with an example in which the performance of the nervous system approaches basic physical limits set by the structure of the inputs in the retina. There are other examples of this near optimal performance, in systems ranging from human vision to bat echolocation to spider thermoreception (for discussion see Rieke et al. 1997). This level of performance requires the nervous system to meet two very different requirements. First, the system must be sufficiently reliable or "quiet" that it does not add significant excess noise. This is especially challenging as the signals propagate through more and more layers of neurons, since the synapses between cells are sometimes observed to be the noisiest components. Second, optimal performance requires that the system make use of very particular algorithms that provide maximal separation of the interesting feature (motion, in the case of H1) from the background of noise. In general these computations must be nonlinear and adaptive, and the theory of optimal signal processing (Potters and Bialek 1994) makes predictions about the nature of these nonlinearities and adaptation that can be tested in independent experiments.

In our broader discussion on the relevance of Poisson firing it is now interesting to quantify to what extent deviations from Poisson behavior help in encoding the stimulus in a spike train. To get an idea of this we do exactly the same analysis as described above, but then on synthetic spike trains that are generated by a modulated Poisson process with the same time dependent rate as the measured spike train. The inset in Fig. 16D shows the post stimulus time histogram for the two step sizes used before in the analysis. From these two, the spike arrival distributions in panel d are computed, and from these again we construct $d'(t)$. Figure 16F shows the end result: The discriminability based on timing of the first spike alone goes from $d' = 0.68$ to $d' = 0.46$, while in the Poisson case the second spike adds only 0.1 to the d' based on the first, so that at 50 ms after the step, the Poisson value for d' is less than half that measured from real spikes. Neural refractoriness was not incorporated in the synthetic train, and it seems likely that that the increased reliability of the real neuron can at least be partly attributed to that.

8.4.2   Taking the fly outside: Counting and timing precision in response to natural stimuli

There is a long tradition of using discrimination tasks, as in the step discrimination experiment of the previous section, to probe the reliability of percpetion in humans and also the reliability of neurons. But such simple tasks are far from the natural ones for which evolution selected these neurons. As a fly flies through the world, angular velocity varies continuously, and this variation has a complicated dynamics. In the past decade, a number of experiments has been done which attempt to approach these more natural conditions. Specifically, experiments with pseudorandom velocity waveforms presented on display oscilloscopes have been used to measure information transmission in H1 by reconstructing the stimulus from the spike train (de Ruyter van Steveninck and Bialek 1988, Bialek et al. 1991, Haag and Borst 1997), and by more direct methods (de Ruyter van Steveninck et al. 1997, Strong et al. 1998). The general conclusion of these studies is that the timing of spikes in the millisecond range does indeed carry significant information, in line with the photoreceptor limits discussed earlier. One would like to know to what extent such conclusions are also relevant

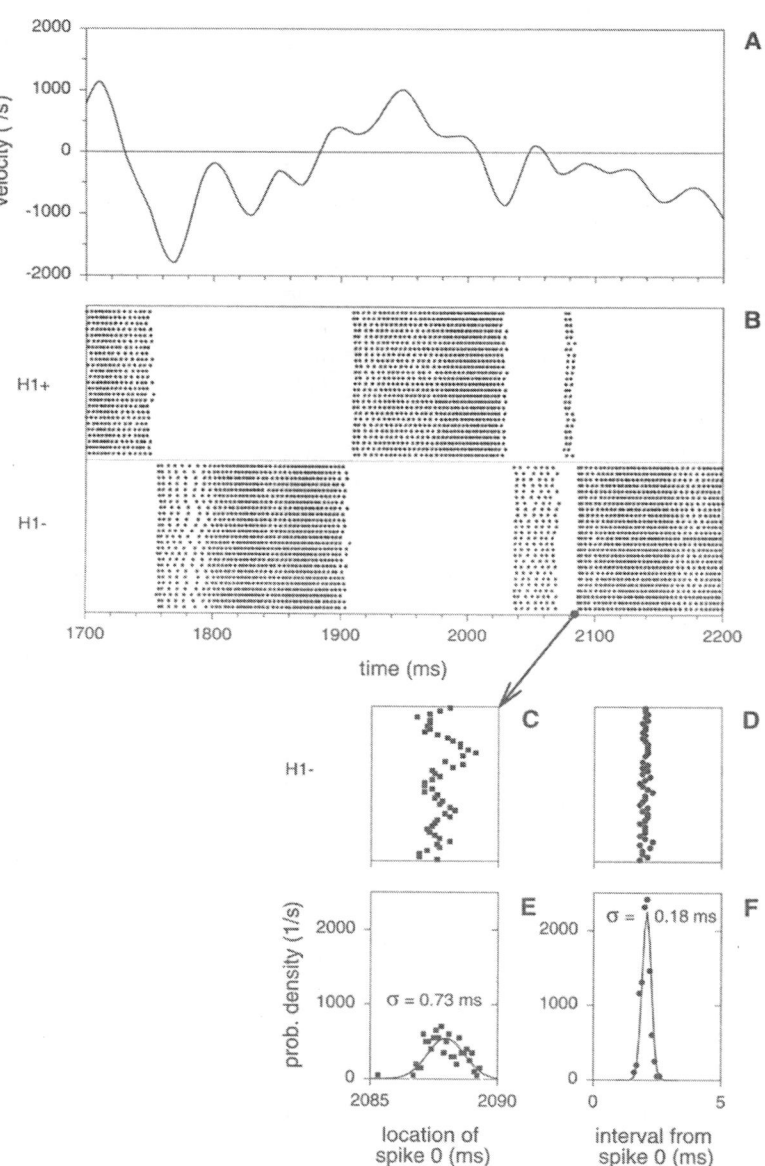

FIGURE 8.17.

FIGURE 8.17. Direct observations of H1 spike timing statistics in response to rotational motion derived from Land and Collett (1974) free flight data (see Fig. 1). The fly was immobilized in a specially designed miniature recording setup, which was fixed to a computer controlled stepper motor. This setup was used in a wooded outdoor environment, in which the fly was rotated repeatedly (200 times in total) along the same motion trajectory. For technical reasons the rotational velocity used in this experiment was scaled down to half of the free flight value. A: A 500 ms segment of the motion trace used in the experiment. B: Top: raster of 25 trials showing occurrences measured from H1. Bottom: 25 trials with spike occurrences from the same cell, but in response to a velocity trace that was the negative of the one shown in A. For ease of reference we call these conditions H1+ and H1- respectively. C: 25 samples of the occurrence time of the first spike fired by H1- following $t=2085$ ms (indicated by the arrow connecting the axis of panel B to panel C). D: Time interval from the spike shown in C to the spike immediately following it. E: Probability density for the timing of the spike shown in C. The spread is characterized by $\sigma = 0.73$ ms, which is defined here as half the width of the peak containing the central 68.3% of the total probability. If the distribution were Gaussian, then this would be equivalent to the standard deviation. Here we prefer this definition instead of one based on computing second moments. The motivation is that there can be an occasional extra spike, or a skipped spike, giving a large outlier which has a disproportionate effect on the width if it is calculated from the second moment. Filled squares represent the experimental histogram, based on 200 observations; the solid line is a Gaussian fit. F: Probability densities for the same interspike interval as shown in D. The definition of σ is the same as the one in E.

to still more natural conditions encountered in fly flight (see for example Warzecha et al. 1998).

Ideally one would perform experiments in the natural habitat of the animal, while it is behaving as freely as possible. Good examples are the study of responses of auditory neurons in *Noctuid* moths to the cries of bats flying overhead (Roeder 1998), and the study of optic nerve responses of *Limulus* lateral eye while the animal is moving under water (Passaglia et al. 1997). Of course, in each specific case concessions are made to be able to record neural signals, and one must decide what is the best compromise between realistic conditions and getting interpretable data.

Here we present data from a setup that allows us to record from a fly while it is rotating on a stepper motor. The rotational motion is mechanically precise, and arbitrary rotation sequences can be programmed. In the case described here the fly was rotated with a time sequence corresponding to the rotations executed by the flies shown in Fig. 1 (based on Land and Collet 1974), except that for stability reasons the amplitude of the entire trace was set to half the value measured from the real flies. The flight trajectories were repeated with their sign inverted, so that for each trajectory we stimulate H1 in complementary ways. It is thus as if we record from the two H1 cells at opposite sides of the head. The setup is portable and the experiment was done outside in the shade of some bushes on a sunny afternoon. Therefore, the light intensities, the stimulated area of the visual field, and the spatial characteristics of the scene are realistic samples of what the animal encounters in nature. The motion trace is somewhat natural, although the rotational velocities are smaller than those measured in free flight, and the measurement was done on a different species of fly.

Important for our analysis is that we can repeat the same motion trace a large number of times, which is of course not really a part of natural behavior. However, it allows us to make quantitative statements about information transmission in the measured spike trains that rely only on the degree of reproducibility of the response to repeated stimuli. Because of this, those statements are independent of any assumptions on how the stimulus is encoded in the spike train, so in that sense they are rather universal.

Fig. 17 presents data from such an experiment. The rotational velocity waveform is shown in panel A. Note that the velocity amplitudes are very large compared to those of the white noise stimuli shown in Fig. 15. The traces in panel B are labeled H1+ and H1−. In reality they were obtained from the same H1 cell, but with a switch of sign in the velocity trace, as described earlier. It is clear that H1+ and H1− alternate their activity quite precisely, and that repeateable patterns of firing occur, such as the pair of spikes in H1+ at about 1900 ms. The edges in spike activity are sharp. Panel C shows the position of the first spike occurring after 2085 ms in H1−. A histogram of arrival times of this spike is shown in Fig. 17E, together with a Gaussian fit with standard deviation 0.73 ms.

In addition to this precision of spike timing with reference to the stimulus, there can also be an internal reference, so that the relative timing of two or more spikes, either from one neuron or among different neurons, may carry information (MacKay and McCulloch, 1952). Panel D and F give an example, where H1 generates a 2 ms interspike interval upon a particularly strong stimulus with a standard deviation of 0.18 ms.

Consequently, interspike intervals may act as special symbols in the code, carrying much more information about the stimulus than what is conveyed by two single spikes in isolation. This was shown indeed to hold for H1 (de Ruyter van Steveninck and Bialek 1988, Brenner et al. 2000). Findings like these should make us cautious. For a complete description of the spike train, timing precision at different levels of resolution may be required, depending on what aspect of the spike train we are talking about. In particular, relative spike timing may have to be much better resolved than absolute timing to recover neural information (see specifically Brenner et al. 2000).

The data presented above show episodes in the stimulus that induce accurately timed events in the spike train, on the millisecond scale. One may worry that such events are very special, and that most spikes are not well defined on the time axis. In other words, we need a "bulk" measure of spike timing precision. The most general way of specifying that is to study the information transmitted by the spike train as a function of time resolution Δt. We do that here by estimating two measures of entropy, the total entropy and the noise entropy, directly from the spike train (cf. de Ruyter van Steveninck et al. 1997, Strong et al. 1998). Loosely speaking, the total entropy measures the size of the neuron's "vocabulary." We calculate it from the distribution, $P(W)$, of neural firing patterns, or words W: $S_{\text{total}} = -\Sigma_W P(W) \log_2[P(W)]$. Here W is a vector of n_W entries, and each entry gives a spike count in a bin of size Δt. All n_W bins taken together form a string of length T. Typical values for Δt are one to a few ms, while T is of order 5-30 ms. $P(W)$ is approximated by the histogram of all firing patterns that the neuron generates in the experiment. The noise entropy characterizes how much the neuron deviates from repeating the same firing pattern at the same phase of a repeated stimulus. If the stimulus is periodic in time with period T_{stim}, then from an experiment with a large number of repetitions we can form histograms of firing patterns $W(t)$ at each time t, $0 \le t \le T_{\text{stim}}$. If the neuron were an ideal noiseless encoder then the response would be the same every time, and for each presentation we would find the same firing pattern $W(t)$ at time t. Then $P(W|t)$ would equal one for $W = W(t)$, and zero otherwise, so that the noise entropy would be zero. In practice, of course, the noise entropy differs from zero, and it will also vary with time t. The time average of the noise entropy is $S_{\text{noise}} = \overline{S_{\text{noise}}(t)} = -\Sigma_W P(W|t) \log_2[P(W|t)]$. The information transmitted by the neuron is the difference of the total and noise entropies, and therefore depends on both T and Δt: $I(T, \Delta t) = S_{\text{total}} - \overline{S_{\text{noise}}}$ (Strong et al. 1998). With enough data we can extrapolate to the limit $T \to \infty$

to get an estimate of the information rate. If we quantify that limiting information rate as a function of the time resolution Δt, we finally arrive at a reasonable bulk measure of the time resolution at which information can be read out from the neuron.

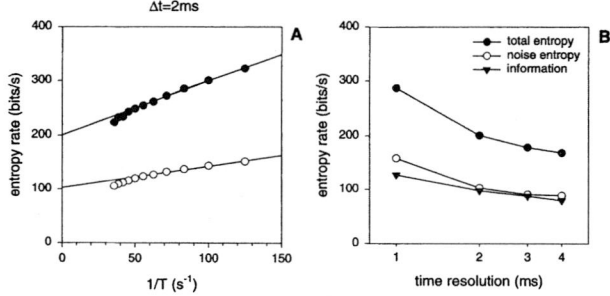

FIGURE 8.18. Information in firing patterns obtained from an experiment with naturalistic motion stimuli (see legends for Figs. 1, 17). A: Rate of total entropy and noise entropy as a function of $1/T$, for time resolution $\Delta t = 2$ ms. The figure shows the corresponding entropy rates, that is the values of total and noise entropy as defined above, divided by T. The fits (solid lines) to the two data sets are extrapolated to zero value of the abscissa, corresponding to $T \to \infty$. The difference of the extrapolated rate for the total entropy and the noise entropy respectively, is the estimate of the information rate at the given time resolution. See Strong et al. (1998) for a more detailed explanation. B: Rates of total entropy, noise entropy and information transmission were computed as explained in the text, and plotted here for different values of time resolution. It is clear that even in the millisecond range the information transmission increases when time resolution becomes finer.

The results are shown in Fig. 18B, which plots the total entropy rate, the noise entropy rate, and the information rate, all as a function of time resolution. The information rate still increases going from $\Delta t = 2$ ms to $\Delta t = 1$ ms, to reach 120 bits per second. The efficiency of encoding, that is the proportion of the total entropy used for transmitting information, is about 50% for $\Delta t \geq 2$ ms, and slightly lower than that for $\Delta t = 1$ ms. Because the information transmission rate must be finite, and the total entropy grows without bound as Δt becomes smaller, the efficiency must go to zero asymptotically as Δt goes to zero. Due to the limited size of the dataset it is not possible to make hard statements about the information transmitted, and thus the efficiency, at time resolutions better than 1 ms. It is clear, however, that spike timing information in the millisecond range, also under natural stimulus conditions, is present in the spike train, and that it could be highly relevant to the fly for getting around.

In the spirit of the discussion in section 8.4.1 we can also ask whether under these more natural conditions, spikes are generated according to a modulated Poisson process. As mentioned in section 8.2.1, the variance

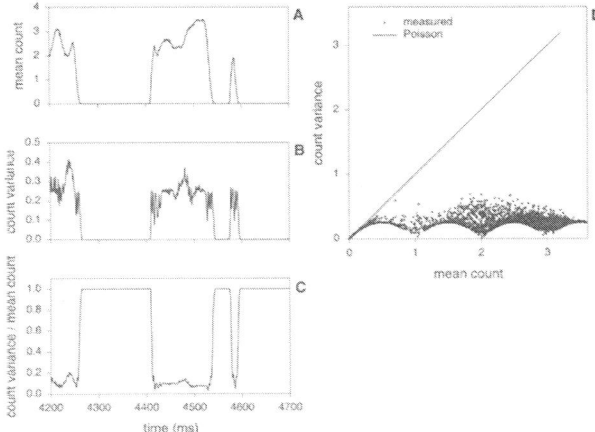

FIGURE 8.19. Example of counting statistics in response to a natural motion stimulus. The data are for the segment represented as H1+ in Fig. 17B. A: Ensemble average count in a 10 ms wide sliding windows. B: Ensemble variance of the count in the same sliding window as in A. C: Ratio of ensemble variance to ensemble average. Where both the variance and the mean are zero the value of the ratio is set to 1. For a Poisson process, all datapoints in this plot should have a value of 1. D: Dots: Scatter plot of simultaneous pairs of the ensemble variance and the ensemble average count. Straight line: Statistics of a Poisson process.

of a Poisson distribution is equal to its mean. Furthermore, if spikes are generated according to a modulated Poisson process, the spike count in a certain window should spread according to a Poisson distribution. Thus, if we compare segments of the responses to repeated identical stimuli, and we compare the variance of the count in a large number of such segments, we can see whether the spike statistics deviate from Poisson (for more details see de Ruyter van Steveninck et al. 1997). In Fig. 19 we show the response segment labeled H1+ in Fig. 17B, and we compute both the mean count and the variance across trials, for a 10 ms wide sliding window. When we plot the ratio of variance to mean we see that, as soon as there is spike activity, the ratio drops to values between 0.1 and 0.2. The comparison of the scatter plot to the Poisson behavior in Fig. 19D makes it clear that there is no strong overall trend for the variance to scale with the mean for these conditions (this, however, may be different for much longer time windows, see Teich and Khanna 1985). Similar results were reported for other systems (Berry et al. 1997, Meister and Berry 1999) and for fly H1 in laboratory conditions (de Ruyter van Steveninck et al. 1997).

These results suggest that when stimuli are dynamic enough, spiking sensory neurons may operate in a regime far removed from the Poissonlike behavior they are often assumed to have. This sub-Poisson behavior can be attributed, at least partly, to the relative refractoriness of spiking cells.

Refractoriness tends to regularize spike trains (Hagiwara 1954), and to increase the information carried by short intervals (de Ruyter van Steveninck and Bialek 1988, Brenner et al. 2000). Therefore, when operating at the same time dependent rate, the real H1 neuron carries much more information about the stimulus than a modulated Poisson train (de Ruyter van Steveninck et al. 1997).

8.5   Discussion and Conclusions

As the fly moves through its environment, its array of photoreceptors contains an implicit representation of motion, and this representation is corrupted by noise. From this the visual brain must extract a time dependent estimate of motion in real time. In some limiting cases, such as stepwise motion and the high frequency limit of white noise motion (cf. section 8.4.1), we can estimate the limits to the precision with which such a running estimate can be made. In those cases we find that the computation performed by the fly's brain up to H1 is efficient in the sense that that H1 retreives a substantial fraction of the motion information implicitly present in the photoreceptor array. To support those conclusions we have presented data on the statistical efficiency of signals at several levels within the blowfly's visual system.

Photoreceptors are stimulated by a Poisson stream of photons and, as long as the modulation is not too strong, they respond linearly to contrast. At low frequencies (up to 50-100 Hz, depending on illumination) the signal to noise ratio with which they encode contrast is close to the limits imposed by the Poisson nature of photon capture. Thus, blowfly photoreceptors are efficient at frequencies up to 50-100 Hz in the sense that they use almost all the information present in the photon stream. At higher frequencies the photoreceptor loses efficiency as latency jitter becomes the dominating noise source.

The LMCs, directly postsynaptic to the photoreceptors, are also efficient (Laughlin et al. 1987; this chapter, section 8.3.4). This means that neural superposition indeed works as was suggested long ago by Braitenberg (1967) and Kirschfeld (1967). As the LMCs receive their signals through an array of chemical synapses, the measured signal to noise ratio of the LMC sets a lower bound to the precision with which these synapses operate. If we hypothesize that vesicle release can be thought of as a modulated Poisson process, then from the measurements we estimate that each synaptic active zone should emit vesicles at a tonic rate of well over a hundred vesicles per second. Based on similar considerations Laughlin et al. (1987) arrive at an even higher number. Direct measurements of vesicle release in this system are unavailable, but recent measurements in goldfish bipolar cells (Lagnado et al. 1999) give a value six times lower than our estimates of tonic rates. This suggests that the hypothesis of Poisson release could be wrong, and that vesicle release is much more tightly regulated. Perhaps the large number of different proteins involved in vesicle docking and release (Kuno 1995), and the delicate anatomical ultrastructure of the synapse (Nicol and Meinertzhagen 1982, Meinertzhagen and Frölich 1983) have a role to play in this type of regulation.

We find that H1 can generate spikes with strongly subPoisson statistics, if it is driven by stimuli representative for at least some of the natural behavior of the fly. Chasing behavior in flies plays an important role in reproduction and territorial defense (Land and Collett 1974) and it presumably taxes the fly's sensory and motor systems to the fullest. It is therefore an interesting limit in which to study the performance of the fly's visual information processing capabilities. However, flies do not often engage in this behavior, and the chases typically last not much longer than a second. One may therefore ask how long the visual system can keep up with these strong dynamic stimuli. As a casual observation, it seems that H1 keeps on reporting about the fly's motion, and we see no signs of habituation when the fly spins around in simulated flight patterns for up to 20 minutes.

A fly, its name notwithstanding, spends most of its time sitting still (Dethier 1976, on page 13 gives a revealing list of how flies spend their time). If the fly's environment does not move, or if movement is steady and slow enough, H1 generates spikes approximately as a Poisson process. But at the same mean firing rate, when stimulated with strong dynamic signals, H1 can be firing far from Poisson, and this makes the encoding more efficient in the sense that the information per spike increases substantially (de Ruyter van Steveninck et al. 1997). Over the years the Poisson process has been the model of choice for describing neural firing statistics, at least to first approximation (see for example Tolhurst et al. 1983, Britten et al. 1993). As H1 can be either close to that limit or far from it, depending on conditions, it is a matter of debate which condition is more relevant. Reference to natural behavior is not conclusive here, because both sitting still and chasing other flies are natural behaviors. We feel that studying neural responses to dynamic stimuli is more interesting and rewarding, both because there is already a long history of characterizing responses to static stimuli, and because one can reasonably assume that well designed dynamic stimuli test the information processing capabilities of the nervous system to the fullest.

From the experiment it has also become clear that H1 can generate spikes that are locked to certain stimulus events with millisecond timing precision. Moreover, interspike intervals can be defined even better, and we saw an example in which an interval was generated with ≈ 0.18 ms accuracy. Overall, the spike train carries information at $\approx 50\%$ efficiency at least down to the millisecond time scale.

At the very least, the combination of these observations should make one cautious in interpreting or modeling neural signals as modulated Poisson processes. As we saw, under some conditions this may be a fair approximation but in others, specifically those that approach conditions of natural flight, it definitely is not. One may wonder whether this latter observation is more generally valid. Here the report by Gur et al. (1997) offers an important clue: The authors studied cells in monkey primary visual cortex, and noticed that these showed marked sub-Poisson statistics when care was taken to exclude episodes with eye movements from the analysis. Thus, from the point of view of specifying neural reproducibility, eye movements add variability. But that of course is by no means necessarily true for the monkey if it knows when it holds its eyes still. It is thus quite possible that sub-Poisson firing is a more general phenomenon, and relevant to natural stimulus conditions; this could be revealed in experiments designed to control carefully for variability in the neural response that is knowable by the animal.

We have encountered an example of this in the H1 data presented in section 8.4.2. Fig. 17C shows the arrival times of spikes with reference to the stimulus, and these jitter by 0.73 ms. Fig. 17D shows the jitter in the interspike interval beginning with the spike depicted in panel C, which at 0.18 ms is much tighter. This implies that the fluctuations in the absolute timing of the two spikes are strongly correlated. If the interval length plays a role in the interpretation of H1 somewhere downstream, then in a sense the fly corrects for the fluctuations by knowing that they are correlated. It would even be more interesting if fluctuations correlated among different neurons, as that would mean that the relative timing between spikes from different cells may carry extra information. Preliminary data from double recording experiments indicate that this is indeed the case for the two H1 neurons in the left and right lobula plates, again under approximately natural stimulus conditions.

Acknowledgments

The work presented here grew out of collaborations and discussions with many people, and it is our great pleasure to thank Naama Brenner, Simon Laughlin, Geoff Lewen, Roland Köberle, Al Schweitzer, and Steve Strong for sharing their insights and helping us out. We thank Simon Laughlin in particular for his thoughtful comments.

8.6 REFERENCES

[1] Barlow HB (1952): The size of ommatidia in apposition eyes. *J Exp Biol 29*, 667-674.

[2] Baylor DA, Lamb TD, Yau KW (1979): Responses of retinal rods to single photons, *J Physiol* (Lond) 288, 613-634.

[3] Berenbaum MR (1995): *Bugs in the System: Insects and Their Impact on Human Affairs.* Addison-Wesley Publishing Co., Reading MA.

[4] Berry M, Warland DK, Meister M (1997): The structure and precision of retinal spike trains. *Proc Natl Acad Sci USA* 94-10, 5411-5416.

[5] Bialek W, Rieke F, de Ruyter van Steveninck RR, Warland D (1991): Reading a neural code. *Science* 252, 1854-1857.

[6] Bracewell RN (1978): *The Fourier Transform and its Applications*, 2nd ed. McGraw-Hill, New York NY.

[7] Braitenberg V (1967) Patterns of projection in the visual system of the fly. I. Retina-lamina projections. *Exp Brain Res* 3, 271-298.

[8] Brenner N, Strong SP, Koberle R, Bialek W, de Ruyter van Steveninck RR (2000) Synergy in a neural code. *Neural Comp* 12, 1531-1552.

[9] Britten KH, Shadlen MH, Newsome WT, Movshon JA (1993) The responses of MT neurons to variable strength stochastic motion stimuli. *Visual Neurosci* 10, 1157-1169.

[10] Bullock TH (1970): The reliability of neurons. *J Gen Physiol* 55, 565-584.

[11] Dethier VG (1976): *The Hungry Fly: A Physiological Study of the Behavior Associated with Feeding.* Harvard University Press, Cambridge MA.

[12] Dodge FA, Knight BW, Toyoda J (1968): Voltage noise in *Limulus* visual cells. *Science* 160, 88-90.

[13] Dubs A, Laughlin SB, Srinivasan MV (1981): Single photon signals in fly photoreceptors and first order interneurones at behavioural threshold. *J Physiol* 317, 317-334.

[14] Exner S (1891): Die Physiologie der facettirten Augen von Krebsen und Insekten. Verlag Franz Deuticke, Vienna. Reprinted as: *The Physiology of the Compound Eyes of Insects and Crustaceans.* Springer Verlag, Heidelberg (1989).

[15] Feller W (1966): *An Introduction to Probability Theory and its Applications*, vol I. John Wiley, New York NY.

[16] Fermi G, Reichardt W (1963): Optomotorische Reaktionen de Fliege Musca Domestica. Abhängigkeit der Reaktion von der Wellenlänge, der Geschwindigkeit, dem Kontrast, und der mittleren Leuchtdichte bewegter periodischer Muster. *Kybernetik* 2, 15-28.

[17] Feynman RP, Leighton R, Sands M (1963): *The Feynman Lectures in Physics*. Addison-Wesley, Reading MA.

[18] Franceschini N, Kirschfeld K (1976): Le contrôle automatique du flux lumineux dans l'oeil composé des Diptères. Propriétés spectrales, statiques et dynamiques du mécanisme. *Biol Cybern* 31, 181-203.

[19] French AS, Järvilehto M (1978): The transmission of information by first and second order neurons in the fly visual system. *J Comp Physiol* 126, 87-96.

[20] Fuortes MGF, Yeandle S (1964): Probability of occurrence of discrete potential waves in the eye of Limulus. *J Gen Physiol* 47, 443-463.

[21] Green DM, Swets JA (1966): *Signal Detection Theory and Psychophysics*. John Wiley, New York NY.

[22] Gur M, Beylin A, Snodderly DM (1997): Response variability of neurons in primary visual cortex (V1) of alert monkeys. *J Neurosci* 17, 2914-2920.

[23] Haag J, Borst A (1997): Encoding of visual motion information and reliability in spiking and graded potential neurons. *J Neurosci* 17, 4809-4819.

[24] Hagiwara S (1954): Analysis of interval fluctuations of the sensory nerve impulse. *Jpn J Physiol* 4, 234-240.

[25] Hardie RC (1985): Functional organization of the fly retina. In: Ottoson D (ed) *Progress in Sensory Physiology* 5. Springer, Berlin, Heidelberg, New York, pp 2-79.

[26] Hardie RC (1988): Effects of antagonists on putative histamine receptors in the first visual neuropile of the housefly *Musca domestica*. *J Exp Biol* 138, 221-241.

[27] van Hateren JH (1984): Waveguide theory applied to optically measured angular sensitivities of fly photoreceptors. *J Comp Physiol A* 154, 761-771.

[28] van Hateren JH (1992): Real and optimal neural images in early vision. *Nature* 360, 68-70.

[29] van Hateren JH (1997): Processing of natural time-series of intensities by the visual system of the blowfly. *Vision Res* 37, 3407-3416.

[30] Hausen K (1981): Monocular and binocular computation of motion in the lobula plate of the fly, *Verh Dtsch Zool Ges* 49-70.

[31] Hausen K, Wehrhahn C (1983): Microsurgical lesion of horizontal cells changes optomotor yaw responses in the blowfly *Calliphora erythrocephala*. *Proc R Soc Lond B* 219, 211-216.

[32] Hausen K, Egelhaaf M (1989): Neural mechanisms of visual course control in insects. In: Stavenga DG, Hardie RC (eds) *Facets of Vision*. Springer, New York, pp 391-424.

[33] Hecht S, Shlaer S, Pirenne MH (1942): Energy, quanta, and vision. *J Gen Physiol* 25, 819-840.

[34] Howard J (1983) Variations in the voltage response to single quanta of light in the photoreceptors of *Locusta migratoria*. *Biophys Struct Mech* 9, 341-348.

[35] Howard J, Blakeslee B, Laughlin SB (1987): The intracellular pupil mechanism and photoreceptor signal: noise ratios in the fly *Lucilia cuprina*. *Proc R Soc Lond B* 231, 415-435.

[36] Juusola M, Uusitalo RO, Weckström M (1995): Transfer of graded potentials at the photoreceptor-interneuron synapse. *J Gen Physiol* 105, 117-148.

[37] Keiper W, Schnakenberg J (1984) Statistical analysis of quantum bump parameters in *Limulus* ventral photoreceptors. *Z Naturforsch* 39c, 781-790.

[38] Kirschfeld K (1967): Die Prokjektion der optischen Umwelt auf das Raster der Rhabdomere im Komplexauge von *Musca*. *Exp Brain Res* 3, 248-270.

[39] Kirschfeld K (1979): The visual system of the fly: physiological optics and functional anatomy as related to behavior. In: Schmitt FO, Worden FG (eds) *Neurosciences: Fourth Study Program*. MIT Press, Cambridge, MA, pp 297-310.

[40] Kirschfeld K, Franceschini N (1969): Ein Mechanismus zur Steuerung des Lichtflusses in den Rhabdomeren des Komplexauges von Musca. *Kybernetik* 6, 13-22.

[41] Krapp HG, Hengstenberg B, Hengstenberg R (1998): Dendritic structure and receptive-field organization of optic flow processing interneurons in the fly. *J Neurophysiol* 79, 1902-1917.

[42] Kuno M (1995) *The Synapse: Function, Plasticity and Neurotrophism.* Oxford University Press, Oxford.

[43] Lagnado L, Gomis A, Job C (1999): Continuous vesicle cycling in the synaptic terminal of retinal bipolar cells. *Neuron* 17, 957-967.

[44] Land MF, Collett TS (1974): Chasing behavior of houseflies (*Fannia canicularis*): A description and analysis. *J Comp Physiol* 89, 331-357.

[45] Laughlin SB (1989): Coding efficiency and design in visual processing. In: Stavenga DG, Hardie RC (eds) *Facets of Vision.* Springer, New York, pp 213-234.

[46] Laughlin SB (1999): Visual motion: Dendritic integration makes sense of the world. *Current Biol* 9, R15-R17.

[47] Laughlin SB, Hardie R (1978): Common strategies for light adaptation in the perpheral visual systems of fly and dragonfly. *J Comp Physiol* 128, 319-340.

[48] Laughlin SB and Lillywhite PG (1982): Intrinsic noise in locust photoreceptors. *J Physiol* 332, 25-45.

[49] Laughlin SB, Howard J, Blakeslee B (1987): Synaptic limitations to contrast coding in the retina of the blowfly *Calliphora. Proc R Soc Lond B* 231, 437-467.

[50] Laughlin SB, de Ruyter van Steveninck RR, Anderson J (1998): The metabolic cost of neural information. *Nature Neuroscience* 1, 36-41.

[51] Lillywhite PG, Laughlin SB (1979) Transducer noise in a photoreceptor. *Nature* 277, 569-572.

[52] MacKay D, McCulloch WS (1952): The limiting information capacity of a neuronal link. *Bull Math Biophys* 14, 127-135.

[53] Meinertzhagen IA, Fröhlich A (1983): The regulation of synapse formation in the fly's visual system. *Trends in Neurosci* 7, 223-228.

[54] Meister M, Berry MJ (1999) *The neural code of the retina.* Neuron 22, 435-450.

[55] Nicol D, Meinertzhagen IA (1982): An analysis of the number and composition of the synaptic populations formed by photoreceptors of the fly. *J Comp Neurol* 207, 29-44.

[56] Passaglia C, Dodge F, Herzog E, Jackson S, Barlow R (1997): Deciphering a neural code for vision. *Proc Natl Acad Sci USA* 94, 12649-12654.

[57] Potters M, Bialek W (1994). Statistical mechanics and visual signal processing. *J Phys I France* 4, 1755-1775.

[58] Reichardt W (1961): Autocorrelation, a principle for the evaluation of sensory information by the central nervous system. In: Rosenblith WA (ed) *Principles of Sensory Communication*. John Wiley, New York, NY, pp. 303-317.

[59] Reichardt W, Poggio T (1976): Visual control of orientation behavior in the fly. Part I: A quantitative analysis. *Q Rev Biophys* 9, 311-375.

[60] Rice SO (1944-45): Mathematical analysis of random noise. *Bell System Technical Journal*, 23, 24. Reprinted in *Wax* (1954), pp. 133-294.

[61] Rieke F, Warland D, de Ruyter van Steveninck R, Bialek W (1997): *Spikes: Exploring the Neural Code*. MIT Press, Cambridge MA.

[62] Roeder KD (1998): *Nerve Cells and Insect Behavior*. Harvard University Press, Cambridge MA.

[63] de Ruyter van Steveninck RR (1986): Real-time performance of a movement-sensitive neuron in the blowfly visual system. Thesis, University of Groningen, The Netherlands.

[64] de Ruyter van Steveninck R, Bialek W (1988) Real-time performance of a movement-sensitive neuron in the blowfly visual system: coding and information transfer in short spike sequences. *Proc R Soc Lond B* 234, 379-414.

[65] de Ruyter van Steveninck R, Bialek W (1995): Reliability and statistical efficiency of a blowfly movement-sensitive neuron. *Phil Trans R Soc Lond B* 348, 321-340.

[66] de Ruyter van Steveninck RR, Laughlin SB (1996a): The rate of information transfer at graded-potential synapses. *Nature* 379, 642-645.

[67] de Ruyter van Steveninck RR, Laughlin SB (1996b): Light adaptation and reliability in blowfly photoreceptors. *Int J Neural Syst* 7, 437-444.

[68] de Ruyter van Steveninck RR, Lewen GD, Strong SP, Koberle R, Bialek W (1997): Reproducibility and variability in neural spike trains. *Science* 275, 1805-1808.

[69] Saleh BEA, Teich MC (1991): *Fundamentals of Photonics* Wiley, New York.

[70] Schrödinger E (1952): *Statistical Thermodymamics*. Cambridge University Press, Cambridge, UK.

[71] Shaw SR (1981): Anatomy and physiology of identified non-spiking cells in the photoreceptor-lamina complex of the compound eye of insects, especially Diptera. In: Roberts A, Bush BMH (eds) *Neurons Without Impulses*. Cambridge Univ Press, Cambridge UK.

[72] Single S, Borst A (1998): Dendritic integration and its role in computing image velocity. *Science* 281, 1848-1850.

[73] Smakman JGJ, van Hateren JH, Stavenga DG (1984): Angular sensitivity of blowfly photoreceptors: Intracellular measurements and wave-optical predictions. *J Comp Physiol* A155, 239-247.

[74] Snyder AW (1979): The physics of vision in compound eyes. In: Autrum H (ed) *Handbook of Sensory Physiology* VII/6A: Comparative physiology and evolution of vision in invertebrates. A: Invertebrate photoreceptors. Springer, Heidelberg, pp 225-314.

[75] Snyder DL, Miller MI (1991): *Random Point Processes in Time and Space* (second edition). Springer, Heidelberg.

[76] Stavenga DG (1974): Visual receptor optics, rhodopsin and pupil in fly retinula cells. Thesis, University of Groningen, Groningen, The Netherlands.

[77] Stavenga DG (1995): Insect retinal pigments. Spectral characteristics and physiological functions. *Progr Ret Eye Res* 15, 231-259.

[78] Stieve H, Bruns M (1983): Bump latency distribution and bump adaptation of *Limulus* ventral nerve photoreceptor in varied extracellular calcium concentrations. *Biophys Struct Mech* 9, 329-339.

[79] Strong SP, Koberle R, de Ruyter van Steveninck RR, Bialek W (1998): Entropy and information in neural spike trains. *Phys Rev Lett* 80, 197-200.

[80] Teich MC, Khanna SM (1985) Pulse number distribution for the spike train in the cat's auditory nerve. *J Acoust Soc Am* 77, 1110-1128.

[81] Tinbergen N (1984): *Curious Naturalist* (revised edition). Univ. of Massachusetts Press, Amherst MA.

[82] Tolhurst DJ, Movshon JA, Dean AF (1983) The statistical reliability of signals in single neurons in cat and monkey visual cortex. *Vision Res* 23, 775-785.

[83] Uusitalo RO, Juusola M, Kouvalainen E, Weckström M (1995): Tonic transmitter release in a graded potential synapse. *J Neurophysiol* 74, 470-473.

[84] Warzecha A-K, Kretzberg J, Egelhaaf M (1998): Temporal precision of the encoding of motion information by visual interneurons. *Curr Biol* 8, 359-368.

[85] Wax N (1954): *Noise and Stochastic Processes*. Dover, New York NY.

[86] Wong F, Knight BW, Dodge FA (1980): Dispersion of latencies in photoreceptors of Limulus and the adapting bump model. *J Gen Physiol* 71, 249-268.

9

Paradigms for Computing with Spiking Neurons

Wolfgang Maass

ABSTRACT In this chapter we define for various neural coding schemes formal models of computation in networks of spiking neurons. The main results about the computational power of these models are surveyed. In particular, we compare their computational power with that of common models for artificial neural networks. Some rigorous theoretical results are presented which show that for temporal coding of inputs and outputs certain functions can be computed in a feedforward network of spiking neurons with fewer neurons than in *any* multi-layer perceptron (i.e., feedforward network of sigmoidal neurons). This chapter also presents a brief survey of the literature on computations in networks of spiking neurons.

9.1 Introduction

Spiking neurons differ in essential aspects from the familiar computational units of common neural network models, such as McCulloch-Pitts neurons or sigmoidal gates. Therefore the question arises how one can *compute* with spiking neurons, or with related computational units in electronic hardware, whose input and output consists of trains of pulses. Furthermore the question arises how the computational power of networks of such units relates to that of common reference models, such as threshold circuits or multi-layer perceptrons. Both questions will be addressed in this chapter.

9.2 A Formal Computational Model for a Network of Spiking Neurons

A neuron model whose dynamics is described in terms of differential equations, such as the Hodgkin-Huxley model, the Fitzhugh-Nagumo model, or the integrate-and-fire model, makes it quite difficult to analyze computations in networks of such neurons. Hence the formulation of the spike-response model (see [11, 10]) has greatly facilitated the investigation of computations in networks of spiking neurons. Although this model has a

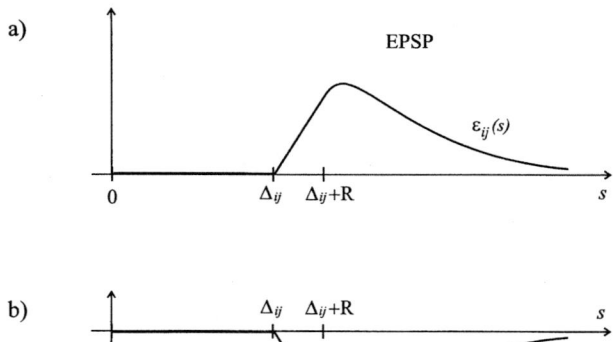

FIGURE 9.1. a) Typical shape of an excitatory postsynaptic potential (EPSP). b) Typical shape of an inhibitory postsynaptic potential (IPSP). The delay Δ_{ij} and the length R of such PSP will turn out to be important for the computations discussed in sections 9.4 and 9.5.

mathematically much simpler formulation, it is able to capture the dynamics of Hodgkin-Huxley neurons quite well [19]. In addition, by choosing suitable response functions one can readily adapt this model to reflect the dynamics of a large variety of different biological neurons.

For the sake of completeness we quickly review the definition of the spike response model. Let I be a set of neurons, and assume that one has specified for each neuron $i \in I$ a set $\Gamma_i \subseteq I$ of immediate predecessors (presynaptic neurons) in the network. The firing times $t_i^{(f)} \in \mathcal{F}_i$ for all neurons $i \in I$ are defined recursively (by simultaneous recursion along the time axis). The neuron i fires whenever the state variable

$$u_i(t) = \sum_{t_i^{(f)} \in \mathcal{F}_i} \eta_i(t - t_i^{(f)}) + \sum_{j \in \Gamma_i} \sum_{t_j^{(f)} \in \mathcal{F}_j} w_{ij}\, \epsilon_{ij}(t - t_j^{(f)}). \qquad (9.1)$$

reaches the firing threshold ϑ of neuron i.

The response functions $\epsilon_{ij}(t - t_j^{(f)})$ model excitatory and inhibitory postsynaptic potentials (EPSPs and IPSPs, see Fig. 9.1) at the soma of neuron i, which result from the firing of a presynaptic neuron j at time $t_j^{(f)}$. The function η_i models the response of neuron i to its own firing, in particular refractory effects. One typically assumes that η_i assumes a strongly negative value for values of $t - t_i^{(f)}$ in the range of a few ms, and then gradually returns to a value near 0 [10]. If the neuron i has not fired for a while, one can ignore the first summand with the refractory terms $\eta_i(t - t_i^{(f)})$ in (9.1).

In order to complete the definition of a network of spiking neurons as a formal computational model one has to specify its network input and output. We assume that subsets of neurons $I_{\text{input}} \subseteq I$ and $I_{\text{output}} \subseteq I$ have been fixed, and that the firing times $\mathcal{F}_i$ for the neurons $i \in I_{\text{input}}$ constitute the network input. Thus we assume that these firing times are determined through some external mechanism, rather than computed according to the previously described rules. The firing times $\mathcal{F}_i$ of the neurons $i \in I_{\text{output}}$

constitute the network output. These firing times are computed in the previously described way (like for all neurons $i \in I - I_{\text{input}}$) with the help of the state variable (9.1).

Thus from a mathematical point of view, a network of spiking neurons computes a function which maps a vector of several time series $\langle \mathcal{F}_i \rangle_{i \in I_{\text{input}}}$ on a vector of several other time series $\langle \mathcal{F}_i \rangle_{i \in I_{\text{output}}}$. In the first part of this chapter we will focus on the mathematically simpler case where these time series $\mathcal{F}_i$ consist of at most one spike each.

9.3   McCulloch-Pitts Neurons versus Spiking Neurons

The simplest computational unit of traditional neural network models is a *McCulloch-Pitts neuron*, also referred to as *threshold gate* or *perceptron*. A McCulloch-Pitts neuron i with real valued weights α_{ij} and threshold ϑ receives as input n binary or real valued numbers $x_1, \ldots, x_n$. Its output has the value

$$
\begin{cases}
1, & \text{if } \sum_{j=1}^{n} \alpha_{ij} \cdot x_j \geq \vartheta \\[2ex]
0, & \text{otherwise.}
\end{cases}
\tag{9.2}
$$

For multilayer networks one usually considers a variation of the threshold gate to which we will refer as a *sigmoidal gate* in the following. The output of a sigmoidal gate is defined with the help of some nondecreasing continuous *activation function* $g : \mathbb{R} \to \mathbb{R}$ with bounded range as

$$
g\Big(\sum_{j=1}^{n} \alpha_{ij} \cdot x_j - \vartheta \Big) .
\tag{9.3}
$$

By using sigmoidal gates instead of threshold gates one can not only compute functions with analog output, but also increase the computational power of neural nets for computing functions with boolean output [24, 6] .

In the following we will compare the computational capabilities of these two computational units of traditional neural network models with that of a spiking neuron. One immediately sees that a spiking neuron i can in principle simulate any given threshold gate (9.2) with positive threshold ϑ for binary input. For that we assume that the response functions $\epsilon_{ij}(x)$ are all identical except for their sign (which we choose to be positive if $\alpha_{ij} > 0$ and negative if $\alpha_{ij} \leq 0$), and that all presynaptic neurons j which fire, fire at the same time $t_j = T_{\text{input}}$. In this case the spiking neuron i fires if and only if

$$
\sum_{j \text{ fires at time } T_{\text{input}}} w_{ij} \cdot \epsilon_{ij} \geq \vartheta ,
\tag{9.4}
$$

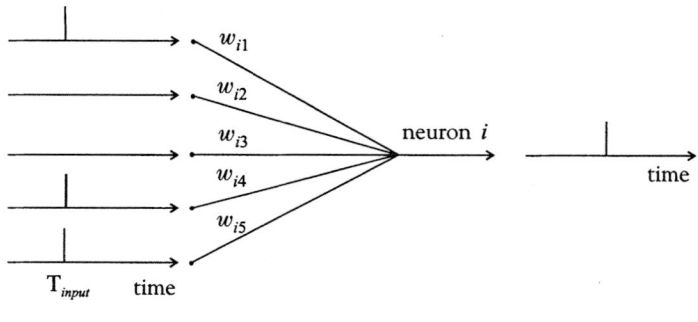

firing times of presynaptic   neuron i fires if

neurons j $$\sum_{j \text{ fires at time } T_{\text{input}}} w_{ij} \cdot \epsilon_{ij} \geq \vartheta$$

FIGURE 9.2. Simulation of a threshold gate by a spiking neuron.

where ϵ_{ij} is the extremal value of $\epsilon_{ij}(s)$ (i.e., $\epsilon_{ij} = \max_s \epsilon_{ij}(s)$ if $\epsilon_{ij}(s)$ represents an EPSP, $\epsilon_{ij} = \min_s \epsilon_{ij}(s)$ if $\epsilon_{ij}(s)$ represents an IPSP), $w_{ij} \geq 0$ is the synaptic weight, and $\vartheta > 0$ is the firing threshold of neuron i; see Fig. 9.2. Then for $w_{ij} := \alpha_{ij}/\epsilon_{ij}$ the spiking neuron i can simulate any given threshold gate defined by (9.2) if the input bits $x_1, \ldots, x_n$ are encoded by the firing or non-firing of presynaptic neurons $j = 1, \ldots, n$ at a common time T_{input}, and if the output bit of the threshold gate is encoded by the firing or non-firing of the spiking neuron i during the relevant time window. A closer look shows that it is substantially more difficult to simulate in the same manner a *multilayer* network of threshold gates (i.e., a threshold circuit) by a network of spiking neurons. The exact firing time of the previously discussed spiking neuron i depends on its concrete input $x_1, \ldots, x_n$. If

$$\sum_{j \text{ fires at time } T_{\text{input}}} w_{ij} \cdot \epsilon_{ij} - \vartheta$$

has a value well above 0, then the state variable

$$u_i(t) = \sum_{j \text{ fires at time } T_{\text{input}}} w_{ij} \cdot \epsilon_{ij}(t - T_{\text{input}})$$

will cross the firing threshold ϑ earlier, yielding an *earlier* firing time of neuron i, compared with an input where

$$\sum_{j \text{ fires at time } T_{\text{input}}} w_{ij} \cdot \epsilon_{ij} - \vartheta$$

is positive but close to 0. Therefore, if one employs several spiking neurons to simulate the threshold gates on the first layer[1] of a threshold circuit, those neurons on the first layer which do fire (corresponding to threshold gates with output 1) will in general fire at slightly different time points. This will occur even if all input neurons j of the network have fired at the

---

[1] We will not count the layer of input neurons in this chapter, and hence refer to the *first hidden layer* as the *first layer* of the network.

same time T_{input}. Therefore the timing of such straightforward simulation of a multi-layer threshold circuit is unstable: even if all firings in one layer occur synchronously, this synchronicity will in general get lost at the next layer. Similar problems arise in a simulation of other types of multilayer boolean circuits by networks of spiking neurons.

Consequently one needs a separate *synchronization mechanism* in order to simulate a multi-layer boolean circuit–or any other common model for digital computation–by a network of spiking neurons with bits $0, 1$ encoded by firing and nonfiring. One can give a mathematically rigorous proof that such synchronization mechanism can in principle be provided by using some auxiliary spiking neurons [20]. This construction exploits the simple fact that the double-negation $\neg \neg b$ of a bit b has the same value as b. Therefore instead of making a spiking neuron i fire in the direct way described by Eq. (9.4), one can make sure that if $\sum_j \alpha_{ij} \cdot x_j \geq \vartheta$, then the spiking neuron i is *not prevented* from firing by auxiliary inhibitory neurons. These auxiliary inhibitory neurons are connected directly to the input neurons whose firing/nonfiring encodes the input bits $x_1, \ldots, x_n$, whereas the driving force for the firing of neuron i comes from input-independent excitatory neurons. With this method one can simulate any given boolean circuit, and in the absence of noise even any Turing machine, by a finite network of spiking neurons [20] (we refer to Judd and Aihara [16] for earlier work in this direction). On this basis one can also implement the constructions of Valiant [46] in a network of spiking neurons.

Before we leave the issue of synchronization we would like to point out that in a practical context one can achieve a near-synchronization of firing times with the help of some common background excitation, for example an excitatory background oscillation $\sin(\omega t)$ that is added to the membrane potential of all spiking neurons i [14]. With proper scaling of amplitudes and firing thresholds one can achieve that for all neurons i the state variable $u_i(t)$ can cross the firing threshold ϑ only when the background oscillation $\sin(\omega t)$ is close to its peak. Our preceding discussion also points to a reason why it may be *less advantageous* for a network of spiking neurons to employ a forced synchronization. The small temporal differences in the firing times of the neurons i that simulate the first layer of a threshold circuit according to Eq. (9.4) contain valuable *additional information* that is destroyed by a forced synchronization: These temporal differences contain information about *how much* larger the weighted sum $\sum_j \alpha_{ij} \cdot x_j$ is compared with ϑ. Thus it appears to be advantageous for a network of spiking neurons to employ instead of a synchronized digital mode an asynchronous or loosely synchronized analog mode where subtle differences in firing times convey additional *analog* information. We will discuss in the next section computational operations which spiking neurons can execute in this quite different computational mode, where *analog* values are encoded in *temporal patterns* of firing times.

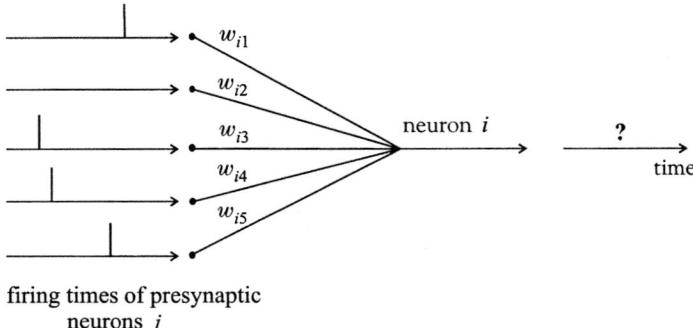

FIGURE 9.3. Typical input for a biological spiking neuron i, where its output cannot be easily described in terms of conventional computational units.

9.4   Computing with Temporal Patterns

9.4.1   Conincidence detection

We will now consider the more typical scenario for a biological neuron, where preceding neurons do not fire in a synchronized manner, see Fig. 9.3. In this case the computational operation of a spiking neuron cannot be easily described with the help of common computational operations or computational models.

We will show that in an asynchronous mode, with analog values encoded by a temporal pattern of firing times, a spiking neuron has in principle not only more computational power than a McCulloch-Pitts neuron, but also more computational power than a sigmoidal gate.

One new feature of a *spiking* neuron–which does not correspond to any feature of the computational units of traditional neural network models – is that it can act as *coincidence detector* for incoming pulses [1]. Hence if the arrival times of the incoming pulses encode *numbers*, a spiking neuron can detect whether some of these numbers have (almost) equal value. On the other hand we will show below that this operation on numbers is a rather "expensive" computational operation from the point of view of traditional neural network models.

We will now make these statements more precise. Assume that $\{1, \ldots, n\} = \Gamma_i$ are the immediate predecessors of a spiking neuron i, that their connections to neuron i all have the same transmission delay Δ_{ij}, and that $w_{ij} = 1$ for all $j \in \Gamma_i$. Furthermore, assume that the response functions $\epsilon_{ij}(s)$ are defined in some common way (see [10]), for example as a difference of two exponentially decaying functions

$$\epsilon_{ij}(s) = \frac{1}{1 - (\tau_s/\tau_m)} \left[\exp\left(-\frac{s - \Delta_{ij}}{\tau_m} \right) - \exp\left(-\frac{s - \Delta_{ij}}{\tau_s} \right) \right] \mathcal{H}(s - \Delta_{ij})$$

a)

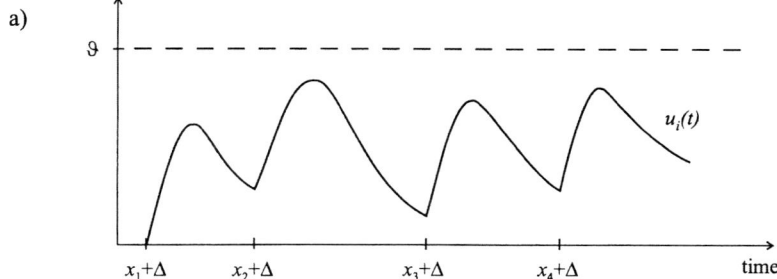

b)

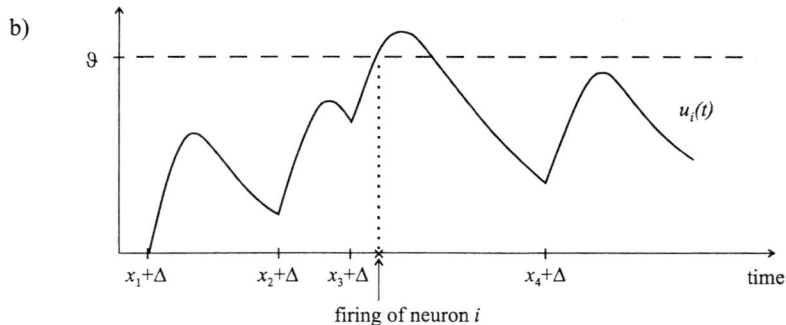

FIGURE 9.4. A) Typical time course of the state variable $u_i(t)$ if $ED_4(x_1, x_2, x_3, x_4) = 0$. B) Time course of $u_i(t)$ in the case where $ED_4(x_1, x_2, x_3, x_4) = 1$ because $|x_3 - x_2| \leq c_1$.

with time constants $0 < \tau_s < \tau_m$ and the heaviside function $\mathcal{H}$. A plot of an EPSP of such shape is shown in Figure 9.1A). It consists of an almost linearly rising phase for small s, exponential decay for large s, and a smooth transition between both phases when it reaches its maximal value in between.

For every given values of the time constants τ_s, τ_m with $\tau_s < \tau_m$ one can find values $0 < c_1 < c_2$ and ϑ so that $u_i(t) < \vartheta$ for any input consisting of an arbitrary number of EPSPs with distance $\geq c_2$, whereas $u_i(t)$ reaches a value $> \vartheta$ for two EPSPs in distance $\leq c_1$. Then the spiking neuron i does not fire if the neurons $j \in \Gamma_i$ fire (each at most once) in temporal distance $\geq c_2$ (see Fig. 9.4A), but it fires whenever two presynaptic neurons $j \in \Gamma_i$ fire in temporal distance $\leq c_1$ (see Fig. 9.4B). Consequently, if for example one encodes n real numbers $x_1, \ldots, x_n$ through the firing times of the n neurons in Γ_i, and decodes the output of neuron i as "1" if it fires and "0" if it does not fire, the neuron i computes the following function $ED_n : \mathbb{R}^n \to \{0, 1\}$:

$$ED_n(x_1, \ldots, x_n) = \begin{cases} 1 & , \quad \text{if there are } j \neq j' \text{ so that } |x_j - x_{j'}| \leq c_1 \\ 0 & , \quad \text{if } |x_j - x_{j'}| \geq c_2 \text{ for all } j \neq j'. \end{cases}$$

Note that this function $ED_n(x_1, \ldots, x_n)$ (where ED stands for "element distinctness") is in fact a partial function, which may output arbitrary values in case that $c_1 < \min\{|x_j - x_{j'}| : j \neq j' \text{ and } j, j' \in \Gamma_i\} < c_2$. Therefore hair-trigger situations can be avoided, and a single spiking neuron can compute this function ED_n even if there is a small amount of noise on its state variable $u_i(t)$.

On the other hand the following results show that the same partial function ED_n requires a substantial number of neurons if computed by neural networks consisting of McCulloch-Pitts neurons (threshold gates) or sigmoidal gates. These lower bounds hold for *arbitrary* feedforward architectures of the neural net, and *any values* of the weights and thresholds of the neurons. The inputs $x_1, \ldots, x_n$ are given to these neural nets in the usual manner as analog input variables.

Theorem 9.1 *Any layered threshold circuit that computes ED_n needs to have at least* $\log(n!) \geq \frac{n}{2} \cdot \log n$ *threshold gates on its first layer.*

The *proof* of Theorem 9.1 relies on a geometrical argument, see [22].

Theorem 9.2 *Any feedforward neural net consisting of arbitrary sigmoidal gates needs to have at least* $\frac{n-4}{2}$ *gates in order to compute ED_n.*

The proof of Theorem 9.2 is more difficult, since the gates of a sigmoidal neural net (defined according to (9.3) with some smooth gain function g) output *analog numbers* rather than *bits*. Therefore a multilayer circuit consisting of sigmoidal gates may have larger computational power than a circuit consisting of threshold gates. The proof procedes in an indirect fashion by showing that any sigmoidal neural net with m gates that computes ED_n can be transformed into another sigmoidal neural net that "shatters" *every* set of $n-1$ different inputs with the help of $m+1$ programmable parameters. According to Sontag [42] this implies that $n-1 \leq 2(m+1)+1$. We refer to [22] for further details.   ∎

9.4.2 RBF-Units in the temporal domain

We have demonstrated in the preceding subsection that for some computational tasks a single spiking neuron has more computational power than a fairly large neural network of the conventional type. We will show in this subsection that the preceeding construction of a spiking neuron that detects coincidences among incoming pulses can be expanded to yield detectors for more complex temporal patterns.

Instead of a common delay Δ between presynaptic neurons $j \in \Gamma_i$ and neuron i (which appears in our formal model as the length of the initial flat part of the response function $\epsilon_{ij}(x)$) one can employ for different j *different* delays Δ_{ij} between neurons j and i. These delays Δ_{ij} represent a new set of parameters that have no counterpart in traditional neural network models[2].

---

[2]Theoretical results about the Vapnik-Chervonenkis dimension (VC-dimension) of neural nets suggest that tuning of delays enhances the flexibility of spiking neurons for computations (i.e., the number of different functions they can compute) even more than

There exists evidence that in some biological neural systems these delays Δ_{ij} can be tuned by "learning algorithms". In addition one can tune the firing threshold ϑ and/or the weights w_{ij} of a spiking neuron to increase its ability to detect specific temporal patterns in the input. In the extreme case one can raise the firing threshold ϑ so high that *all* pulses from presynaptic neurons have to arrive nearly simultaneously at the soma of i to make it fire. In this case the spiking neuron can act in the temporal domain like an RBF-unit (i.e., radial basis function unit) in traditional neural network models: it will fire only if all presynaptic neurons $j \in \Gamma_i$ fire at times t_j so that for some constant T_{input} one has $t_j \approx T_{\text{input}} - \Delta_{ij}$ for all $j \in \Gamma_i$, where the vector $(\Delta_{ij})_{j \in \Gamma_i}$ of transmission delays plays now the role of the *center* of an RBF-unit. This possibility of using spiking neurons as RBF-like computational units in the temporal domain was first observed by Hopfield [14]. In the same article Hopfield demonstrates an advantageous consequence of employing a *logarithmic* encoding $x_j = \log y_j$ of external sensory input variables y_j through firing times $t_j = T_{\text{input}} - x_j$. Since spiking neurons have the ability to detect temporal patterns irrespective of a common additive constant in their arrival times, they can with the help of logarithmic encoding ignore constant *factors* λ in sensory input variables $\langle \lambda \cdot y_j \rangle_{j \in \Gamma_i}$. It has been argued that this useful mechanisms may be related to the amazing ability of biological organisms to classify patterns over a very large scale of intensities, such as for example visual patterns under drastically different lighting conditions.

In [35] the previous construction of an RBF-unit for temporal patterns has been extended to a an RBF-network with the help of lateral inhibition between RBF-units. Alternatively one can add linear gates on the second layer of an RBF-network of spiking neurons with the help of the construction described in the following section.

9.4.3 Computing a weighted sum in temporal coding

A characteristic feature of the previously discussed computation of the function $ED_{\neg}$ and the simulation of an RBF-unit is the *asymmetry* between coding schemes used for *input* and *output*. Whereas the input consisted of a vector of analog numbers, encoded through temporal delays, the output of the spiking neuron was just binary, encoded through firing or nonfiring of that neuron. Obviously for multilayer or recurrent computations with spiking neurons it is desirable to have mechanisms that enable a layer of spiking neurons to *output* a vector of analog numbers encoded in the same way as the input. For that purpose one needs mechanisms for *shifting* the firing time of a spiking neuron i in dependence of the firing times t_j of presynaptic neurons, in a manner that can be controlled through the

---

tuning the weights [28].

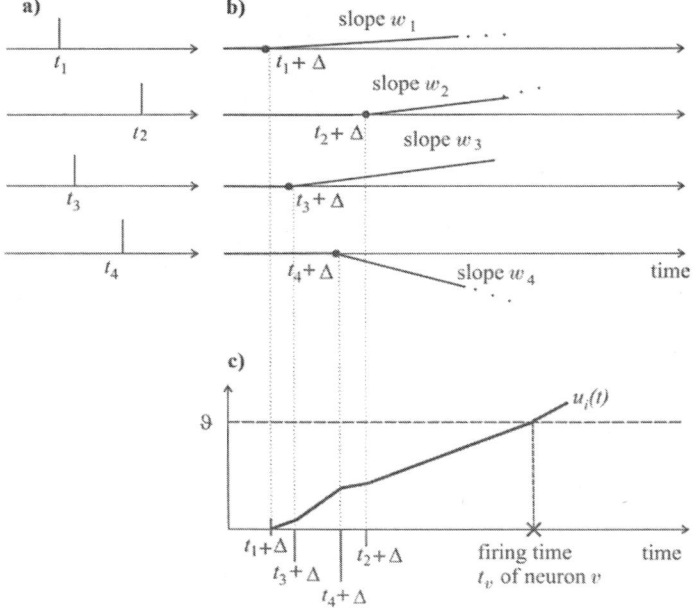

FIGURE 9.5. Mechanisms for computing a weighted sum in temporal coding according to Eq. (9.6). A) Firing times t_j of presynaptic neurons j. B) Initial linear segments of the weighted response functions $w_{ij} \cdot \epsilon_{ij}(t - t_j)$ at the soma of neuron i. C) State variable $u_i(t)$ and resulting firing time t_i of neuron i.

internal parameters w_{ij} and Δ_{ij}. As an example for that we will now describe a simple mechanism for computing for arbitrary parameters $\alpha_{ij} \in \mathbb{R}$ and inputs $x_j \in [0, 1]$ the weighted sum $\sum_j \alpha_{ij} \cdot x_j$ through the firing time of neuron i.

We assume that each response function $\epsilon_{ij}(s)$ has a shape as shown in Fig. 9.1: $\epsilon_{ij}(s)$ has value 0 for $s \leq \Delta_{ij}$ and then rises approximately lineary (in the case of an EPSP) or descends approximately lineary (in the case of an IPSP) with slope $\lambda_{ij} \in \mathbb{R}$ for an interval of length at least $R > 0$. Assume that the presynaptic neurons $j \in \Gamma_i$ fire at times $t_j = T_{\text{input}} - x_j$. If the state variable $u_i(t) = \sum_{j \in T_i} w_{ij} \cdot \epsilon_{ij}(t - t_j)$ of neuron i reaches the threshold ϑ at a time t_i when the response functions $\epsilon_{ij}(t - t_j)$ are all in their initial linear phase of length $\geq R$, then t_i is determined by the equation

$$\sum_{j \in \Gamma_i} w_{ij} \cdot \epsilon_{ij}(t_i - t_j) = \sum_{j \in \Gamma_i} w_{ij} \cdot \lambda_{ij} \cdot (t_i - t_j - \Delta_{ij}) = \vartheta . \qquad (9.5)$$

Obviously (9.5) implies that

$$t_i = \frac{\vartheta}{\sum_{j \in \Gamma_i} w_{ij} \cdot \lambda_{ij}} + \frac{\sum_{j \in \Gamma_i} w_{ij} \cdot \lambda_{ij} \cdot (t_j + \Delta_{ij})}{\sum_{j \in \Gamma_i} w_{ij} \cdot \lambda_{ij}} . \tag{9.6}$$

Then by writing λ for $\sum_{j \in \Gamma_i} w_{ij} \cdot \lambda_{ij}$ and expressing t_j as $T_{\text{input}} - x_j$ we get

$$t_i = \frac{\vartheta}{\lambda} + \sum_{j \in \Gamma_i} \frac{w_{ij} \cdot \lambda_{ij}}{\lambda} \cdot (T_{\text{input}} - x_j + \Delta_{ij}),$$

or equivalently

$$t_i = T_{\text{output}} - \sum_{j \in \Gamma_i} \alpha_{ij} \cdot x_j \tag{9.7}$$

for some input-independent constant

$$T_{\text{output}} := \frac{\vartheta}{\lambda} + \sum_{j \in \Gamma_i} \frac{w_{ij} \cdot \lambda_{ij}}{\lambda} (T_{\text{input}} + \Delta_{ij}),$$

and formal "weights" α_{ij} defined by $\alpha_{ij} := \frac{w_{ij} \cdot \lambda_{ij}}{\lambda}$. These "weights" α_{ij} are automatically normalized: By the definition of λ they satisfy $\sum_{j \in \Gamma_i} \alpha_{ij} = 1$. Such automatic weight normalization may be desirable in some situations [13]. One can circumvent it by employing an auxiliary input neuron (see [21]). In this way one can compute an arbitrary given weighted sum $\sum_{j \in \Gamma_i} \alpha_{ij} \cdot x_j$ in *temporal* coding by a *spiking* neuron. Note that in this construction the analog output $\sum_{j \in \Gamma_i} \alpha_{ij} \cdot x_j$ is encoded in exactly the same way as the analog inputs x_j.

9.4.4 *Universal approximation of continuous functions with spiking neurons in the temporal domain*

We will show in this subsection that on the basis of the computational mechanism described in the preceding subsection one can build networks of spiking neurons that can approximate arbitrary given bounded continuous functions in the temporal domain. We first observe that one can expand the previously described mechanism for computing a weighted sum $\sum_{j \in \Gamma_i} \alpha_{ij} \cdot x_j$ in the temporal domain to yield for *temporal coding* also a simulation of an arbitrary given sigmoidal neuron with the piecewise linear gain function

$$\text{sat}(x) = \begin{cases} x & , \quad \text{if} \quad 0 \leq x \leq 1 \\ 0 & , \quad \text{if} \quad x \leq 0 \\ 1 & , \quad \text{if} \quad x \geq 1 \end{cases}$$

In this case we want that neuron i responds to firing of its presynaptic neurons at times $t_j = T_{\text{input}} - x_j$ by firing at time

$$t_i = T_{\text{output}} - \text{sat}(\sum_{j \in \Gamma_i} \alpha_{ij} \cdot x_j).$$

For that purpose one just needs auxiliary mechanisms that support an approximation of the given sigmoidal neuron in the saturated regions of its gain function sat, i.e. for $x \leq 0$ and $x \geq 1$. Translated into the temporal domain this requires that the spiking neuron i does not fire before some fixed time T (simulating $\mathrm{sat}(x) = 1$ for $x \geq 1$) and by the latest at some fixed time $T_{\mathrm{output}} > T$ (simulating $\mathrm{sat}(x) = 0$ for $x \leq 0$). This can easily be achieved with the help of auxiliary spiking neurons. Computer simulations suggest that in a practical situation such auxiliary neurons may not even be necessary [26].

According to the preceding construction one can simulate any sigmoidal neuron with the piecewise linear gain function sat by spiking neurons with analog inputs *and* outputs encoded by temporal delays of spikes. Since inputs and outputs employ the same coding scheme, the outputs from a first layer of spiking neurons (that simulate a first layer of sigmoidal gates) can be used as inputs for another layer of spiking neurons, simulating another layer of sigmoidal gates. Hence on the basis of the assumption that the initial segments of response functions $\epsilon_{ij}(s)$ are linear one can show with a rigorous mathematical proof [21]:

Theorem 9.3 *Any feedforward or recurrent analog neural net (for example any multilayer perceptron), consisting of s sigmoidal neurons that employ the gain function* sat, *can be simulated arbitrarily closely by a network of $s + c$ spiking neurons (where c is a small constant) with analog inputs and outputs encoded by temporal delays of spikes. This holds even if the spiking neurons are subject to noise.* ∎

Theorem 9.2 and 9.3 together exhibit an interesting *asymmetry* regarding the computational power of standard sigmoidal neural nets (multilayer perceptrons) and networks of spiking neurons: Whereas any sigmoidal neural net can be simulated by an insignificantly larger network of spiking neurons (with temporal coding), certain networks of spiking neurons can only be simulated by substantially larger sigmoidal neural nets.

It is well known that feedforward sigmoidal neural nets with gain function sat can approximate any given continuous function $F : [0,1]^n \rightarrow [0,1]^m$ with any desired degree of precision. Hence Theorem 9.3 implies:

Corollary 9.4 *Any given continuous function $F : [0,1]^n \rightarrow [0,1]^m$ can be approximated arbitrarily closely by a network of spiking neurons with inputs and outputs encoded by temporal delays.*

Remarks:
a) The construction that yields the proof of Theorem 9.3 shows that a network of spiking neurons can *change* the function: $F : [0,1]^n \rightarrow [0,1]^m$ that it computes in the same way as a traditional neural net: by changing the synaptic "weights" w_{ij} that scale the slopes of the initial segments of postsynaptic pulses. The delays Δ_{ij} between neurons need not be changed for that purpose (but they *could* be used to modulate the effective "weights" of the simulated sigmoidal neural net by additive constants). From that point of view this construction is complementary to the simulation of RBF-units by spiking neurons described in subsection 9.4.2: There the "program" of the encoded function was encoded exclusively in the delays Δ_{ij}.

b) It turns out that the network of spiking neurons constructed for the proof of Theorem 9.3 computes approximately the same function in rate-coding *and* in temporal coding.

9.4.5 Other computations with temporal patterns in networks of spiking neurons

The previously described method for emulating classical artificial neural networks in the temporal domain with spiking neurons can also be applied to Hopfield nets [26], Kohonen's self-organizing map [39] and RBF-networks [35]. The latter construction refines Hopfield's construction of an RBF-unit in the temporal domain. It simulates RBF-units by neurons that output an analog number (encoded in its firing time), rather than a single bit (encoded by firing/nonfiring). They implement a competition among different RBF-units through lateral inhibition. Furthermore they show through computer simulations that a variation of the Hebb-rule for spiking neurons with temporal coding, that has been experimentally observed for biological neurons (see [32] and [31]), yields good performance for unsupervised learning of temporal input patterns. It is of interest for applications that their RBF-network also exhibits some robustness with regard to warping of temporal input patterns.

Simon Thorpe and his collaborators have independently explored several options for computing with information encoded in the timing of spikes ([43, 44, 40]). The goal of their constructions is that the output neurons of the net respond in a given way to the firing *order* of the input neurons. This is a special case of the computations with spiking neurons considered in sections 9.4.3 and 9.4.4. Obviously each firing order is naturally encoded in the vector $\underline{x}$ of delays of these spikes. Hence any classification task for spike orders can be viewed as a special case of a classification task for delay vectors $\underline{x}$. Theorem 9.3 shows that networks of spiking neurons can apply to this task the full classification power of multilayer perceptrons.

Finally we would like to mention that Watanabe and Aihara [48] have explored *chaos* in the temporal pattern of firing in a network of spiking neurons.

9.5   Computing with a Space-Rate Code

The second model for fast analog computation with spiking neurons that we will discuss is more suitable for neural systems consisting of a large number of components which are not very reliable. In fact this model, which was recently developed in collaboration with Thomas Natschläger [27], *relies* on the assumption that individual synapses are "unreliable". It takes into account evidence from Dobrunz and Stevens [7] and others, which shows that individual synaptic release sites are highly stochastic: They release a vesicle (filled with neurotransmitter) upon the arrival of a spike from the presynaptic neuron u with a certain probability (called *release probability*). This release probability varies among different synapses between values less than 0.1 and more than 0.9.

This computational model is based on a space-rate encoding (also referred to as population coding) of analog variables, i.e., an analog variable $x \in [0, 1]$ is encoded by the percentage of neurons in a population that fire within a short time window (say, of length 5 ms).

Although there exists substantial empirical evidence that many cortical systems encode relevant analog variables by such space-rate code, it has remained unclear how networks of spiking neurons can *compute* in terms of such a code. Some of the difficulties become apparent if one just wants to understand for example how the trivial linear function $f(x) = x/2$ can be computed by such a network if the input $x \in [0, 1]$ is encoded by a space-rate code in a pool U of neurons and the output $f(x) \in [0, 1/2]$ is supposed to be encoded by a space-rate code in another pool V of neurons. If one assumes that all neurons in V have the same firing threshold and that reliable synaptic connections from all neurons in U to all neurons in V exist with approximately equal weights, a firing of a percentage x of neurons in U during a short time interval will typically trigger *almost none* or *almost all* neurons in V to fire, since they all receive about the same input from U.

Several mechanisms have already been suggested that could in principle achieve a *smooth* graded response in terms of a space-rate code in V instead of a binary "all or none" firing, such as strongly varying firing thresholds or different numbers of synaptic connections from U for different neurons $v \in V$ [49]. Both of these options would fail to spread average activity over all neurons in V, and hence would make the computation less robust against failures of individual neurons.

We assume that n pools $U_1, \ldots, U_n$ consisting of N neurons each are given, and that all neurons in these pools have synaptic connections to all neurons in another pool V of N neurons.[3] We assume that for each pool U_j all neurons in U_j are excitatory, or all neurons in U_j are inhibitory. We will first investigate the question which functions $\langle x_1, \ldots, x_n \rangle \rightarrow y$ can be computed by such a network if x_j is the firing probability of a neuron in pool U_j during a short time interval I_{input} and y is the firing probability of a neuron in pool V during a slightly later time interval I_{output}.

Consider an idealized mathematical model where all neurons which fire in the pool U_j fire synchronously at time T_{input}, and the probability that a neuron $v \in V$ fires (at time $\rightarrow$) can be described by the probability that the sum h_v of the amplitudes of EPSPs and IPSPs resulting from firing of neurons in the pools $U_1, \ldots, U_n$ exceeds the firing threshold θ (which is assumed to be the same for all neurons $v \in V$). We assume in this section that the firing rates of neurons in pool V are relatively low, so that the impact of their refractory period can be neglected. We investigate refractory effects in section 9.6. The random variable (r.v.) h_v is the sum of random variables h_{vu} for all neurons $u \in \bigcup_{j=1}^n U_j$, where h_{vu} models the contribution of neuron u to h_v. We assume that h_{vu} is nonzero only if neuron $u \in U_j$ fires at time T_{input} (which occurs with probability x_j)[4] and if the synapse between u and v releases one or several vesicles (which occurs with probability r_{vu} whenever u fires). If both events occur then the value of h_{vu} is chosen according to some probability density function ϕ_{vu}. The functions ϕ_{vu}, as well as the parameters r_{vu}, are allowed to vary arbitrarily for different pairs u, v of neurons. For each neuron $v \in V$ we consider the sum $h_v = \sum_{j=1}^n \sum_{u \in U_j} h_{vu}$ of the r.v.'s h_{vu} and we assume that v fires at time $\rightarrow$ if and only if $h_v \geq \theta$. Although the r.v.'s h_{vu} may have quite different distributions (for example due to different ϕ_{vu} and r_{vu}), their stochastic independence allows us to approximate the firing probability $\mathrm{P}\{h_v \geq \theta\}$ through a normal distribution Φ. The Berry-Esseen Theorem [36] implies that

$$|\mathrm{P}\{h_v \geq \theta\} - (1 - \Phi(\theta; \mu_v, \sigma_v))| \leq 0.7915 \frac{\rho_v}{\sigma_v^3}, \qquad (9.8)$$

where $\Phi(\cdot; \mu_v, \sigma_v)$ denotes the normal distribution function with mean μ_v and variance σ_v^2. The three moments occurring in (9.8) can be related to the r.v.'s h_{vu} through the equations $\mu_v = \sum_{j=1}^n \sum_{u \in U_j} \mathrm{E}[h_{vu}]$, $\sigma_v^2 = \sum_{j=1}^n \sum_{u \in U_j} \mathrm{Var}[h_{vu}]$, and $\rho_v = \sum_{j=1}^n \sum_{u \in U_j} \mathrm{E}[|h_{vu} - \mathrm{E}[h_{vu}]|^3]$. According to the definition of the r.v. h_{vu} we have $\mathrm{E}[h_{vu}] = x_j r_{vu} \bar{a}_{vu}$ and

---

[3]Our results remain valid if one considers instead connections by fixed random graphs with lower density between pools U_j and V.

[4]This holds if the pool size is large enough such that we can treat x_j (y) as the probability that a neuron $u \in U_j$ ($v \in V$) will fire once during a certain input (output) interval of length Δ.

$E[h^2_{vu}] = x_j r_{vu} \hat{a}_{vu}$ where $\bar{a}_{vu} = \int a\phi_{vu}(a)da$ denotes the mean EPSP (IPSP) amplitude and $\hat{a}_{vu} = \int a\phi_{vu}(a)da$ denotes the second moment. Hence we can assign to μ_v and σ_v in (9.8) the values

$$\mu_v = \sum_{j=1}^{n} \sum_{u \in U_j} x_j r_{vu} \bar{a}_{vu}, \tag{9.9}$$

$$\sigma_v^2 = \sum_{j=1}^{n} \sum_{u \in U_j} (x_j r_{vu} \hat{a}_{vu} - x_j^2 r_{vu}^2 \bar{a}_{vu}^2). \tag{9.10}$$

A closer look reveals that the right-hand side of (9.8) scales like $N^{-1/2}$.[5] Hence, Eq. (9.8) implies that for large N we can approximate the firing probability $P\{h_v \geq \theta\}$ by the term $1 - \Phi(\theta; \mu_v, \sigma_v)$, which smoothly grows with μ_v. The gain of this sigmoidal function depends on the size of σ_v. In particular, if synaptic transmission were reliable, this function would degenerate to a step function. With the definition of the formal weights $w_{vj} := \sum_{u \in U_j} r_{vu} \bar{a}_{vu}$ we have $\mu_v = \sum_{j=1}^{n} w_{vj} x_j$, and hence $1 - \Phi(\theta; \mu_v, \sigma_v)$ smoothly grows with the weighted sum $\sum_{j=1}^{n} w_{vj} x_j$ of the inputs x_j.

So far we have just considered the probability $P\{h_v \geq \theta\}$ that a single neuron $v \in V$ will fire, but we are really interested in the expected fraction y of neurons in pool V which will fire in response to a firing of a fraction x_j of neurons in the pools U_j for $j = 1, \ldots, n$. According to Eq. (9.8) one can approximate y for sufficiently large pool sizes N by

$$y = \frac{1}{N} \sum_{v \in V} P\{h_v \geq \theta\} = \frac{1}{N} \sum_{v \in V} 1 - \Phi(\theta; \mu_v, \sigma_v).$$

Hence y is approximated by an average of the N sigmoidal functions $1 - \Phi(\theta; \mu_v, \sigma_v)$. If the weights w_{vj} have similar values for different $v \in V$ one can expect that y grows smoothly with the weighted sum $\bar{\mu} = \sum_{j=1}^{n} w_j x_j$, where we write $w_j = \sum_{v \in V} w_{vj}/N$ for the "effective weights" w_j between the pools of neurons U_j and V.

In order to test these theoretical predictions for an idealized mathematical model we have carried out computer simulations of a more detailed model consisting of more realistic models for spiking neurons and time intervals I_{input} and I_{output} of length $\Delta = 5\,\text{ms}$ for space-rate coding (see Fig. 9.6).

9.5.1   Multilayer computations

The preceding arguments show that approximate computations of functions of the form $\langle x_1, \ldots, x_n \rangle \rightarrow y = \sigma(\sum_{j=1}^{n} w_j x_j)$, with inputs and output in

---

[5]More precisely, the right-hand side of (9.8) scales like $N^{-1/2}$ if for all N the average value of the terms $E[|h_{vu} - E[h_{vu}]|^3]$ is uniformly bounded from above and the average value of the terms $\text{Var}[h_{vu}]$ is uniformly bounded from below by a constant > 0 for $j \in \{1, \ldots, n\}$ and $u \in \bigcup_{j=1}^{n} U_j$. The latter can only be achieved for inputs where $x_j > 0$ for some j, since otherwise $\text{Var}[h_{vu}] = 0$ for all $u \in U_j$ and all $j \in \{1, \ldots, n\}$. But in the case $x_j = 0$ for all $j \in \{1, \ldots, n\}$ both $P\{h_v \geq \theta\}$ and $1 - \Phi(\theta; \mu_v, \sigma_v)$ have value 0 if $\theta > 0$ and hence the left hand side of (9.8) has value 0.

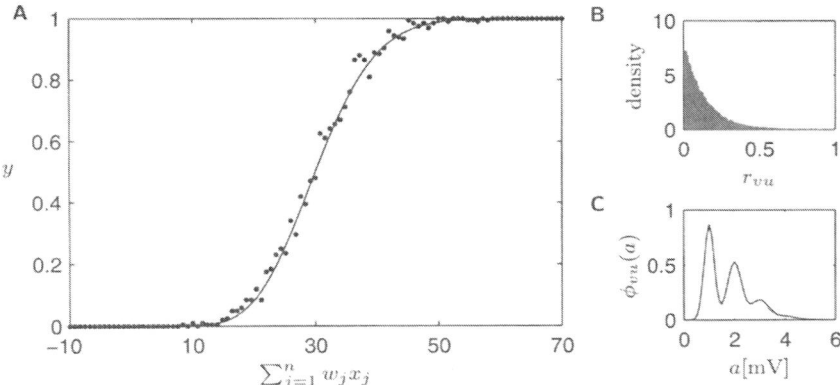

FIGURE 9.6. A) Computer simulation of the model described in section 9.5 with a time interval Δ of length $\Delta = 5\,\text{ms}$ for space-rate coding and neurons modeled by the spike response model of Gerstner [10]. We have chosen $n = 6$, a pool size $N = 200$ and $\langle w_1, \ldots, w_6 \rangle = \langle 10, -20, -30, 40, 50, 60 \rangle$ for the effective weights. Each dot is the result of a simulation with an input $\langle x_1, \ldots, x_6 \rangle$ selected randomly from $[0, 1]^6$ in such a way that $\sum_{j=1}^n w_j x_j$ covers the range $[-10, 70]$ almost uniformly. The y-axis shows the fraction y of neurons in pool V that fire during a 5 ms time interval in response to the firing of a fraction x_j of neurons in pool U_j for $j = 1, \ldots, 6$ during an earlier time interval of length 5 ms. B) Distribution of non-failure probabilities r_{vu} for synapses between the pools U_4 and V underlying this simulation. C) Example of a probability density function ϕ_{vu} of EPSP amplitudes as used for this simulations. This corresponds to a synapse with 5 release sites and a release probability of 0.3.

space-rate code, can be carried out within 10 ms by a network of spiking neurons. Hence the universal approximation theorem for multilayer perceptrons implies that *arbitrary continuous functions* $f : [0, 1]^n \rightarrow [0, 1]^m$ can be approximated with a computation time of not more than 20 ms by a network of spiking neurons with 3 layers. Thus our model provides a possible theoretical explanation for the empirically observed very fast multilayer computations in biological neural systems that were mentioned in the introduction.

Results of computer simulations of the computation of a specific function f in space-rate coding that requires a multilayer network because it interpolates the boolean function XOR shown in Fig. 9.7.

9.6   Analog Computation on Time Series in a Space-Rate Code

We have shown that biological networks of spiking neurons with space-rate coding have at least the computational power of multilayer perceptrons. In this section we will demonstrate that they have strictly more computational power. This becomes clear if one considers computations on *time series*, rather than on *static batch inputs*.

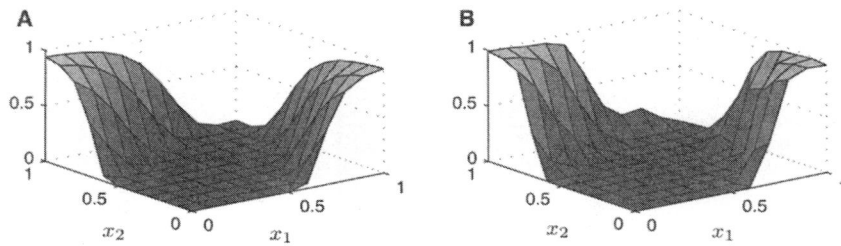

FIGURE 9.7. A) Plot of a function $f(x_1, x_2) : [0,1]^2 \rightarrow [0,1]$ which interpolates XOR. Since f interpolates XOR it cannot be computed by a single sigmoid unit. B) Computation of f by a 3-layer network in space-rate coding with spike response model neurons ($N = 200$) according to our model (for details see [27]).

We now analyze the behavior of our computational model if the firing probabilities in the pools U_i change with time. Writing $x_i(t)$ ($y(t)$) for the probability that a neuron in pool U_i (V) fires during the t-th time window of length Δ (e.g., for Δ in the range between 1 and 5 ms), our computational model from section 9.5 maps a vector of n analog time series $\{x_i(t)\}_{t \in \mathbb{N}}$ onto an output time series $\{y(t)\}_{t \in \mathbb{N}}$ (where ($\mathbb{N} = \{0, 1, 2, \ldots\}$).

As an example consider a network which consists of one presynaptic pool U_1 connected to the output pool V with the same type of synapses as discussed before. In addition there are *feedback connections* between individual neurons $v \in V$. The results of simulations reported in Fig. 9.8 show that this network computes an interesting map in the time series domain: The space-rate code in pool V represents a sigmoidal function σ applied to the output of a *bandpass filter*. Fig. 9.8B shows the response of such network to a sine wave with some bursts of activity added on top (Fig. 9.8A). Fig. 9.8C shows the output of another network of spiking neurons (which approximates a *lowpass* filter) to the same input (shown in Fig. 9.8A).

One can prove that in principle any linear filter with finite or infinite impulse response can be implemented with any desired degree of precision by recurrent networks of spiking neurons with space-rate coding [27].

9.7 Computing with Firing Rates

Traditionally a link between sigmoidal neural networks and biological neural systems is established by interpreting the firing rate (i.e., the spike count over time) of a spiking neuron as an analog number between 0 and 1. In this interpretation one gets a plausible correspondence between the dependence of the output value $g(\sum_{j \in \Gamma_i} w_{ij} x_j)$ of a sigmoidal gate on its

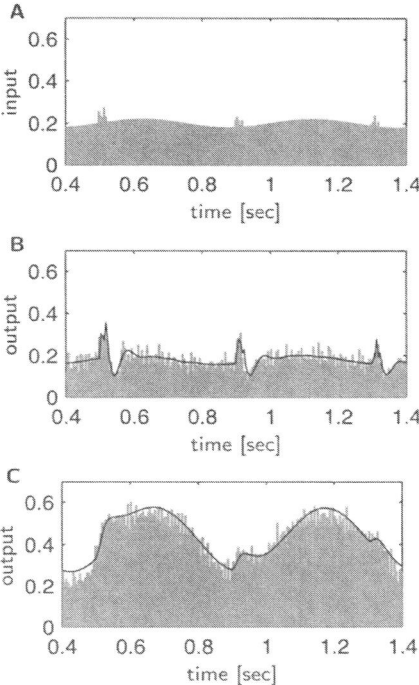

FIGURE 9.8. Response of two different networks to the same input (panel A). Panel B shows the response of a network which approximates a bandpass filter, whereas panel C shows the response of a network which approximates a lowpass filter. The gray shaded bars in panels B and C show the actual measured fraction of neurons which fire in pool V of the network during a time interval of length 5 ms in response to the input activity in pool U_1 shown in panel A. The solid lines in panels B and C are plots of the theoretically predicted output (see [27]).

input values x_j on one hand, and the dependence of the firing rate of a spiking neuron i on the firing rates of presynaptic neurons $j \in \Gamma_i$ on the other hand.

There exists ample biological evidence that information about a stimulus is in many biological neural systems encoded in the firing rates of neurons. However recent empirical results from neurophysiology have raised doubts whether the firing rate of a biological neuron i does in fact depend on the firing rates x_j of presynaptic neurons $j \in \Gamma_i$ in a way that can be described by an expression of the form $g(\sum_{j \in \Gamma_i} w_{ij} x_j)$. Results of Abbott et al. [4] and others about the dynamic behavior of biological synapses show that for some neural systems above a "limiting frequency" of about 10 Hz the amplitudes of postsynaptic potentials are inversely proportional to the firing rate x_j of the presynaptic neuron $j \in \Gamma_i$. These results suggest that instead of a fixed parameter w_{ij} one has to model the "strength" of a biological synapse for rate coding by a quantity $w_{ij}(x_j)$ that *depends* on the firing rate x_j of the presynaptic neuron, and that this quantity $w_{ij}(x_j)$ is proportional to $\frac{1}{x_j}$. But then the weighted sum $\sum_{j \in \Gamma_i} w_{ij}(x_i) \cdot x_j$, which models the average membrane potential at the soma of a spiking neuron i, does no longer depend on the firing rates x_j of those presynaptic neurons j that fire above the limiting frequency. We refer to [30] for a survey of computational implications of synaptic dynamics.

9.8   Computing with Firing Rates and Temporal Correlations

We will discuss in this section computations that employ a quite different type of "temporal coding". Communication by spike trains offers a direct way to encode transient *relations* between different neurons: through coincidences (or near coincidences) in their firing times. Hence computations with spiking neurons may in principle also involve complex operations on *relations between computational objects*, a computational mode which has no parallel in traditional neural network models–or any other common computational model. This type of temporal coding need not necessarily take place on the microscopic level of coding by single spikes, but can also take place on a macroscopic level of statistical correlations between firing times of different neurons. Milner had conjectured already in 1974 that visual input might be encoded in the visual cortex in a way where "cells fired by the same figure fire together but not in synchrony with cells fired by other figures" ([33]). This conjecture has subsequently been elaborated by von der Malsburg [47] and has been supported more recently by experimental data from several labs (see for example Eckhorn et al., 1988 [8]; Gray et al., 1989 [12]; Kreiter and Singer, 1996 [18]; Vaadia et al., 1995 [45]).

A variety of models have been proposed in order to shed light on the possible *organization* of *computing with firing correlations* in networks of spiking neurons. We have already shown in the preceding sections that spiking neurons are well suited for *detecting* firing correlations among preceeding neurons. They also can *induce* firing correlations in other neurons k by sending the same output spike train to several other neurons k. But the question remains what exactly can be computed with firing correlations in a *network* of spiking neurons.

In [9] a computational model was introduced whose computational units are modifications of integrate-and-fire neurons that receive two types of input: *feeding input* and *linking input*. Both types of inputs are processed in this model by leaky integrators with different time constants, and are then multiplied to produce the potential $u_i(t)$ of an integrate-and-fire neuron i. In networks of such computational units the feeding input typically is provided by feedforward connections, starting from the stimulus, whereas the linking input comes through feedback connections from higher layers of the network. These higher layers may represent information stored in an associative memory like in a Hopfield net. Computer simulations have shown that this model is quite successful in reproducing firing patterns in response to specific stimuli that match quite well firing patterns that have been experimentally observed in the visual cortex of cat and monkey. So far no theoretical results have been derived for this model.

A related model, but on a more abstract level without spiking neurons was proposed in [17] (see also [37] for a survey of this and related models). That model also involves two types of input, called RF and CF, where RF corresponds to "feeding input" and CF corresponds to "linking input". No computational model has been specified for the generation of the CF-values. A computational unit in their model outputs a continuous value $2 \cdot g(\frac{r}{2} \cdot (1 + e^{2r \cdot c})) - 1$ that ranges between -1 and 1. In this formula r is a weighted sum of RF-input, s is a weighted sum of CF-input to that unit, and g is a sigmoidal gain function. In this computational unit the RF-input r determines the sign of the output. Furthermore $r = 0$ implies that the output has value 0, independently of the value of the CF-input c. However for $r \neq 0$ the size of the output is increased through the influence of the CF-input c if c has the same sign as r, and decreased otherwise. Computer simulations of large networks of such computational units have produced effects which are expected on the basis of psychological studies of visual perception in humans[37].

Other models aim directly at simulating effects of computing with firing correlations on an even more abstract conceptual level, see for example [von der Malsburg, 1981; Shastri and Ajjanagadde, 1993]. In [41] a formal calculus is developed for exploiting the possibility to compute with *relations* encoded by temporal coincidences.

No rigorous results are available so far which show that the previously described models have more computational power than conventional neural network models. In principle every function that is computed on any of the previously discussed models can also be computed by a conventional sigmoidal neural net, i.e., by an abstract model for a network of spiking neurons that encode all information in their firing rates. This follows from the simple fact that a sigmoidal neural net has the "universal approximation property"– i.e., it can approximate any given continuous function. Thus the question about a possible increase in computational power through the use of firing correlations boils down to a *quantitative* rather than *qualitative* question: How much hardware, computation time, etc. can a neural network save by computing with firing correlations in addition to firing rates?

We will sketch a new model from Maass, 1997 [23] that provides some first results in this direction. We write $\nu(i)$ for the output of a sigmoidal gate i, which is assumed to range over $[0, 1]$. One may interpret $\nu(i)$ as the firing rate of a spiking neuron i. We now introduce for certain sets S of neurons a new variable $c(S)$, also ranging over $[0, 1]$, whose value models in an abstract way the current amount of *temporal correlation* in the firing times of the neurons $i \in S$. For example, for some time internal A of 5 msec one could demand that $c(S) = 0$ if

$$\frac{Pr[\text{ all } j \in S \text{ fire during } A]}{\Pi_{j \in S} \ Pr[j \text{ fires during } A]} \leq 1,$$

and that $c(S)$ approaches 1 when this quotient approaches infinity. Thus we have $c(S) = 0$ if all neurons $j \in S$ fire stochastically independently.

One then needs computational rules that extend the standard rule

$$\nu(i) = g(\sum_{j \in \Gamma_i} w_{ij} \cdot \nu(j))$$

for sigmoidal gates so that they also involve the new variables $c(S)$ in a meaningful way. In particular, one wants to have that the firing rate $\nu(i)$ is increased if one or several subsets $S \subseteq \Gamma_i$ of preceding neurons fire with temporal correlation $c(S) > 0$. This motivates the first rule of our model:

$$\nu(i) = g(\sum_{j \in \Gamma_i} w_{ij} \cdot \nu(j) + \sum_{S \subseteq \Gamma_i} w_{iS} \cdot c(S) \cdot \Pi_{j \in S} \nu(j) + \vartheta). \qquad (9.11)$$

The products $c(S) \cdot \Pi_{j \in S} \nu(j)$ in the second summand of (9.11) reflect the fact that statistical correlations in the firing times of the neurons $j \in S$ can only increase the firing rate of neuron i by a significant amount if the firing rates of all neurons $j \in S$ are sufficiently high. These products also arise naturally if $c(S)$ is interpreted as being proportional to

$$\frac{Pr[\text{ all } j \in S \text{ fire during } A]}{\Pi_{j \in S} \ Pr[j \text{ fires during } A]}, \qquad (9.12)$$

and $\nu(j)$ is proportional to $Pr[j$ fires during $A]$. Hence multiplying (9.12) with $\Pi_{j \in S} \nu(j) \approx \Pi_{j \in S} Pr[j$ fires during $A]$ yields a term proportional to $Pr[\text{all } j \in S$ fire during $A]$. This term is the one that really determines by how much the firing rate of neuron i may increase through correlated firing of neurons in S: If the neurons $j \in S$ fire almost simultaneously, this will move the state variable $u_i(t)$ of neuron i to a larger peak value compared with a situation where the neurons $j \in S$ fire in a temporally dispersed manner.

In order to complete the definition of our model for computing with firing rates $\nu(i)$ and firing correlations $c(S)$ one also has to specify how the correlation variable $c(S)$ is computed for a set S of "hidden" units i. Two effects have to be modeled:

(a) $c(S)$ increases if all neurons $i \in S$ receive common input from some other neuron k.

(b) $c(S)$ increases if there is a set S' of other neurons with significant correlation (i.e., $c(S') > 0$) so that each neuron $i \in S$ has some neuron $i' \in S'$ as predecessor (i.e., $\forall\, i \in S\, \exists\, i' \in S'(i' \in \Gamma_i)$).

These two effects give rise to the two terms in the following rule:

$$c(S) = g(\sum_k w_{Sk} \cdot o(k) + \sum_{S'} w_{SS'} \cdot c(S') \cdot \Pi_{i' \in S'}\, o(i') + \vartheta_S). \qquad (9.13)$$

¿From the point of view of computational complexity it is interesting to note that in a network of spiking neurons no additional units are needed to compute the value of $c(S)$ according to (9.13). The new parameters $w_{Sk}, w_{SS'}$ can be chosen so that they encode the relevant information about the connectivity structure of the net, for example $w_{Sk} = 0$ if not $\forall\, i \in S\, (k \in \Gamma_i)$ and $w_{SS'} = 0$ if not $\forall\, i \in S\, \exists i' \in S'\, (i' \in \Gamma_i)$. Then the rule (9.13) models the previously described effects (a) and (b).

The rules (9.11) and (9.12) involve besides the familiar "synaptic weights" w_{ij} also new types of parameters w_{iS}, w_{Sk}, and $w_{SS'}$. The parameter w_{iS} scales the influence that correlated firing of the presynaptic neurons $j \in S$ has on the firing rate of neuron i. Thus for a biological neuron this parameter w_{iS} not only depends on the connectivity structure of the net, but also on the geometric and biochemical structure of the dendritic tree of neuron i and on the locations of the synapses from the neurons $j \in S$ on this dendritic tree. For example correlated firing of neurons $j \in S$ has a larger impact if these neurons either have synapses that are clustered together on a single subtree of the dendritic tree of i that contains voltage-gated channels (hot spots), or if they have synapses onto disjoint subtrees of the dendritic tree (thus avoiding sublinear summation of their EPSP's in the Hodgkin-Huxley model. Taking into account that very frequently pairs of biological neurons are not connected just by one synapse, but by multiple synapses that may lie on different branches of their dendritic tree, one sees that in the context of computing with firing correlations the "program" of the computation can be encoded through these additional parameters w_{iS} in much more subtle and richer ways than just through the "synaptic weights" w_{ij}. Corresponding remarks apply to the other new parameters w_{Sk} and $w_{SS'}$ that arise in the context of computing with firing correlations.

One should add that the interpretation of $c(S)$ becomes more difficult in case that one considers correlation variables $c(S')$ for a family of sets S' whose intersection contains more than a single neuron. For example, if $|S'| \geq 2$ and $S' \subsetneq S$ then $c(S)$ should be interpreted as the impact of correlated firing of neurons in S *beyond* the impact that correlated firing of the neurons in S' already has. One can escape this technical difficulty by considering for example in a simplified setting only correlation variables $c(S)$ for sets S of size 2.

The following result shows that a computational unit i that computes its output $\nu(i)$ according to rule (9.11) has more computational power than a sigmoidal gate (or even a small network of sigmoidal gates) that receives the same numerical variables $\nu(j), c(S)$ as input. This arises from the fact that the computational role (2.10) involves a *product* of input variables.

Consider the boolean function $F : \{0,1\}^{n+\binom{n}{2}} \to \{0,1\}$ that outputs 1 for n boolean input variables $\nu(j)$, $j \in \{1,\ldots,n\}$, and $\binom{n}{2}$ boolean input variables $c(S)$ for all subsets $S \subseteq \{1,\ldots,n\}$ of size 2 *if and only if* $c(S) = 1$ and $o(j_1) = o(j_2) = 1$ for some subset $S = \{j_1,j_2\}$. It is obvious from equation (9.11) that if one takes as gain function the Heaviside function $\mathcal{H}$, then a *single* computational unit i of the type described by equation (9.11) can compute the function F_n. On the other hand the following result shows that a substantial number of threshold gates or sigmoidal gates are needed to compute the same function F_n. Its proof can be found in [23].

Theorem 9.5 *The function F_n can be computed by a single neuron that carries out computations with firing rates and firing correlations according to rule (9.11).*

On the other hand any feedforward threshold circuit that computes the function F_n needs to have on the order of $n^2/\log n$ gates. Any feedforward circuit consisting of sigmoidal gates[6] needs to have at least proportional to n many gates to compute F_n. ∎

9.9   Networks of Spiking Neurons for Storing and Retrieving Information

Synfire chains [2] are models for networks of spiking neurons that are well suited for storing and retrieving information from a network of spiking neurons. A synfire chain is a chain of pools of neurons with a rich (diverging/converging) pattern of excitatory feedforward connection from each pool to the next pool, that has a similar effect as complete connectivity between successive pools: an almost synchronous firing of most neurons in one pool in a synfire chain triggers an almost synchronous firing of most neurons in the next pool in the chain. Neurons may belong to different synfire chains, which has the consequence that the activation of one synfire chain may trigger the activation of another synfire chain (see [5]). In this way a pointer from one memory item (implemented by one synfire chain) to another memory item (implemented by another synfire chain) can be realized by a network of spiking neurons. A remarkable property of synfire chains is that the temporal delay between the activation time of the first pool and the kth pool in a synfire chain has a very small variance, even

---

[6]with piecewise rational activation functions

for large values of k. This is due to the temporal averaging of EPSPs from neurons in the preceding pool that is carried out through rich connectivity between successive pools.

An *analog* version of synfire chains results from the model for computing with space-rate coding discussed in section 9.5. If one takes synaptic unreliability into account for a synfire chain, one can achieve that the percentage of firing neurons of a later pool in the synfire chain becomes a smooth function of the percentage of firing neurons in the first pool. This variation of the synfire chain model predicts that precisely timed firing patterns of a fixed set of 2 or 3 neurons in different pools of a biological neural system occur more often than can be expected by chance, but not *every* time for the same stimulus. This prediction is consistent with experimental data [3].

Other types of networks of spiking neurons that are useful for storing and retrieving information are various implementations of attractor neural networks with recurrent networks of spiking neurons, see for example [Maass and Natschläger, 1997].

9.10 Computing on Spike Trains

We still know very little about the power of networks of biological neurons for computations on spike trains, for example for spike trains that encode a time series of analog numbers as for example the spike trains from neuron $H1$ in the blowfly (see [38]). One problem is that the previously discussed formal models for networks of spiking neurons are not really adequate for modeling computations by biological neural systems on *spike trains*, because they are based on the assumption that synaptic weights w_{ij} are *static* during a computation. Some consequences of the inherent temporal dynamics of biological synapses on the computational power of networks of spiking neurons are discussed in [30, 29, 34]. In particular, it is shown in [29] that theoretically a single layer of spiking neurons with dynamic synapses can approximate any nonlinear filter given by an arbitrary Volterra series.

9.11 Conclusions

This chapter has shown that networks of spiking neurons present a quite interesting new class of computational models. In particular, they can carry out analog computation not only under a rate code, but also under temporal codes where the timing of spikes carries analog information. We have presented theoretical evidence which suggests that through the use of temporal coding a network of spiking neurons may gain for certain computational tasks more computational power than a traditional neural network of comparable size.

The models for networks of spiking neurons that we have discussed in this chapter originated in the investigation of biological neurons. However it is obvious that many of the computational ideas and architectures presented in this chapter are of a more general nature and can just as well be applied to implementations of pulsed neural nets in electronic hardware, such as those surveyed in [25].

9.12 REFERENCES

[1] Abeles, M. (1982). Role of the cortical neuron: integrator or coincidence detector? *Israel J. Med. Sci.*,18:83–92.

[2] Abeles, M. (1991). *Corticonics*. Cambridge University Press, Cambridge.

[3] Abeles, M., Bergmann, H., Margalit, E., and Vaadia, E. (1993). Spatiotemporal firing patterns in the frontal cortex of behaving monkeys. *J. Neurophysiol.*, 70(4), 1629–1638.

[4] Abbott, L. F., Sen, K., Varela, J. A., and Nelson, S. B. (1997). Synaptic depression and cortical gain control. *Science*, 275:220–222.

[5] Bienenstock, E. (1995). A model of neocortex. *Network*, 6:179–224.

[6] DasGupta, B. and Schnitger G. (1996). Analog versus discrete neural networks. *Neural Computation*, 8(4), 805–818.

[7] Dobrunz, L. and Stevens, C. (1997). Heterogenous release probabilities in hippocampal neurons. *Neuron*, 18:995–1008.

[8] Eckhorn, R., Bauer, R., Jordan, W., Brosch, M., Kruse, W., Munk, M., and Reitboeck, H. J. (1988). Coherent oscillations: A mechanism of feature linking in the visual cortex? Multiple electrode and correlation analysis in the cat. *Biological Cybernetics*, 60:121–130.

[9] Eckhorn, R., Reitboeck, H. J., Arndt, M., and Dicke, P. (1990). Feature linking via synchronization among distributed assemblies: simulations of results from cat visual cortex. *Neural Computation*, 2:293–307.

[10] Gerstner, W. (1999) Spiking neurons. In: *Pulsed Neural Networks*, W. Maass and C. Bishop, eds., MIT Press (Cambridge), 3–53.

[11] Gerstner, W. and van Hemmen, J. L. (1992). Associative memory in a network of "spiking" neurons. *Network: Computation in Neural Systems*, 3:139–164.

[12] Gray, C. M., König, P., Engel, A. K., and Singer, W. (1989). Oscillatory responses in cat visual cortex exhibit inter-columnar synchronization which reflects global stimulus properties. *Nature*, 338:334–337.

[13] Haefliger, P., Mahowald, M., and Watts, L. (1997). A spike based learning neuron in analog VLSI. *Advances in Neural Information Processing Systems, vol. 9*, MIT Press, (Cambridge), 692–698.

[14] Hopfield, J. J. (1995). Pattern recognition computation using action potential timing for stimulus representation. *Nature*, 376:33–36.

[15] Hopfield, J. J., Herz, A. V. M. (1995). Rapid local synchronization of action potentials: Toward computation with coupled integrate-and-fire neurons. *Proc. Natl. Acad. Sci. USA*, 92:6655–6662.

[16] Judd, K. T. and Aihara, K. (1993). Pulse propagation networks: A neural network model that uses temporal coding by action potentials. *Neural Networks*, 6:203–215.

[17] Kay, J. and Phillips, W. A. (1997). Activation functions, computational goals, and learning rules for local processors with contextual guidance. *Neural Computation*, 9(4):895–910.

[18] Kreiter, A. K. and Singer, W. (1996). Stimulus-dependent synchronization of neuronal responses in the visual cortex of the awake macaque monkey. *The Journal of Neuroscience*, 16(7):2381–2396.

[19] Kistler, W., Gerstner, W., and van Hemmen, J. L. (1997). Reduction of Hodkin-Huxley equations to a single-variable threshold model. *Neural Computation*, 9:1015–1045.

[20] Maass, W. (1996). Lower bounds for the computational power of networks of spiking neurons. *Neural Computation*, 8(1):1–40.

[21] Maass, W. (1997). Fast sigmoidal networks via spiking neurons. *Neural Computation*, 9:279–304.

[22] Maass, W. (1997). Networks of spiking neurons: The third generation of neural network models. *Neural Networks*, 10(9):1659–1671. Extended abstract (with a different title) appeared in: *Advances in Neural Information Processing Systems, vol. 9*, MIT Press, (Cambridge), 211–217.

[23] Maass, W. (1998). A simple model for neural computation with firing rates and firing correlations. *Network: Computation in Neural Systems*, 9:1–17.

[24] Maass, W., Schnitger, G., and Sontag, E. (1991). On the computational power of sigmoid versus boolean threshold circuits. *Proc. of the 32nd Annual IEEE Symposium on Foundations of Computer Science 1991*, 767–776; extended version appeared in: *Theoretical Advances in Neural Computation and Learning*, V. P. Roychowdhury, K. Y. Siu, A. Orlitsky, eds., Kluwer Academic Publishers (Boston, 1994), 127–151.

[25] Maass, W., and Bishop, C., eds. (1999) *Pulsed Neural Networks*, MIT Press, (Cambridge).

[26] Maass, W. and Natschläger, T. (1997). Networks of spiking neurons can emulate arbitrary Hopfield nets in temporal coding. *Network: Computation in Neural Systems*, 8(4):355–372.

[27] Maass, W. and Natschläger, T. (2000). A model for fast analog computation based on unreliable synapses. *Neural Computation*, 12(7):1679–1704.

[28] Maass, W. and Schmitt, M. (1997). On the complexity of learning for spiking neurons with temporal coding. *Proc. of the Tenth Annual Conference on Computational Learning Theory*, ACM, New York, 54–61; journal version in *Information and Computation* 153, 26–46, 1999.

[29] Maass, W. and Sontag, E. D. (2000). Neural systems as nonlinear filters. *Neural Computation*, 12(8), 1743–1772.

[30] Maass, W. and Zador, A. (1998). Dynamic stochastic synapses as computational units. *Advances in Neural Information Processing Systems*, vol. 10, MIT-Press (Cambridge), 194–200; journal version: *Neural Computation*, 11(4), 1999, 903–918.

[31] Markram, H., Lübke, J., Frotscher, M., and Sakman, B. (1997). Regulation of synaptic efficacy by coincidence of postsynaptic APs and EPSPs. *Science*, 275:213 – 215.

[32] Markram, H. and Sakmann, B. (1995). Action potentials propagating back into dendrites triggers changes in efficacy of single-axon synapses between layer V pyramidal neurons. *Society for Neuroscience Abstracts*, 21:2007.

[33] Milner, P. M. (1974). A model for visual shape recognition. *Psychological Review*, 81(6):521–535.

[34] Natschläger, T., Maass, W., Sontag, E.D., and Zador, A., (2001). Processing of time series by neural circuits with biologically realistic synaptic dynamics. In *Advances in Neural Information Processing Systems 2000 (NIPS '2000)*, volume 13, Cambridge, 2001. MIT Press, to appear.

[35] Natschläger, T. and Ruf, B. (1998). Spatial and temporal pattern analysis via spiking neurons. *Network: Computation in Neural Systems*, 9(3):319-332.

[36] Petrov, V. V. (1995). *Limit Theorems of Probability Theory*. Oxford University Press.

[37] Phillips, W. A. and Singer, W. (1997). In search of common foundations for cortical computation. *Behavioral and Brain Sciences*, 20(4):657–722.

[38] Rieke, F., Warland, D., de Ruyter van Steveninck, R. R., and Bialek, W. (1997). *Spikes – Exploring the Neural Code*. MIT Press, (Cambridge), MA.

[39] Ruf, B., and Schmitt, M. (1998). Self-organizing maps of spiking neurons using temporal coding. In *Computational Neuroscience: Trends in Research*, J. M. Bower, ed., Plenum Press, New York, 509–514.

[40] Samuelides, M., Thorpe, S., and Veneau, E. (1997). Implementing hebbian learning in a rank-based neural network. *Proc. 7th Int. Conference on Artificial Neural Networks - ICANN'97* in Lausanne, Switzerland, Springer, Berlin, 145–150.

[41] Shastri, L. and Ajjanagadde, V. (1993). From simple associations to systematic reasoning: a connectionist representation of rules, variables and dynamic bindings using temporal synchrony. *Behavioural and Brain Sciences*, 16:417–494.

[42] Sontag, E. D. (1997). Shattering all sets of "k" points in "general position" requires $(k - 1)/2$ parameters. *Neural Computation*, 9(2):337–348.

[43] Thorpe, S. J. and Imbert, M. (1989). Biological constraints on connectionist models. *Connectionism in Perspective*. R. Pfeifer, Z. Schreter, F. Fogelman-Soulié, L. Steels, eds. Elsevier, Amsterdam, (1989), 63–92.

[44] Thorpe, S. J. and Gautrais, J. (1997). Rapid visual processing using spike asynchrony. *Advances in Neural Information Processing Systems*, vol. 9, MIT Press, (Cambridge), MA, 901–907.

[45] Vaadia, E., Aertsen, A., and Nelken, I. (1995). Dynamics of neuronal interactions cannot be explained by neuronal transients. *Proc. Royal Soc. of London B*, 261:407–410.

[46] Valiant, L. G. (1994). *Circuits of the Mind*, Oxford University Press, Oxford.

[47] von der Malsburg, C. (1981). The correlation theory of brain function. *Internal Report 81-2 of the Dept. of Neurobiology of the Max Planck Institute for Biophysical Chemistry in Göttingen*, Germany. Reprinted in *Models of Neural Networks II*, Domany et al., eds., Springer, 1994, 95–119.

[48] Watanabe, M., and Aihara, K. (1997). Chaos in neural networks composed of coincidence detector neurons. *Neural Networks*, 10(8), 1353-1359.

[49] Wilson, H. R. and Cowan, J. D. (1972). Excitatory and inhibitory interactions in localized populations of model neurons. *Biophysics Journal*, 12:1–24.

Index